To Sandip Sinharay —

Wim J. van der Linden

Computerized Adaptive Testing:
Theory and Practice

Computerized Adaptive Testing: Theory and Practice

Edited by

Wim J. van der Linden

and

Cees A.W. Glas
University of Twente,
The Netherlands

In collaboration with Interuniversitair Centrum voor
Onderwijskundig Onderzoek (ICO)

KLUWER ACADEMIC PUBLISHERS
DORDRECHT / BOSTON / LONDON

Library of Congress Cataloging-in-Publication Data

ISBN 0-7923-6425-2

Published by Kluwer Academic Publishers,
P.O. Box 17, 3300 AA Dordrecht, The Netherlands.

Sold and distributed in North, Central and South America
by Kluwer Academic Publishers,
101 Philip Drive, Norwell, MA 02061, U.S.A.

In all other countries, sold and distributed
by Kluwer Academic Publishers,
P.O. Box 322, 3300 AH Dordrecht, The Netherlands.

Printed on acid-free paper

Reprinted with corrections 2001

Printed in the Netherlands.

TABLE OF CONTENTS

Preface   iii

PART 1: ITEM SELECTION AND ABILITY ESTIMATION

1. Item selection and ability estimation in adaptive testing   1
   Wim J. van der Linden & Peter J. Pashley
2. Constrained adaptive testing with shadow tests   27
   Wim J. van der Linden
3. Principles of multidimensional adaptive testing   53
   Daniel O. Segall

PART 2: APPLICATIONS IN LARGE-SCALE TESTING PROGRAMS

4. The GRE computer adaptive test: Operational issues   75
   Graig N. Mills & Manfred Steffen
5. MATHCAT: A flexible testing system in mathematics education for adults   101
   Alfred J. Verschoor & Gerard J. J. M. Straetmans
6. Computer-adaptive sequential testing   117
   Richard M. Luecht & Ronald J. Nungester

PART 3: ITEM POOL DEVELOPMENT AND MAINTENANCE

7. Innovative item types for computerized testing   129
   Cynthia G. Parshall, Tim Davey & Peter J. Pashley
8. Designing item pools for computerized adaptive testing   149
   Bernard P. Veldkamp & Wim J. van der Linden
9. Methods of controlling the exposure of items in CAT   163
   Martha L. Stocking & Charles Lewis

PART 4: ITEM CALIBRATION AND MODEL FIT

10. Item calibration and parameter drift   183
    Cees A. W. Glas
11. Detecting person misfit in adaptive testing using statistical   201
    process control techniques
    Edith M. L. A. van Krimpen-Stoop & Rob R. Meijer
12. The Assessment of differential item functioning in   221
    computer adaptive tests
    Rebecca Zwick

PART 5: TESTLET-BASED ADAPTIVE TESTING

13. Testlet response theory: An analog for the 3PL model   245
    useful in testlet-based adaptive testing
    Howard Wainer, Eric T. Bradlow & Zuru Du
14. MML and EAP estimation in testlet-based adaptive   271
    testing
    Cees A. W. Glas, Howard Wainer & Eric T. Bradlow
15. Testlet-based adaptive mastery testing   289
    Hans J. Vos & Cees A. W. Glas

Author Index   311
Subject Index   319

Preface

For a long time, educational testing has focused mainly on paper-and-pencil tests and performance assessments. Since the late 1980s, when the large-scale dissemination of personal computers (PCs) in education began, these testing formats have been rapidly extended to formats suitable for delivery by computer. Computer-based delivery of tests has several advantages. For example, it allows for testing on demand, that is, whenever and wherever an examinee is ready to take the test. Also, the enormous power of modern PCs as well as their ability to control multiple media can be used to create innovative item formats and more realistic testing environments. Furthermore, computers can be used to increase the statistical accuracy of test scores using computerized adaptive testing (CAT). Instead of giving each examinee the same fixed test, CAT item selection adapts to the ability level of individual examinees. After each response the examinee's ability estimate is updated and the subsequent item is selected to have optimal properties at the new estimate.

The idea of adapting the selection of the items to the examinee is certainly not new. In the Binet-Simon (1905) intelligence test, the items were classified according to mental age, and the examiner was instructed to infer the mental age of the examinee from the earlier responses to the items and to adapt the selection of the subsequent items to his or her estimate until the correct age could be identified with sufficient certainty. In fact, the idea of adaptive testing is even as old as the practice of oral examinations. Good oral examiners have always known to tailor their questions to their impression of the examinees' knowledge level.

The development of item response theory (IRT) in the middle of the last century has provided a sound psychometric footing for CAT. The key feature of IRT is its modeling of response behavior with distinct parameters for the examinee's ability and the characteristics of the items. Due to this parameter separation, the question of optimal item parameter values for the estimation of examinee ability became relevant. The main answer to this question was given by Birnbaum (1968) who proved that, unless guessing is possible, the optimal item is the one with the highest value for the item discrimination parameter and a value for the difficulty parameter equal to the ability of the examinee.

The further development and fine tuning of the psychometric techniques needed to implement CAT took several decades. Because the first computers were slow and did not allow for ability estimation in real time, early research was almost exclusively directed at finding approximations or alternative formats that could be implemented in a traditional paper-and-pencil environment. Examples include the two-stage testing format (Cronbach & Gleser, 1965), Bayesian item selection with an approximation to the posterior distribution of the ability parameter (Owen, 1969), the up-and-down method of item selection (Lord, 1970), the Robbins-Monro algorithm (Lord, 1971a), the flexilevel test (Lord, 1971b), the stradaptive test (Weiss, 1973), and pyramidal adaptive testing (Larkin & Weiss, 1975).

With the advent of more powerful computers, application of CAT in large-scale high-stakes testing programs became feasible. A pioneer in this field was the US Department of Defense with its Armed Services Vocational Aptitude Battery (ASVAB). After a developmental phase, which began in 1979, the first CAT version of the ASVAB became operational in the mid 1980s. An informative account of the development of the CAT-ASVAB is given in Sands, Waters, and McBride (1997). However, the migration from paper-and-pencil testing to CAT truly began when the National Council of State Boards of Nursing launched a CAT version of its licensing exam (NCLEX/CAT) and was followed with a CAT version of the Graduate Record Examination (GRE). Ever since, many other large-scale testing programs have followed. It seems safe to state that at the moment the majority of large-scale testing programs either has already been computerized or are in the process of becoming so.

Some of the early reasons to switch to computerized test administration were: (1) CAT makes it possible for students to schedule tests at their convenience; (2) tests are taken in a more comfortable setting and with fewer people around than in large-scale paper-and-pencil administrations; (3) electronic processing of test data and reporting of scores is faster; and (4) wider ranges of questions and test content can be put to use (Educational Testing Service, 1996). In the current CAT programs, these advantages have certainly been realized and appreciated by the examinees. When offered the choice between a paper-and-pencil and a CAT version of the same test, typically most examinees choose the CAT version.

However, the first experiences with actual CAT have also given rise to a host of new questions. For example, in high-stakes testing programs, item security quickly became a problem. The capability of examinees to memorize test items as well as their tendency to share them with future examinees appeared to be much higher than anticipated.

As a consequence, the need arose for effective methods to control for item-exposure as well as to detect items that have been compromised. Also, the question of how to align test content with the test specifications and balance content across test administrations appeared to be more complicated than anticipated. This question has led to a search for new testing algorithms as well as the introduction of a variety of new forms such as testlet-based adaptive testing. Furthermore, items now have to be calibrated on line, and the feasibility of efficient methods of item calibration, using background information on the examinee and employing optimal design techniques, are currently investigated. These examples highlight only a few practical issues met when the first CAT programs were implemented in practice. A more comprehensive review of such issues is given in Mills and Stocking (1996).

The goal of this volume is to present a snapshot of the latest results from CAT research and development efforts conducted to answer these new questions. The book thus serves the same goal as an earlier volume edited by Weiss (1983), which still offers an excellent summary of the theoretic developments in an earlier period of CAT research. Because several of the research questions addressed in this volume have solutions that rely heavily on new types of psychometric modeling or the application of new statistical techniques, this volume is occasionally on the technical side. The book is not intended as a first introduction to CAT. Though broad in its coverage of modern research on CAT, neither is it intended to offer a comprehensive treatment of all the problems met when designing and implementing a new CAT program. A reference of proven quality that does serve both goals is the introductory volume edited by Wainer (1990). Finally, newer forms of CAT are intricately related to developments in computerized test item technology. Chapter 7 in this volume offers a comprehensive review of these developments. For a more in-depth introduction to examples of this technology, the reader is referred to a recent volume edited by Drasgow and Olson-Buchanan (1999).

The volume is organized into five sections. The first section shows some recent advances in item selection and ability estimation in CAT. The chapter by van der Linden and Pashley discusses both Bayesian and likelihood-based item selection criteria and ability estimators. It shows how both can be enhanced using prior information and allowing for random effects in the values of the item parameters. The chapter by van der Linden addresses the problem of balancing CAT administrations with respect to test content. Several existing methods are reviewed and a new method based on the assembly of shadow tests before item selection is developed. The problem of CAT with an item pool measuring multiple abilities is discussed by Segall. His approach is based on a multidimensional logistic IRT model with Bayesian parameter estimation and item selection.

The second section discusses three applications of CAT in large-scale testing programs. The two chapters by Mills and Steffen, and by Luecht and Nungester, describe applications in high-stakes programs, whereas the chapter by Verschoor and Straetmans discusses an application to a low-stakes program. Mills and Steffen describe the introduction of the CAT version of the Graduate Record Examination (GRE) program and discuss its research into the problems of item overlap, maintaining item pool quality over time, and scoring incomplete tests. Verschoor and Streatmans discuss the development of MATHCAT, an adaptive testing program to support placement of students in mathematics courses in an adult education program in the Netherlands. They describe the types of items, the testing algorithms as well as results from a statistical evaluation of the program. Luecht and Nungester explain the design methodology used for developing computer-adaptive versions of the United States Medical Licensing Examination^{TM} (USMLE^{TM}). The methodology entails an integrated system for test development, test administration, and data management.

The third section addresses the topic of item pool development and maintenance in CAT. The chapter by Parshall, Davey, and Pashley presents a taxonomy of innovative item types. The taxonomy has five dimensions: item format, response action, media inclusion, level of interactivity, and scoring algorithm. For each of these dimensions several new types of items are discussed and references to applications are given. In the next chapter, Veldkamp and van der Linden present an algorithmic approach to designing item pools for CAT. They show how the costs of item production can be minimized by translating the test objectives into a discrete optimization problem that provides a blueprint for the item pool. The blueprint can be used to guide the item production process. An important problem in CAT is how to maintain the integrity of the item pool in an operational CAT program. Stocking and Lewis review various methods of item-exposure based on the Sympson-Hetter algorithm. In addition, they present research into the problem of conditioning item-exposure control on estimated instead of true abilities of the examinees.

The topic of item calibration and model fit is addressed in the fourth section. In the first chapter, Glas reviews marginal maximum-likelihood (MML) item calibration. Then he addresses the issue of item-parameter drift, which may occur, for example, when the transition from item pretesting to operational CAT is made. He shows how item-parameter drift can be detected using statistical tests based on the Lagrange multiplier statistic or the cumulative sum statistic. Detection of aberrant response behavior is addressed in the next chapter by van Krimpen-Stoop and Meijer. These authors distinguish among various forms of aberrant response behavior and review several test statistics sensitive to each of these forms. These statistics are also derived from statistical quality control theory. The last problem of model fit discussed in this section is the problem of differential item functioning (DIF) in adaptive testing. Zwick reviews several methods for DIF detection based on the Mantel-Heanszel statistic.

The final section contains three chapters on testlet-based adaptive testing. If adaptive testing is based on pools of testlets rather than items to incorporate content constraints into the test, the price to be paid is loss of conditional independence among responses to items within the same testlet. Wainer, Bradlow, and Du solve this problem by positing a 3PL model with a random testlet effect. In the next chapter, Glas, Wainer, and Bradlow show how the parameters in this model can be estimated using the MML method and how the model can be implemented in a CAT program. The final chapter by Vos and Glas optimizes mastery testing rules for application in adaptive testing from a testlet-based item pool. The chapter also studies the loss that would occur if adaptive mastery testing were based on a model that ignores conditional dependence between responses within testlets.

This book is the result of contributions by many people whose roles we gratefully acknowledge. First, we would like to express our gratitude to the Dutch *Interuniversitair Centrum voor Onderwijsonderzoek* (ICO). The plan to produce a series of edited volumes on active research programs in ICO that contain their latest results and show their international cooperation is due to their current directors. In particular we thank Ton J. M. de Jong for his support during the production of this volume as well as his subtle way of asking us about our progress. Second, we deeply appreciate the cooperation of all contributors and their willingness to report on their research and developmental work in this volume. In spite of the current tendency to use journals rather than books as a primary outlet for new research, these contributors allowed us to edit a volume with chapters that are all based on original work. Third, at various stages in the production of this book we received secretarial support from Lorette A. M. Bosch and Sandra G. B. M. Dorgelo. Their efforts to meet our deadlines and prepare the manuscripts for import in Scientific Workplace were beyond the call of duty. Fourth, without forgetting the many good recommendations and advice we received from many of our colleagues during the editorial process, we thankfully acknowledge the helpful comments on a selection of chapters from Ronald K. Hambleton, University of Massachusetts. Finally, our thanks are due to *Applied Measurement in Education, Ap-*

plied Psychological Measurement, and *Psychometrika* for their permission to reproduce portions of figures in Chapter 1 and 2 and to the *Defense Manpower Data Center, Educational Testing Service,* and *Law School Admission Council* to use their data in the empirical examples in Chapter 2.

Wim J. van der Linden & Cees A. W. Glas

University of Twente

References

Binet, A. & Simon, Th. A. (1905). Méthodes nouvelles pour le diagnostic du niveau intellectual des anormaux. *l'Anneé Psychologie, 11,* 191-336.

Birnbaum, A. (1968). Some latent trait models and their use in inferring an examinee's ability. In F. M. Lord & M. R. Novick, *Statistical theories of mental test scores* (pp. 397-479). Reading, MA: Addison-Wesley.

Cronbach, L. J. & Gleser, G. C. (1965). *Psychological test and personnel decisions* (2nd Ed). Urbana: University of Illinois Press.

Dragsow, F. & Olson-Buchanan, J. B. (1999). *Innovations in computerized assessment.* Mahwah, NJ: Lawrence Erlbaum Associates.

Educational Testing Service (1994). *Computer-based tests: Can they be fair to everyone?* Princeton, NJ: Educational Testing Service.

Larkin, K. C. & Weiss, D. J. (1975). *An empirical comparison of two-stage and pyramidal adaptive ability testing* (Research Report, 75-1). Minneapolis: Psychometrics Methods Program, Department of Psychology, University of Minnesota.

Lord, F. M. (1970). Some test theory for tailored testing. In W. H. Holtzman (Ed.), *Computer assisted instruction, testing, and guidance* (pp. 139-183). New York: Harper and Row.

Lord, F. M. (1971a). Robbins-Monro procedures for tailored testing. *Educational and Psychological Measurement, 31,* 2-31.

Lord, F. M. (1971b). The self-scoring flexilevel test. *Journal of Educational Measurement, 8,* 147-151.

Mills, C. N., & Stocking, M. L. (1996). Practical issues in computerized adaptive testing. *Applied Psychological Measurement, 9,* 287-304.

Owen, R. J. (1969). *A Bayesian approach to tailored testing* (Research Report 69-92). Princeton, NJ: Educational Testing Service.

Sands, W. A., Waters, B. K., & McBride, J. R. (Eds.) (1997). *Computerized adaptive testing: From inquiry to operation.* Washington DC: American Psychological Association.

Weiss, D. J. (1973). *The stratified adaptive computerized ability test* (Research Report 73-3). Minneapolis: University of Minnesota, Department of Psychology.

Weiss, D. J. (Ed.) (1983). *New horizons in testing: Latent trait test theory and computerized adaptive testing.* New York: Academic Press.

Wainer, H. (Ed.) (1990). *Computerized adaptive testing: A primer.* Hilsdale, NJ: Lawrence Erlbaum Associates.

Chapter 1
Item Selection and Ability Estimation in Adaptive Testing

Wim J. van der Linden & Peter J. Pashley
University of Twente, The Netherlands
Law School Admission Council, USA

1. Introduction

This century has seen a progression in the refinement and use of standardized linear tests. The first administered College Board exam occurred in 1901 and the first Scholastic Aptitude Test (SAT) was given in 1926. Since then, progressively more sophisticated standardized linear tests have been developed for a multitude of assessment purposes, such as college placement, professional licensure, higher-education admissions, and tracking educational standing or progress. Standardized linear tests are now administered around the world. For example, the Test Of English as a Foreign Language (TOEFL) has been delivered in approximately 88 countries.

Seminal psychometric texts, such as those authored by Gulliksen (1950), Lord (1980), Lord and Novick (1968) and Rasch (1960), have provided increasingly sophisticated means for selecting items for linear test forms, evaluating them, and deriving ability estimates using them. While there are still some unknowns and controversies in the realm of assessment using linear test forms, tried-and-true prescriptions for quality item selection and ability estimation abound. The same cannot yet be said for adaptive testing. To the contrary, the theory and practice of item selection and ability estimation for computerized adaptive testing (CAT) is still evolving.

Why has the science of item selection and ability estimation for CAT environments lagged behind that for linear testing? First of all, the basic statistical theory underlying adapting a test to an examinee's ability was only developed relatively recently. (Lord's 1971 investigation of flexi-level testing is often credited as one of the pioneering works in this field.) But more importantly, a CAT environment involves many more delivery and measurement complexities as compared to a linear testing format.

To illustrate these differences, consider the current development and scoring of one paper-and-pencil Law School Admission Test (LSAT). To begin, newly written items are subjectively rated for difficulty and

1

W.J. van der Linden and C.A.W. Glas (eds.),
Computerized Adaptive Testing: Theory and Practice, 1–25.
© 2000 *Kluwer Academic Publishers. Printed in the Netherlands.*

placed on pretest sections by test specialists. Items that statistically survive the pretest stage are eligible for final form assembly. A preliminary test form is assembled using automated test assembly algorithms, and is then checked and typically modified by test specialists. The form is then pre-equated. Finally, the form is given operationally, to about 25,000 examinees on average, and most likely disclosed. Resulting number-right scores are then placed on a common LSAT scale by psychometricians using IRT scaling and true-score equating. The time lag between operational administrations and score reporting is usually about three weeks.

In contrast, within a CAT environment item selection and ability estimation occur in real time. As a result, computer algorithms must perform the roles of both test specialists and psychometricians. Because the test adapts to the examinee, the task of item selection and ability estimation is significantly harder. In other words, procedures are needed to solve a very complex measurement problem. These procedures must at the same time be robust enough to be relied upon with little or no human intervention.

Consider another, perhaps more subtle, difference between linear and CAT formats. As indicated above with the LSAT example, item selection and ability estimation associated with linear tests are usually conducted separately, though sometimes using similar technology, such as item response theory. Within a CAT format, item selection and ability estimation proceed hand in hand. Efficiencies in ability estimation are heavily related to the selection of appropriate items for an individual. In a circular fashion, the appropriateness of items for an individual depends in large part on the quality of interim ability estimates.

To start the exposition of these interrelated technologies, this chapter discusses what could be thought of as baseline procedures for the selection of items and the estimation of abilities within a CAT environment. In other words, basic procedures appropriate for unconstrained, unidimensional CATs that adapt to an examinee's ability level one item at a time for the purposes of efficiently obtaining an accurate ability estimate. Constrained, multidimensional, and testlet-based CATs, and CATs appropriate for mastery testing, are discussed in other chapters in this volume (Glas, Wainer & Bradlow; Segall; van der Linden). Also note that in this chapter, item parameters are assumed to have been provided, with or without significant estimation error. A discussion of item calibration is given elsewhere in this volume (Glas).

Classical procedures are covered first. Quite often these procedures were strongly influenced by a common assumption or a specific circumstance. The common assumption was that what works well for linear tests probably works well for CATs. Selecting items based on maximal information is an example of this early thinking. The specific circumstance was that these procedures were developed during a time when fast PCs were not available. For example, approximations, such as Owen's (1969) approximate Bayes procedure, were often advocated to make CATs feasible to administer with slow PCs.

More modern procedures, better suited to adaptive testing using fast PCs, are then discussed. Most of these procedures have a Bayesian flavor to them. Indeed, adaptive testing seems to naturally fit into an empirical or sequential Bayesian framework. For example, the posterior distribution of θ estimated from $k-1$ items can readily be both used to select the kth item and as the prior for the derivation of the next posterior distribution. However, problems with some Bayesian approaches are also highlighted.

When designing a CAT, a test developer must decide how initial and interim ability estimates will be calculated, how items will be selected based on those estimates, and how the final ability estimate will be derived. This chapter provides state-of-the-art alternatives that could guide the development of these core procedures for efficient and robust item selection and ability estimation.

2. Classical Procedures

2.1. NOTATION AND SOME STATISTICAL CONCEPTS

To discuss the classical procedures of ability estimation and item selection in CAT, the following notation and concepts are needed. The items in the pool are denoted by $i = 1, ..., I$, whereas the rank of the items in the adaptive test is denoted by $k = 1, ..., K$. Thus, i_k is the index of the item in the pool administered as the kth item in the test. The theory in this chapter will be presented for the case of selecting the kth item in the test. The previous $k - 1$ items form the set $S_k = \{i_i, ..., i_{k-1}\}$; they have responses that are represented by realizations of the response variables $U_{i_1} = u_{i_1}, ..., U_{i_{k-1}} = u_{i_{k-1}}$. The set of items in the pool remaining after $k - 1$ items have been selected is $R_k = \{1, ..., I\}\backslash S_{k-1}$. Item k is selected from his set.

For the sake of generality, the item pool is assumed to be calibrated by the three-parameter logistic (3PL) model. That is, the probability of a correct response on item i is given as:

$$p_i(\theta) \equiv \Pr(U_i = 1 \mid \theta) \equiv c_i + (1 - c_i)\frac{\exp[a_i(\theta - b_i)]}{1 + \exp[a_i(\theta - b_i)]}, \qquad (1)$$

where $\theta \in (-\infty, \infty)$ is the parameter representing the ability of the examinee and $b_i \in (-\infty, \infty)$, $a_i \in [0, \infty)$ and $c_i \in [0, 1]$ represent the difficulty, discriminating power, and the guessing probability on item i, respectively. One of the classical item-selection criteria discussed below is based on the three-parameter normal-ogive model:

$$p_i(\theta) \equiv c_i + (1 - c_i)\Phi[a_i(\theta - b_i)], \tag{2}$$

where Φ is the normal cumulative distribution function.

The likelihood function associated with the responses on the first $k - 1$ items is:

$$L(\theta \mid u_{i_1}...u_{i_{k-1}}) \equiv \prod_{j=1}^{k-1} \frac{\{\exp[a_{i_j}(\theta - b_{i_j})]\}^{u_{i_j}}}{1 + \exp[a_{i_j}(\theta - b_{i_j})]}. \tag{3}$$

The second-order derivative of the loglikelihood reflects the curvature of the observed likelihood function at θ relative to the scale chosen for this parameter. The negative of this derivative is generally known as the observed information measure

$$J_{u_{i_1}...u_{i_{k-1}}}(\theta) \equiv -\frac{\partial}{\partial\theta^2} \ln L(\theta \mid u_{i_1}, ..., u_{i_{k-1}}). \tag{4}$$

The expected valued of the observed information measure over the response variables is Fisher's expected information measure:

$$I_{U_{i_1}...U_{i_{k-1}}}(\theta) \equiv E[J_{U_{i_1}...U_{i_{k-1}}}(\theta)]. \tag{5}$$

For the model in (1), the expected information measure reduces to:

$$I_{U_{i_1}...U_{i_{k-1}}}(\theta) = \sum_{j=1}^{k-1} \frac{[p'_{i_j}(\theta)]^2}{p_{i_j}(\theta)[1 - p_{i_j}(\theta)]}, \tag{6}$$

with

$$p'_{i_j}(\theta) \equiv \frac{\partial}{\partial\theta} p_{i_j}(\theta). \tag{7}$$

In a Bayesian approach, a prior for the unknown value of the ability parameter, $g(\theta)$, is assumed. Together, the likelihood and prior yield the posterior distribution of θ:

$$g(\theta \mid u_{i_1}...u_{i_{k-1}}) = \frac{L(\theta \mid u_{i_1}...u_{i_{k-1}})g(\theta)}{\int L(\theta \mid u_{i_1}...u_{i_{k-1}})g(\theta)d\theta}. \tag{8}$$

Typically, this density is assumed to be uniform or, if the examinees can be taken to be exchangeable, to represent an empirical estimate of the ability distribution in the population of examinees. The population distribution is often modeled to be normal. For the response models in (1) and (2), a normal prior does not yield a (small-sample) normal posterior.

It is common practice in CAT to assume that the values of the item parameters have been estimated with enough precision to treat the estimates as the true parameter values. Under this assumption, the two-parameter logistic (2PL) and one-parameter logistic (1PL) or Rasch model, obtained from (1) by setting $c_i = 1$ and $a_i = 0$, subsequently, belong to the exponential family. Because the information measures in (4) and (5) are identical for models belonging to the exponential family (e.g., Andersen, 1980, sect. 3.3), the distinction between the two measures has only practical meaning for the 3PL model. This fact is relevant for some of the Bayesian criteria later in this chapter.

2.2. ABILITY ESTIMATORS

The ability estimator after the responses to the first $k - 1$ items is denoted as $\widehat{\theta}_{u_{i_1}, \ldots, u_{i_{k-1}}}$ (for brevity it is sometimes denoted as $\widehat{\theta}_{k-1}$). Several ability estimators have been used in CAT. In the past, the maximum-likelihood (ML) estimator was the most popular choice. The estimator is defined as the maximizer of the likelihood function in (3) over the range of possible θ values:

$$\widehat{\theta}_{u_{i_1} \ldots u_{i_{k-1}}}^{\mathrm{ML}} \equiv \arg\max_{\theta} \left\{ L(\theta \mid u_{i_1} \ldots u_{i_{k-1}}) : \theta \in (-\infty, \infty) \right\}. \qquad (9)$$

An alternative to (9) is Warm's (1989) weighted likelihood estimator (WLE). The estimator is the maximizer of the likelihood in (3) weighted by a function $w_{k-1}(\theta)$:

$$\widehat{\theta}_{u_{i_1} \ldots u_{i_{k-1}}}^{\mathrm{WLE}} \equiv \arg\max_{\theta} \left\{ w_{k-1}(\theta) L(\theta \mid u_{i_1} \ldots u_{i_{k-1}}) : \theta \in (-\infty, \infty) \right\}, \qquad (10)$$

The weight function is defined to satisfy:

$$\frac{\partial w_{k-1}(\theta)}{\partial \theta^2} \equiv \frac{H_{k-1}(\theta)}{2 I_{k-1}(\theta)}, \qquad (11)$$

where

$$H_{k-1}(\theta) \equiv \sum_{j=1}^{k-1} \frac{[p'_{i_j}(\theta)][p''_{i_j}(\theta)]}{p_{i_j}(\theta)[1 - p_{i_j}(\theta)]}, \qquad (12)$$

$$p''_{i_j}(\theta) \equiv \frac{\partial^2 p_{i_j}(\theta)}{\partial \theta^2}, \qquad (13)$$

and $I_{k-1}(\theta) \equiv I_{U_{i_1} \ldots U_{i_{k-1}}}(\theta)$ is defined in (5). For a linear test, the WLE is attractive because of its unbiasedness to order n^{-1}.

In a more Bayesian fashion, a point estimator of the value of θ can be based on its posterior distribution in (8). Posterior-based estimators used in adaptive testing are the Bayes Modal (BM) or Maximum A Posteriori (MAP) estimator and the Expected A Posteriori (EAP) estimator. The former is defined as the maximizer of the posterior of θ:

$$\widehat{\theta}^{\text{MAP}}_{u_{i_1}...u_{i_{k-1}}} \equiv \arg\max_{\theta} \left\{ g(\theta \mid u_{i_1}...u_{i_{k-1}}) : \theta \in (-\infty, \infty) \right\}; \qquad (14)$$

the latter as its expected value:

$$\widehat{\theta}^{\text{EAP}}_{u_{i_1}...u_{i_{k-1}}} \equiv \int \theta g(\theta \mid u_{i_1}...u_{i_{k-1}}) d\theta. \qquad (15)$$

The MAP estimator was introduced in IRT in Lord (1986) and Mislevy (1986). Use of the EAP estimator in adaptive testing is discussed extensively in Bock and Mislevy (1988).

A more principled Bayesian approach is to refrain from using a point estimate and consider the full posterior as the estimator of the ability of the examinee. This estimator does not only show the most plausible value of θ but also the plausibility of any other value. If the point estimators in (14) and (15) are used, it is common to summarize this uncertainty in the form of the variance of the posterior distribution of θ:

$$\text{Var}(\theta \mid u_{i_1}...u_{i_{k-1}}) \equiv \int [\theta - E(\theta \mid u_{i_1}...u_{i_{k-1}})]^2 g(\theta \mid u_{i_1}...u_{i_{k-1}}) d\theta. \quad (16)$$

For the 3PL model, a unique maximum for the likelihood function in (3) does not always exist (Samejima, 1973). Also, for response patterns with all items correct or all incorrect, no finite ML estimates exist. However, for linear tests, the ML estimator is consistent and asymptotically efficient. For adaptive tests, both the large- and small-sample properties of the ML estimator depend on such factors as the distribution of the items in the pool and the item-selection criterion used. Large-sample theory for the ML estimator for an infinite item pool and one of the popular item-selection criteria will be reviewed later in this chapter.

For a uniform prior, the posterior distribution in (8) becomes proportional to the likelihood function over the support of the prior, and the maximizers in (9) and (14) are equal. Hence, for this case, the MAP estimator shares all the above properties of the ML estimator. For nonuniform prior distributions, the small-sample properties of the MAP estimator depend not only on the likelihood but also on the shape of the prior distribution. Depending on the choice of prior distribution, the posterior distribution may be multimodal. If so, unless precaution is taken, MAP estimation may result in a local maximum.

For a proper prior distribution, the EAP estimator always exists. Also, unlike the previous estimators, it is easy to calculate. No iterative procedures are required; only one round of numerical integration suffices. This feature use to be important in the early days of computerized adaptive testing but has become less critical now that cheap, plentiful computing power is available.

2.3. CHOICE OF ESTIMATOR

The practice of ability estimation in linear testing has been molded by the availability of a popular computer program (BILOG; see Mislevy & Bock, 1983). In adaptive testing, such a *de facto* standard is missing. Most testing programs run their operations using their own software. In developing their software, most of them have taken an eclectic approach to ability estimation. The reason for this practice is that, unlike linear testing, in adaptive testing three different stages of ability estimation can be distinguished: (1) ability estimation to start the item-selection procedure; (2) ability estimation during the test to adapt the selection of the items to the examinee's ability; and (3) ability estimation at the end of the test to report a score to the examinee. Each of these stages involves its own requirements and problems.

2.3.1. *Initial Ability Estimation*
As already noted, the method of ML estimation does not produce finite estimates for response patterns with all items correct or all incorrect. Because the probability of such patterns is substantial, ML estimation can not be used to guide item selection at the beginning of the test. Several measures have been proposed to resolve this problem. First, it has been proposed to fix the ability estimate at a small or large value until finite estimates are obtained. Second, ability estimation is sometimes postponed until a larger set of items has been answered. Third, the problem has been an important motive to use a Bayesian methods such as the EAP estimator. Fourth, if empirical information on background variables for the examinees is available, initial ability estimates can be inferred from this information. A method for calculating these estimates is discussed later in this chapter.

None of solutions is entirely satisfactory, though. The first two solutions involve an arbitrary choice of ability values and items, respectively. The third solution involves the choice of a prior distribution which, in the absence of response data, completely dominates the choice of the first item. If the prior distribution is located away from the true ability of the examinee, it becomes counterproductive and can easily produce a longer initial string of correct or incorrect responses than necessary. (Bayesian methods are often said to produce a smaller posterior variance after each new datum, but this statement is not true; see Gelman, Carlin, Stern, & Rubin, 1995, sect. 2.2. Initial ability estimation in adaptive testing with a prior at the wrong location is a good counterexample.) As for the fourth solution, though there are no technical objections to using empirical priors (see the discussion later in this chapter), examinees may nevertheless try blaming their scores on the test to the easy or hard items they got at the beginning due to their background characteristics. Fortunately, the problem of inferring an initial ability estimate is only acute for short tests, for example, 10-item tests in a battery. For longer tests, of more than 20-30 items, say, the ability estimator generally does have enough time to recover from a bad start.

2.3.2. *Interim Ability Estimation*

Ideally, the next estimates should converge quickly to the true ability value. In principle, any combination of ability estimator and item-selection criterion that does this job for the item pool could be used. Though some of these combinations look more "natural" than others (e.g., ML estimation with maximum-information item selection and Bayesian estimation with item selection based on the posterior distribution), practice of CAT has not been impressed by this argument and has often taken a more eclectic approach. For example, a popular choice has been the EAP estimator in combination with maximum-information item selection.

Because ability estimation and item selection have to take place in real time, their numerical aspects use to be important. For example, in the first practical attempts at adaptive testing, Owen's item selection procedure was an important practical alternative to a fully Bayesian procedure because it did not involve any time-consuming, iterative calculations. However, for modern powerful PCs, computational limitations to CAT, for all practical purposes, no longer exist.

2.3.3. *Final Ability Estimation*

Though final ability estimates should have optimal statistical properties, their primary function is no longer to guide item selection, but to score the test and provide the examinee with a meaningful summary of his or her performance. For this reason, final estimates are sometimes transformed to an equated number-correct score on a released linear version of the test. The equations typically used for this procedure are the test characteristic function (e.g., Lord, 1980, sect. 4.4) and the equipercentile transformation that equates the ability estimates on the CAT into number-correct scores on a paper-and-pencil version of the test (Segall, 1997).The former is known once the items are calibrated; the latter has to be estimated in a separate empirical study. To avoid the necessity of explaining complicated ML scoring methods to examinees, Stocking (1966) proposed a modification to the likelihood equation such that its solution is a monotonic function of the number-correct score. Finally, adaptive testing with observed number-correct scoring is also possible through appropriate constraints on the item selection (van der Linden, this volume).

The answer to the question of what method of ability estimation is best is intricately related to other aspects of the CAT. First of all, the choice of item-selection criterion is critical. Other aspects that have an impact on ability estimates are the composition of the item pool, whether or not the estimation procedure uses examinee background information, the choice of the method to control the exposure rates of items, and the presence of content constraints on item selection. The issue will be returned to at the end of this chapter where some of these aspects have been discussed in more detail.

2.4. Classical Item-Selection Criteria

2.4.1. *Maximum-Information Criterion*

Birnbaum (1968) introduced the test information function as the main criterion for linear test assembly. The test information function is the expected information measure in (5) defined as a function of the ability parameter. Birnbaum's motivation for this function was the fact that, for increasing test length, the variance of the ML estimator is known to converge to the reciprocal of (5). In addition, the measure in (5) is easy to calculate and additive in the items.

Though no asymptotic motivation existed for the use of (5) as item-selection criterion in CAT, the maximum-information criterion was immediately adopted as a popular choice. The criterion selects the kth item to maximize (5) at $\theta = \widehat{\theta}_{u_{i_1},...,u_{i_{k-1}}}$. Formally, it can be presented as

$$i_k \equiv \arg\max_j \left\{ I_{U_1, ..., U_{k-1}, U_j} \left(\widehat{\theta}_{u_{i_1},...,u_{i_{k-1}}} \right) : j \in R_k \right\}. \tag{17}$$

Because of the additivity of the information function, the criterion boils down to

$$i_k \equiv \arg\max_j \left\{ I_{U_j} \left(\widehat{\theta}_{u_{i_1},...,u_{i_{k-1}}} \right) : j \in R_k \right\}. \tag{18}$$

Observe that, though the ML estimator is often advocated as the natural choice, the choice of estimator of θ in (18) is open. Also, the maximum-information criterion is often used in the form of a previously calculated information table for a selection of θ values (for an example, see Thissen & Mislevy, 1990, Table 5.2).

Recently, an asymptotic motivation for the use of (18) in adaptive testing has been provided by Chang and Ying (in press). These authors show that for this criterion the ML estimator of θ converges to the true value with a sampling variance approaching the reciprocal of (5). The results hold only for an infinite item pool with values for the discrimination parameter in the item pool bounded away from 0 and ∞, and values for the guessing parameter bounded away from 1. Also, for the 3PL model, a slight modification of the likelihood equation is necessary to prevent from multiple roots. Because these conditions are mild, the results are believed to provide a useful approximation to adaptive testing from a well-designed item pool. As shown in Warm (1989), the WLE in (10), outperforms the ML estimator in adaptive testing. The results by Chang and Ying are therefore expected to hold also for the combination of (18) with the WLE.

2.4.2. Owen's Approximate Bayes Procedure

Owen (1969; see also 1975) was the first to introduce a Bayesian approach to adaptive testing. His proposal followed the format of an empirical or sequential Bayes procedure where at each stage the previous posterior distribution of the unknown parameter serves as the new prior.

Owen's proposal was formulated for the three-parameter normal-ogive model in (2) rather than its logistic counterpart. His criterion was to chose the kth item such that

$$\left| b_{i_k} - E(\theta \mid u_{i_1}...u_{i_{k-1}}) \right| < \delta \tag{19}$$

for a small value of $\delta \geq 0$, where $E(\theta \mid u_{i_1}...u_{i_{k-1}})$ is the EAP estimator defined in (15). After the item is administered, the likelihood is updated and combined with the previous posterior to calculate a new posterior. The same criterion is then applied to select a new item. The procedure is repeated until the posterior variance in (16) is below the level of uncertainty on θ the test administrator is willing to tolerate. The last posterior mean is the ability estimate reported to the examinee.

In Owen's procedure, the selection of the first item is guided by the choice of a normal density for the prior, $g(\theta)$. However, the class of normal priors is not the conjugate for the normal-ogive model in (2); that is, they do not yield a normal posterior distribution. Because it was impossible to calculate the true posterior in real time, Owen provided closed-form approximations to the true posterior mean and variance

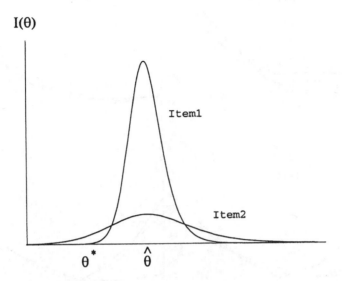

Figure 1. Attenuation paradox in item selection in CAT

and suggested to used these with a normal posterior. The approximation for the mean was motivated by its convergence to the true value
of θ in mean square for $k \to \infty$ (Owen, 1975, Theorem 2).

Note that in (19), b_i is the only item parameter that determines
the selection of the kth item. No further attempt is made to optimize
item selection. However, Owen did made a reference to the criterion of
minimal preposterior risk (see below), but refrained from pursuing this
option because of its computational complexity.

3. Modern Procedures

Ideally, item-selection criteria in adaptive testing should allow for two
different types of possible errors: (1) errors in the first ability estimates
in the tests and (2) errors in the item parameter estimates.

Because the errors in the first ability estimates in the test are generally large, item-selection criteria that ignore them tend to favor items
with optimal measurement properties at the wrong value of θ. This
problem, which has been known as the attenuation paradox in test
theory for a long time (Lord & Novick, 1968, sect. 16.5), has been
largely ignored in adaptive testing. For the maximum-information criterion in (18), the "paradox" is illustrated in Figure 1, where the item
that performs best at the current ability estimate does worse at the

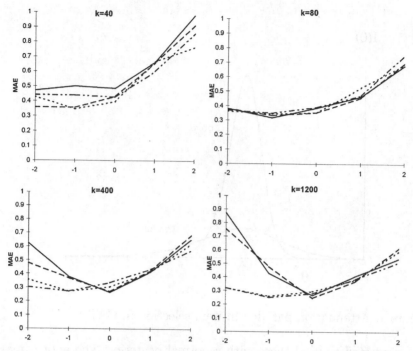

Figure 2. Mean absolute error (MAE) in ability estimation from item pools with k=40, 80, 400, and 1200 items (size of calibration samples: 250: solid; 500: dashed; 1200: dotted; 2500: dashed-dotted)

true ability value. The classical solution for a linear test was to maintain high values for the discrimination parameter, but space the values for the difficulty parameter (Birnbaum, 1968, sect. 20.5). This solution goes against the nature of adaptive testing.

Ignoring errors in the estimates of the item parameter values is a strategy without serious consequences if the calibration sample is large. However, the first large-scale CAT applications showed that to maintain item pool integrity, the pools had to be replaced much more often than anticipated. Because the costs of replacement are high, the current trend is to minimize the size of the calibration sample. A potential problem for CAT from a pool of items with errors in their parameter values, however, is capitalization on chance. Because the items are selected to be *optimal* according to the item-selection criterion, the test will tend to have both items with extreme true values and large estimation errors in their item parameters. Figure 2 illustrates the effect of capitalization on chance on the ability estimates for an adaptive test of 20 items from item pools of varying sizes. Note that, unlike for a pool with true values for the item parameters, for the smaller calibration samples the error in the ability estimates at the lower end scale goes up if the item pool becomes larger. This counterintuitive result is due only to capitalization on chance (for other examples of this phenomenon, see van der Linden and Glas, 2000).

Recently, new item-selection criteria have been introduced to fix the above problems. Though not been tested in practice yet, these criteria have shown favorable statistical properties in extended computer simulation studies. Also, with the possible exception of one criterion, they can be used in real time on the current PCs.

3.1. Maximum Global-Information Criterion

To deal with large estimation error in the beginning of the test, Chang and Ying (1996) suggested to replace Fisher's information in (17) by a measure based on Kullback-Leibler information. Generally, Kullback-Leibler information measures the "distance" between two likelihoods. The larger the Kullback-Leibler information, the easier it is to discriminate between two likelihoods, or equivalently, between the values of the parameters that index them (Lehmann & Casella, 1998, sect. 1.7).

For the response model in (1), Kullback-Leibler information in the response variable associated with the kth item in the test with respect to true ability value (θ_0) of the examinee and his/her current ability estimate $\left(\widehat{\theta}_{k-1}\right)$ is

$$K_{i_k}\left(\widehat{\theta}_{k-1}, \theta_0\right) \equiv E[\log \frac{L(\theta_0 \mid U_{i_k})}{L(\widehat{\theta}_{k-1} \mid U_{i_k})}]. \qquad (20)$$

The expectation in (20) is taken over response variable U_{i_k}. The measure can therefore be calculated as:

$$\begin{aligned} K_{i_k}\left(\widehat{\theta}_{k-1}, \theta_0\right) \;=\; & p_{i_k}(\theta_0) \log \frac{p_{i_k}(\theta_0)}{p_{i_k}(\widehat{\theta}_{k-1})} \\ & +[1 - p_{i_k}(\theta_0)] \log \frac{1 - p_{i_k}(\theta_0)}{1 - p_{i_k}(\widehat{\theta}_{k-1})}. \end{aligned} \qquad (21)$$

Because of conditional independence between the response variables, information in the variables for the first k items in the test can be written as:

$$K_k\left(\widehat{\theta}_{k-1}, \theta_0\right) \equiv E[\log \frac{L(\theta_0 \mid U_{i_1}, ..., U_{i_k})}{L(\widehat{\theta}_{k-1} \mid U_{i_1}, ..., U_{i_k})}] = \sum_{h=1}^{k} K_{i_h}\left(\widehat{\theta}_{k-1}, \theta_0\right). \quad (22)$$

Kullback-Leibler information tells us how well the response variables discriminate between the current ability estimate, $\widehat{\theta}_{k-1}$, and the true ability value, θ_0. Because the true value θ_0 is unknown, Chang and

Ying propose to replace (20) by its integral over an interval about the current ability estimate, $[\widehat{\theta}_{k-1} - \delta_k, \widehat{\theta}_{k-1} + \delta_k]$, with δ_k chosen to be a decreasing function of the rank number of the item in the adaptive test. The kth item in the test is selected according to:

$$i_k \equiv \arg\max_j \left\{ \int_{\widehat{\theta}_{k-1}-\delta_k}^{\widehat{\theta}_{k-1}+\delta_k} K_j(\widehat{\theta}_{k-1}, \theta)d\theta : j \in R_k \right\}. \qquad (23)$$

Evaluation of the criterion will be postponed until all further criteria in this section have been reviewed.

3.2. LIKELIHOOD-WEIGHTED INFORMATION CRITERION

Rather than integrating the unknown parameter θ out, as in (23), the integral could have been taken over a measure for the plausibility of all possible values of θ. This idea has been advocated by Veerkamp and Berger (1997). Though they presented it for the Fisher information measure, it can easily be extended to the Kullback-Leibler measure.

In a frequentistic framework, the likelihood function associated with the responses $U_{i_1=u_{i_1}}, ..., U_{i_{k-1}} = u_{i_{k-1}}$ expresses the plausibility of the various values of θ given the data. Veerkamp and Berger's propose to weigh Fisher's information with the likelihood function and select the kth item according to

$$i_k \equiv \arg\max_j \left\{ \int_{-\infty}^{\infty} L(\theta \mid u_{i_1}, ..., u_{i_{k-1}}) I_{i_k}(\theta)d\theta : j \in R_k \right\}. \qquad (24)$$

If maximum-likelihood estimation of ability is used, the criterion in (24) places most weight on θ values close to the current ability estimate. In the beginning of the test, the likelihood function is flat and values away from $\widehat{\theta}_{k-1}$ received substantial weight. Towards the end of the test the likelihood function tends to become more peaked, and all the weight will go to values close to $\widehat{\theta}_{k-1}$.

Veerkamp and Berger (1997) also specify an interval information criterion that, like (23), assumes integration over a finite interval of θ values about the current ability estimate. However, rather than defining the size of the interval as a function of a new parameter δ_k, they suggest using a confidence interval for θ. The same suggestion would be possible for the criterion in (23).

3.3. Fully Bayesian Criteria

All Bayesian criteria for item selection involve some form of weighting based on the posterior distribution of θ. Because the posterior distribution is a combination of the likelihood function and a prior distribution, the basic difference with the previous criterion is the assumption of a prior distribution. The question of how to choose a subjective prior for θ will not be addressed in this chapter. Instead, the next section discusses how an empirical prior can be estimated from data on background variables for the examinee. The purpose of this section is only to review several Bayesian criteria for item selection proposed in van der Linden (1998).

Analogous to (24), a posterior-weighted information criterion can be defined:

$$i_k \equiv \arg\max_j \left\{ \int I_{U_j}(\theta)g(\theta \mid u_{i_1}, ..., u_{i_{k-1}})d\theta : j \in R_k \right\}. \qquad (25)$$

The criterion puts more weight on item information near the location of the posterior distribution. If the MAP or EAP estimators in (14) and (15) are used, the criterion favors items with the mass of their information at these estimates. However, in doing so, the criterion uses the specific shape of the posterior distribution to discriminate between candidate items.

Note that the criterion in (25) is still based on Fisher's expected information in (5). Though the distinction between expected and observed information makes practical sense only for the 3PL model, a more Bayesian choice would be to use observed information in (4). Also, note that it is possible to combine (25) with the Kullback-Leibler measure (Chang & Ying, 1996).

The subsequent criteria are all based on preposterior analysis. They predict the responses on the remaining items in the pool, $i \in R_k$, after $k - 1$ items have been administered and then choose the next item according to the updates of a posterior quantity for these responses. A key element in this analysis is the predictive posterior distribution for the response on item i, which has probability function

$$p_i(u_i \mid u_{i_1}, ..., u_{i_{k-1}}) = \int p_i(u_i \mid \theta)g(\theta \mid u_{i_1}, ..., u_{i_{k-1}})d\theta. \qquad (26)$$

Suppose item $i \in R_k$ would be selected. The examinee responds correctly to this item with probability $p_i(1 \mid u_{i_1}, ..., u_{i_{k-1}})$. A correct response would lead to an update of the following quantities: (1) the full posterior distribution of θ; (2) the point estimate of the ability value of the examinee, $\widehat{\theta}_k$; (3) observed information at $\widehat{\theta}_k$; and (4) the posterior variance of θ. An incorrect response has probability $p_i(0 \mid u_{i_1}, ..., u_{i_{k-1}})$ and would to a similar set of updates. It should be noticed that observed information is updated not only for the response on item i. Because of the update to point estimate $\widehat{\theta}_k$, the information measure must be re-evaluated at this estimate for all previous response as well.

The first item-selection criterion based on preposterior analysis is the maximum expected information criterion. The criterion maximizes observed information over the predicted responses on the kth item. Formally, it can be represented as

$$i_k \equiv \arg\max_j \Big\{ p_j(0 \mid u_{i_1}, ..., u_{i_{k-1}}) J_{u_{i_1},...,u_{i_{k-1}},U_j=0}\big(\widehat{\theta}_{u_{i_1},...,u_{i_{k-1}},U_j=0}\big)$$
$$+ p_j(1 \mid u_{i_1}, ..., u_{i_{k-1}}) J_{u_{i_1},...,u_{i_{k-1}},U_j=1}\big(\widehat{\theta}_{u_{i_1},...,u_{i_{k-1}},U_j=1}\big)$$
$$: j \in R_k \Big\}. \tag{27}$$

If in (27) observed information is replaced by the posterior variance, the minimum expected posterior variance criterion is obtained:

$$i_k \equiv \arg\min_j \big\{ p_j(0 \mid u_{i_1}, ..., u_{i_{k-1}}) \mathrm{Var}(\theta \mid u_{i_1}, ..., u_{i_{k-1}}, U_j = 0)$$
$$+ p_j(1 \mid u_{i_1}, ..., u_{i_{k-1}}) \mathrm{Var}(\theta \mid u_{i_1}, ..., u_{i_{k-1}}, U_j = 1)$$
$$: j \in R_k \big\}. \tag{28}$$

The expression in (28) is known to be the preposterior risk under a quadratic loss function for the estimator. Owen (1975) referred to this criterion as a numerically more complicated alternative to his criterion in (19).

It is possible to combine the best elements of the ideas underlying the criteria in (25) and (28) by first weighting observed information using the posterior distribution of θ and then taking the expectation over the predicted responses. The new criterion is

$$i_k \equiv \arg\max_j \big\{ p_j(0 \mid u_{i_1}, ..., u_{i_{k-1}}) \tag{29}$$

$$\cdot \int J_{u_{i_1},...,u_{i_{k-1}},U_j=0}(\theta) g(\theta \mid u_{i_1}, ..., u_{i_{k-1}}, U_j = 0) d\theta$$

$$\cdot \int J_{u_{i_1},...,u_{i_{k-1}},U_j=1}(\theta) g(\theta \mid u_{i_1}, ..., u_{i_{k-1}}, U_j = 1) d\theta : j \in R_k \big\}.$$

It is also possible to generalize the criteria in (26)-(28) to a larger span of prediction. For example, if the responses are predicted for the next two items, the generalization involves replacing the posterior predictive probability function in the above criteria by

$$p_i(u_i \mid u_{i_1}, ..., u_{i_{k-1}})p_h(u_h \mid u_{i_1}, ..., u_{i_{k-1}}, u_i), \qquad (30)$$

modifying the other posterior updates accordingly. Though optimization is over pairs of candidates for items k and $k+1$, better adaptation to the examinee's ability is obtained is if only item k is administered and the other item returned to the pool, whereupon the procedure is repeated. Combinatorial problems inherent to application of the procedure with large item pools can be avoided by using a trimmed version of the pool, with unlikely candidates for item k or $k+1$ left out.

3.4. BAYESIAN CRITERIA WITH AN EMPIRICAL PRIOR

As indicated earlier, an informative prior located at the true value of θ is essential when choosing a Bayesian ability estimator. For a large variety of item selection-criteria, an informative prior would not only yield finite initial ability estimates, but also improve item selection and speed up convergence of the estimates during the test. If background information on the examinee exist, for example, on previous achievements or performances on a related test, an obvious idea is to base the prior on it. Technically, no objections against this idea exist; in adaptive testing, the selection criterion is ignorable if the interest is only in an ML estimate of θ or in a Bayesian estimator (Mislevy & Wu, 1988). Nevertheless, if policy considerations preclude the use of test scores based on prior information, a practical strategy may be to base item selection on updates of the empirical prior, but to calculate the final ability estimate only from the likelihood.

Procedures for adaptive testing with a prior distribution regressed on background variables for the 2PL model are described in van der Linden (1999). Let the predictor variables be denoted by X_p, $p = 0, ..., P$. The regression of θ on the predictor variables can be modeled as

$$\theta = \beta_0 + \beta_1 X_1 + ... + \beta_P X_P + \varepsilon, \qquad (31)$$

with

$$\varepsilon \sim N(0, \sigma^2). \qquad (32)$$

Substitution of (30) into the response model in (1) gives

$$p_i(\theta) = \frac{\exp[a_i(\beta_0 + \beta_1 X_1 + ... + \beta_P X_P + \varepsilon - b_i)]}{1 + \exp[a_i(\beta_0 + \beta_1 X_1 + ... + \beta_P X_P + \varepsilon - b_i)]}. \qquad (33)$$

For known values for the item parameters, the model in (31) is a logistic regression model with examinee scores on ε missing. The values of the parameters β_p and σ can be estimated from data using the EM algorithm. The estimation procedure boils down to iteratively solving two recursive relationships given in van der Linden (1999, Eqs. 16-17). These equations are easily solved for a set of pretest data. They also allow for an easy periodical update of the parameter estimates from response data when the adaptive test is operational.

If the item selection is based on point estimates of ability, the regressed value of θ on the predictor variables,

$$\widehat{\theta}_0 = \beta_0 + \beta_1 x_1 + ... \beta_P x_P, \tag{34}$$

can be used as the prior ability estimate for which the initial item is selected. If the items are selected using a full prior distribution for θ, the choice of prior following (32)-(33) is

$$g(\theta) \equiv N(\widehat{\theta}_0, \sigma). \tag{35}$$

Empirical initialization of adaptive tests does have more than just statistical advantages. If initialization is based on a fixed item, the first items in the test are always chosen from the same subset in the pool and become quickly overexposed (Stocking & Lewis, this volume). On the other hand, empirical initialization of the test entails a variable entry point to the pool and hence a more even exposure of the items.

3.5. BAYESIAN CRITERIA WITH RANDOM ITEM PARAMETERS

If the calibration sample is small, errors in the estimates of the values of the item parameters should not be ignored but rather dealt with explicitly when estimating the value of θ. A Bayesian approach would not fix the item parameters at point estimates but leave them random, using their posterior distribution given previous response data in the ability estimation procedure. An empirical Bayes procedure for ability estimation from linear test data based on this framework was presented in Tsutakawa and Johnson (1990). Their procedure can easily be modified for application in CAT.

The modification is as follows: Let $\mathbf{y}$ be the matrix with response data from all previous examinees. For brevity, the parameters (a_i, b_i, c_i) for the items in the pool are collected into a vector $\boldsymbol{\xi}$. Suppose a new examinee has answered $k-1$ items. Given a prior for $\boldsymbol{\xi}$, the derivation of the posterior distribution $g(\boldsymbol{\xi} \mid u_{i_1}, ..., u_{i_{k-1}}, \mathbf{y})$ is standard.

Using the assumptions in Tsutakawa and Johnson (1990), the posterior distribution of θ after item $k-1$ can be updated as:

$$g(\theta \mid u_{i_1}, ..., u_{i_{k-1}}, \mathbf{y}) = \frac{g(\theta) \int p(u_{i_{k-1}} \mid \theta, \boldsymbol{\xi}) g(\boldsymbol{\xi} \mid u_{i_1}, ..., u_{i_{k-2}}, \mathbf{y}) d\boldsymbol{\xi}}{p(u_{i_{k-1}} \mid u_{i_1}, ..., u_{i_{k-2}}, \mathbf{y})}. \tag{36}$$

Key in this expression is the replacement of the likelihood for response u_i by one averaged over the posterior distribution for the item parameters given all previous data. Such averaging is the Bayesian way of accounting for posterior uncertainty in unknown parameters. Given the posterior of θ, the posterior predictive probability function for the response on item i can be derived as:

$$p_i(u_i \mid u_{i_1}, ..., u_{i_{k-1}}, \mathbf{y}) \equiv \int p_i(u_i \mid \theta) g(\theta \mid u_{i_1}, ..., u_{i_{k-1}}, \mathbf{y}) d\theta. \qquad (37)$$

Once (37) has been calculated, it can be used in one of the criteria in (25) or (27)-(29).

In spite of our current computational power, direct update of the posterior distribution of the item parameters, $g(\boldsymbol{\xi} \mid u_{i_1}, ..., u_{i_{k-1}}, \mathbf{y})$, is prohibitive, due to the evaluation of complex multiple integrals. However, in practice, it make sense to update the posterior only periodically, each time first screening the set of interim response patterns for possible aberrant behavior by the examinees as well as compromise of the items. When testing the current examinee, the posterior can then be fixed at the posterior $g(\boldsymbol{\xi} \mid \mathbf{y})$ obtained after the last update. The resulting expression in (36)-(37) can easily be calculated in real time by an appropriate method of numerical integration.

Simplifying assumptions for the update of $g(\boldsymbol{\xi} \mid \mathbf{y})$ are given in Tsutakawa and Johnson (1990). Also, numerical approximations based on Markov chain Monte Carlo approaches are now generally available. Finally, the computational burden is gradually decreased if a mixed approach is followed, fixing item parameters at their posterior means as soon as their posterior uncertainties become sufficiently small.

3.6. MISCELLANEOUS CRITERIA

The item-selection criteria presented thus far were statistically motivated. An item-selection procedure that addresses both a statistical and a more practical goal is the method of multistage α-stratified adaptive testing proposed in Chang and Ying (1999).

This method was introduced primarily to reduce the effect of ability estimation error on item selection. As illustrated in Figure 1, if the errors are large, an item with a lower discrimination parameter value is likely to be more efficient than one with a higher value. These authors therefore propose to stratify the pool according to the values of the discrimination parameter for the items and restrict item selection to strata with increasing values during the test. In each stratum, items are selected according to the criterion of minimum distance between the value of the difficulty parameter and the current ability estimate.

The procedure also provides a remedy to the problem of uneven item exposure in CAT. If the sizes of the strata in the pool and the numbers of items per stratum in the test are chosen appropriately, the items in the pool can have a fairly uniform probability of being selected.

To prevent from capitalization on chance (see Figure 2), it may be effective to cross-validate item parameter estimation during the adaptive test. A practical way to cross-validate is to split the calibration sample into two parts and estimate the item parameters for each part separately. One set of estimates is used to select the item, the other to update the ability estimates after the examinee has taken the item. Item selection then still tends to capitalize on the errors in the estimates in the first set, but the effects on ability estimation are neutralized by using the second set of estimates. Conditions under which this neutralization offsets the loss in precision due to calibration from a smaller sample are studied in van der Linden and Glas (submitted).

A final suggestion for item selection in adaptive testing is offered in Wainer, Lewis, Kaplan and Brasswell (1992). As selection criterion they use the posterior variance between the subgroups that scored the item in the pretest correctly and incorrectly . Results from an empirical study of this criterion are given in Schnipke and Green (1995).

3.7. EVALUATION OF ITEM-SELECTION CRITERIA AND ABILITY ESTIMATORS

The question of which combination of item-selection criterion and ability estimation is best is too complicated for analytic treatment. Current statistical theory only provides us with asymptotic conclusions. A well-known result from Bayesian statistics is that for $k \to \infty$, the posterior distribution $g(\theta \mid u_{i_1}, ..., u_{i_{k-1}})$ converges to degeneration at the true value of θ. Hence, it can be concluded that all posterior-based ability estimation and item selection procedures reviewed in this chapter produce identical asymptotic results. Also, the result by Chang and Ying (in press) referred to earlier shows that for maximum-information item selection, the ML estimator converges to the true value of θ as well. The WLE in (10) is expected to show the same behavior. Because the information measure and the posterior variance are known to be asymptotic reciprocals, asymptotic equality is extended to the likelihood-based ability estimation and item selection procedures in this chapter as well.

However, particularly in CAT, not asymptotic but small-sample comparison of estimators and criteria should be of interest. For such comparisons we have to resort to simulation studies.

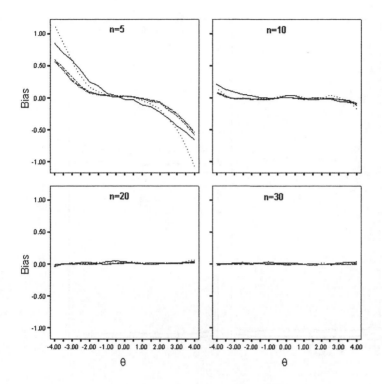

Figure 3. Bias functions for five item-selection criteria after n=5, 10, 20, 30 items (Maximum-Information with MLE: solid; Maximum-Posterior Weighted Information: dotted; Maximum Expected Information: dashed-dotted; Maximum Expected Posterior Variance: dashed ;Maximum Expected Posterior Weighted Information: finely dotted)

Relevant studies have been reported on in Chang and Ying (1999), van der Linden (1998), Veerkamp and Berger (1997), Wang, Hanson & Lau (1999), Wang and Vispoel (1998), Weiss ((1982), Weiss and McBride (1984) and Warm (1989), among others. Sample results for the bias and mean squared error (MSE) functions for five different combinations of ability estimators and item-selection criteria are given in Figure 3 and 4. All five combinations show the same slight inward bias for $n=10$, which disappears completely for $n=20$ and 30. Note that the bias for the ML estimators in Figure 3 has a direction opposite to the one in the estimator for a linear test (e.g., Warm, 1989). This result is due to a feedback mechanism created by the combination of the contributions of the items to the bias in the estimator and the maximum-information criterion (van der Linden, 1998).

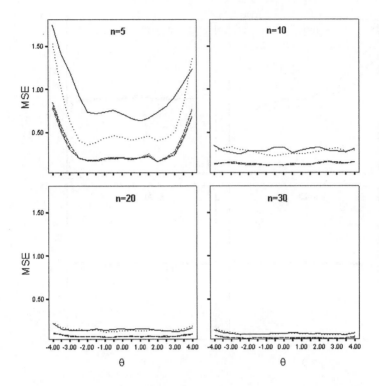

Figure 4. MSE functions for five item-selection criteria after n=5, 10, 20, 30 items (Maximum-Information with MLE: solid; Maximum-Posterior Weighted Information: dotted; Maximum Expected Information: dashed-dotted; Maximum Expected Posterior Variance: dashed ;Maximum Expected Posterior Weighted Information: finely dotted)

The MSE functions for the adaptive tests in Figure 4 are flat for all three test lengths, whereas the MSE functions for linear tests are typically U-shaped with the dip at the θ values where the items are located. Flat MSE functions are the major benefit of adaptive testing. In Figure 4, the best MSE functions were obtained for the criteria in (27)-(29). Note that each of these criteria was based on a preposterior analysis. Hence, a critical element in the success of an item-selection criterion seems to be its use of posterior predictive probability functions to predict the item responses on the remaining items in the pool. As revealed by the comparison between the MSE functions for the maximum-information and maximum posterior-weighted information criterion in Figure 4, simply using the posterior distribution of θ appears to have little effect.

Weiss (1982) reported basically the same results for the maximum-information criterion and Owen's criterion in (19). In Wang and Vispoel's (1998) study, the behavior of the ML, EAP and MAP estimators in combination with the maximum-information criterion were compared with Owen's criterion. For a 30-item test from a real item pool the three Bayesian procedures behaved comparably, whereas the ML estimator produced a worse standard error but a better bias function. Wang, Hanson and Lau (1999) report several conclusions for modifications of the ML and Bayesian estimators intended to remove their bias. A sobering result was given by Sympson, Weiss and Ree (see Weiss, 1982, p. 478) who, in a real-life application of the maximum-information and Owen's selection criterion, found that approximately 85% of the items selected by the two criteria were the same. However, the result may largely be due to the choice of a common initial item for all examinees.

4. Concluding Remarks

As noted in the introduction section of this chapter, methods for item selection and ability estimation within a CAT environment are not yet as refined as those currently employed for linear testing. Hopefully, though, this chapter has provided evidence that significant progress has been made in this regard. Modern methods have begun to emerge that directly address the peculiarities of adaptive testing, rather than relying on simpler solutions previously applied to linear testing situations. Recent analytical studies that have provided theoretical frameworks within which different procedures may be evaluated have been especially good to see. In addition, the constraints on timely numerical computations imposed by older and slower PCs have all but disappeared. For example, while an examinee is answering an item, modern CAT programs often select two items for delivery next—one if the examinee answers the current item correctly, and another if he or she answers it incorrectly.

The studies discussed in this chapter only relate to a small part of the conditions that may prevail in an adaptive testing program. Clearly, programs can differ in the type of item-selection criterion and ability estimator they use. However, they can also vary in numerous other ways, such as the length of the test and whether the length is fixed or variable; the size and composition of the item pools; the availability of relevant background information on the examinees; the size and composition of the calibration samples; the ability to update item parameter estimates using operational test data; the use of measures to control item exposure rates; and the content constraints imposed on the item selection process. Important trade-offs exist among several of these factors, which also interact in their effect on the statistical behavior of the final ability estimates.

Given the complexities of a CAT environment and the variety of approaches (some untested) that are available, how should one proceed? One method would be to delineate all the relevant factors that could be investigated and then undertake an extensive simulation study-a daunting task at best. Perhaps a better strategy would be to study a few feasible arrangements in order to identify a suitable, though not necessarily optimal, solution for a specific practical application.

References

Andersen, E. B. (1980). *Discrete statistical models with social sciences applications.* Amsterdam: North-Holland.

Birnbaum, A. (1968). Some latent trait models and their use in inferring an examinee's ability. In F. M. Lord & M. R. Novick, *Statistical theories of mental test scores* (pp. 397-479). Reading, MA: Addison-Wesley.

Bock, R. D., & Mislevy, R. J. (1988). Adaptive EAP estimation of ability in a microcomputer environment. *Applied Psychological Measurement, 6,* 431-444.

Chang, H.-H., & Ying, Z. (1996). A global information approach to computerized adaptive testing. *Applied Psychological Measurement, 20,* 213-229.

Chang, H.-H., & Ying, Z. α-stratified multistage computerized adaptive testing. *Applied Psychological Measurement, 23,* 211-222.

Chang, H.-H., & Ying, Z. (in press). Nonlinear designs for logistic item response models with application in computerized adaptive tests. *The Annals of Statistics.*

Gelman, A., Carlin, J. B, Stern, H. S., & Rubin, D. B. (1995). *Bayesian data analysis.* London: Chapman & Hall.

Gulliksen, H. (1950). *Theory of mental tests.* Hillsdale, NJ: Erlbaum.

Lehmann, E. L., & Casella, G. (1998). *Theory of point estimation.* New York: Springer Verlag

Lord, F. M. (1971). The self-scoring flexilevel test. *Journal of Educational Measurement, 8,* 147-151.

Lord, F. M. (1980). *Applications of item response theory to practical testing problems.* Hillsdale, NJ: Erlbaum.

Lord, F. M. (1986). Maximum likelihood and Bayesian parameter estimation in item response theory. *Journal of Educational Measurement, 23,* 157-162.

Lord, F. M., & Novick, M. R. (1968). *Statistical theories of mental test scores.* Reading, MA: Addison-Wesley.

Mislevy, R. J. (1986). Bayes modal estimation in item response models. *Psychometrika, 51,* 177-195.

Mislevy, R. J., & Bock, R. D. (1983). *BILOG: Item and test scoring with binary logistic models* [Computer program and manual]. Mooresville, IN: Scientific Software.

Mislevy, R. J., & Wu, P.-K. (1988). *Inferring examinee ability when some items response are missing* (Research Report 88-48-ONR). Princeton, NJ: Educational Testing Service.

Owen, R. J. (1969). *A Bayesian approach to tailored testing* (Research Report 69-92). Princeton, NJ: Educational Testing Service

Owen, R. J. (1975). A Bayesian sequential procedure for quantal response in the context of adaptive mental testing. *Journal of the American Statistical Association, 70*, 351-356.

Rasch, G. *Probabilistic models for some intelligence and attainment tests.* Copenhagen: Denmarks Paedogogiske Institut.

Samejima, F. (1973). A comment on Birnbaum's three-parameter logistic model in latent trait theory. *Psychometrika, 38*, 221-233.

Samejima, F. (1993). The bias function of the maximum-likelihood estimate of ability for the dichotomous response level. *Psychometrika, 58*, 195-210.

Schnipke, D. L., & Green, B. F. (1995).A comparison of item selection routines in linear and adaptive testing. *Journal of Educational Measurement, 32*, 227-242.

Segall, D. O. (1997). Equating the CAT-ASVAB. In W. A. Sands, B. K. Waters and J. R. McBride (Eds.), *Computerized adaptive testing: From inquiry to operation* (pp. 181-198). Washington, DC: American Psychological Association.

Stocking, M. L. (1996). An alternative method for scoring adaptive tests. *Journal of Educational and Bahavioral Statistics, 21*, 365-389.

Thissen, D., & Mislevy, R. J. (1990). Testing algorithms. In H. Wainer (Ed.), *Computerized adaptive testing: A primer* (pp. 103-134). Hillsdale, NJ: Lawrence Erlbaum.

Tsutakawa, R. K., & Johnson, C. (1990). The effect of uncertainty on item parameter estimation on ability estimates. *Psychometrika, 55*, 371-390.

van der Linden, W. J. (1998). Bayesian item-selection criteria for adaptive testing. *Psychometrika, 62*, 201-216.

van der Linden, W. J. (1999). A procedure for empirical initialization of the trait estimator in adaptive testing. *Applied Psychological Measurement, 23*, 21-29.

van der Linden, W. J., & Glas, C. A. W. (2000). Capitalization on item calibration in adaptive testing. *Applied Measurement in Education, 13*, 35-53.

van der Linden, W. J., & Glas, C. A. W. (submitted). Cross validating item parameter estimation in adaptive testing.

Veerkamp, W. J. J. , & Berger, M. P. F. Item-selection criteria for adaptive testing. *Journal of Educational and Behavioral Statistics, 22*, 203-226.

Wainer, H., Lewis, C., Kaplan, B., & Braswell, J. (1991). Building algebra testlets: A comparison of hierarchical and linear structures. *Journal of Educational Measurement, 28*, 311-323.

Wang, T., Hanson, B. A., Lau, C.-M. A. (1999). Reducing bias in CAT trait estimation: A comparison of approaches. *Applied Psychological Measurement, 23*, 263-278.

Wang, T. &, Vispoel, W. P. (1998). Properties of ability estimation methods in computerized adaptive testing. *Journal of Educational Measurement, 35*, 109-135.

Warm, T. A. (1989). Weighted likelihood estimation of ability in item response theory with tests of finite length. *Psychometrika, 54*, 427-450.

Weiss, D. J. (1982). Improving measurement quality and efficiency with adaptive testing. *Applied Psychological Measurement, 4*, 473-285.

Weiss, D. J., & McBride, J. R. (1984). Bias and information of Bayesian adaptive testing. *Applied Psychological Measurement, 8*, 273-285.

Chapter 2
Constrained Adaptive Testing with Shadow Tests

Wim J. van der Linden*
University of Twente, The Netherlands

1. Introduction

The intuitive principle underlying the first attempts at adaptive testing was that a test has better measurement properties if the difficulties of its items match the ability of the examinee. Items that are too easy or difficult have predictable responses and cannot provide much information about the ability of the examinee. The first to provide a formalization of this principle was Birnbaum (1968). The information measure he used was Fisher's well-known information in the sample. For the common dichotomous item response theory (IRT) models, the measure is defined as

$$I(\theta) = \sum_{i=1}^{n} I_i(\theta) = \sum_{i=1}^{n} \frac{(P'(\theta))^2}{P(\theta)\left[1 - P(\theta)\right]},\tag{1}$$

where $P_i(\theta)$ is the probability that a person with ability θ gives a correct response to item $i = 1 \ldots n$.

For the one-parameter logistic (1PL) model, the information measure takes a maximum value if for each item in the test the value of the difficulty parameter $b_i = \theta$. The same relation holds for the 2PL model, though the maximum is now monotonically increasing in the value of the discrimination parameter of the items, a_i. The empirical applications discussed later in this chapter are all based on response data fitting the 3PL model:

$$P_i(\theta_j) \equiv c_i + (1 - c_i)\frac{e^{a_i(\theta_j - b_i)}}{1 + e^{a_i(\theta_j - b_i)}}.\tag{2}$$

For this model, the optimal value of the item-difficulty parameter is larger than the ability of the examinee due to the possibility of guessing on the items. The difference between the optimal value and the ability

* Part of the research in this chapter received funding from the Law School Admission Council (LSAC). The opinions and conclusions contained in this chapter are those of the author and do not necessarily reflect the position or policy of LSAC.

W.J. van der Linden and C.A.W. Glas (eds.),
Computerized Adaptive Testing: Theory and Practice, 27–52.
© 2000 *Kluwer Academic Publishers. Printed in the Netherlands.*

of the examinee is known to be a monotonically increasing function of the guessing parameter, c_i.

Both test theoreticians and practitioners immediately adopted the information measure in (1) as their favorite criterion of test assembly. The fact that item information additively contributes to the test precision has greatly enhanced its popularity. Though other criteria of item selection have been introduced later (for a review see van der Linden and Pashley, this volume), the most frequently used criterion in computerized adaptive testing (CAT) has also been the one based on the information measure in (1).

Though adaptive testing research was initially motivated by the intention to make testing statistically more informative, the first real-life testing programs to make the transition to CAT quickly discovered that adaptive testing operating only on this principle would lead to unrealistic results. For example, if items are selected only to maximize the information in the ability estimator, test content may easily become unbalanced for some ability levels. If examinees happened to learn about this feature, they may change their test preparations and the item calibration results would no longer be valid. Likewise, without any further provisions, adaptive tests with maximum information can also be unbalanced with respect to such attributes as their possible orientation towards gender or minority groups and become unfair for certain groups of examinees. Furthermore, even a simple attribute such as the serial position of the correct answer for the items could become a problem if the adaptive test administrations produced highly disproportionate use of particular answer keys. Lower ability examinees might benefit from patterned guessing and some of the more able examinees might become anxious and begin second-guessing their answers to previous items. Examinees may get alerted by this fact and start bothering if their answers to the previous questions were correct.

More examples of nonstatistical specifications for adaptive tests are easy to provide. In fact, what most testing programs want if they make the transition from linear to adaptive testing, is test administrations that have exactly the same "look and feel" as their old linear forms but that are much shorter because of a better adaptation to the ability levels of the individual examinees. The point is that adaptive testing will only be accepted if the statistical principle of adapting the item selections to the ability estimates for examinees is implemented in conjunction with serious consideration of many other nonstatistical test specifications.

Formally, each specification that an adaptive test has to meet imposes a constraint on the selection of the items from the pool. As a consequence, a CAT algorithm that combines maximization of statistical information with the realization of several nonstatistical specifications can be viewed as an algorithm for *constrained sequential optimization*. The *objective function* to be optimized is the statistical information in the test items at the current ability estimate. All other specifications are the *constraints* subject to which the optimization has to take place.

The goal of this chapter is to develop this point of view further and present a general method of constrained sequential optimization for application in adaptive testing. This method has proven to be successful in several applications. The basic principle underlying the method is to implement all constraints through a series of *shadow tests* assembled to be optimal at the updated ability estimates of the examinee. The items to be administered are selected from these shadow tests rather than directly from the item pool. Use of the method will be illustrated for item pools from several well-known large-scale testing programs.

2. Review of Existing Methods for Constrained CAT

2.1. ITEM-POOL PARTITIONING

An adaptation of the maximum-information criterion to make item selection balanced with respect to test content was presented in Kingsbury and Zara (1991). Their proposal was to partition the item pool according to the item attributes. While testing, the numbers of items selected from each class in the partition are recorded. To maintain content balance, the algorithm is forced to follow a minimax principle selecting the next item from the class for which the largest number of items is lacking. A further modification was proposed to prevent items from being readministered to examinees that have taken the same test earlier. Finally, to reduce the exposure of the most informative items in the pool, these authors suggested not to select the most informative item from the current class in the partition but to pick one at random from among the k best items in the class. The last adaptation has also been used in an early version of the CAT-ASVAB (Hetter & Sympson, 1997).

2.2. WEIGHTED-DEVIATION METHOD

A more general approach is the weighted deviation method (WDM) by Stocking and Swanson (1993). In their approach, all content specifications for the CAT are formulated as a series of upper and lower bounds on the numbers of items to be selected from the various content classes. The objective of maximum information is reformulated as a series of upper and lower bounds on the values of the test information function. A weighted sum of the deviations from all bounds is taken as the objective function, with the weights reflecting the desirability of the individual specifications. The items in the adaptive test are selected sequentially to minimize the objective function.

2.3. TESTLET-BASED ADAPTIVE TESTING

Some of the first to address the necessity of combining content specification and statistical criteria in item selection for CAT were Wainer and Kiely (1987). Their solution was to change the size of the units in the item pool. Rather than discrete items, the pool instead comprised "testlets"; that is, bundles of items related to sets of content specifications that can be selected only as intact units. Testlets are preassembled to have a fixed item order. For example, they may be organized according to a hierarchical branching scheme with examinees proceeding with items at a higher (lower) difficulty level if their ability estimate goes up (down). Examinees may proceed from one testlet to another in a hierarchical fashion, a linear fashion, or a combination thereof. New psychometric theory for testlet-based adaptive testing is offered by Wainer, Bradlow, and Du (this volume) and Glas, Wainer, and Bradlow (this volume).

2.4. MULTI-STAGE TESTING

The idea of testlet-based adaptive testing is closely related to the older format of multi-stage testing (Lord, 1980). In multi-stage testing, examinees proceed through a sequence of subtests, moving to a more difficult subtest if they do well but to an easier one if their previous performances are low. Though the earlier literature discussed a paper-and-pencil version of this format with nonstatistical scoring of the ability of the examinees after each subtest, the advent of computers in testing practice has made an implementation with statistical estimation of ability after each subtest possible. Adema (1990) van der Linden and Adema (1998) offer 0-1 linear programming models for the design of multi-stage testing systems based on the maximum-information criterion that allow for a large variety of constraints on the composition of the subtests. The possibility to include such constraints into multi-stage testing systems has rekindled the interest in this testing format (Luecht & Nungester, this volume; 1998).

2.5. EVALUATION OF EXISTING APPROACHES

The above approaches differ in an important way. The first two approaches implement the constraints through a modification of the item selection algorithm. The last two approaches build all constraints directly into the units in the pool from which the test is administered. This distinction has consequences with respect to:

1. the degree of adaptation possible during the test;
2. the extent of item coding necessary;
3. the possibility of expert review of actual test content;
4. the nature of constraint realization; and
5. the possibility of constraint violation.

Both the item-pool partitioning and WDM approach allow for an update of the ability estimate after each item. Thus, they offer the maximum degree of adaptation possible. However, to be successful, both approaches assume coding of all relevant item attributes. If a potentially important attribute is overlooked, test content may still become unbalanced. The WDM minimizes only a weighted sum of deviations from the constraints. Some of its constraints can therefore be violated even with complete coding of the items. In either approach, both the selection of the items and the realization of the constraints is sequential. Though sequential item selection allows for optimal adaptation, sequential realization of constraints is less than ideal. Algorithms with this feature tend to favor items with an attractive value for the objective function early in the test. However, the choice of some of these items may turn out to be suboptimal later in the test. If so, the result is less than optimal adaptation to the ability estimates and/or the impossibility to complete the test without constraint violation.

The testlet-based and multi-stage approach have the option of expert review of all intact testing material prior to administration. Explicit coding of relevant item attributes is not always necessary. However, these approaches lack full adaptation of item selection to the ability estimates (multi-stage testing and linear testlet-based CAT) or allow only for partial adaptation (testlet-based CAT with hierarchical branching). Also, the task of assembling a pool of testlets or a multi-stage testing system such that any path an examinee takes satisfies all constraints involves a huge combinatorial problem that can quickly become too complicated to be solved by intuitive methods. The result may be a suboptimal branching system and/or constraint violation. However, as already noted, formal methods for assembling multi-stage testing systems do exist. Use of these methods does assume explicit coding of all relevant item attributes. Both in the testlet-based CAT and multi-stage testing approaches, groups of items are selected sequentially. Hence, adaptation of item selection to ability estimation is suboptimal. On the other hand, the constraints are realized simultaneously when assembling the testlets or subtests prior to the test administration, provided formal methods and adequate item coding are used to assemble the pool of testlets or the multi-stage testing system.

This evaluation of the existing methods for constrained CAT reveals an important dilemma. An algorithm with optimal properties would need *to select its items sequentially to allow for optimal adaptation but realize all constraints simultaneously to prevent violation of certain constraints or suboptimal adaptation later in the test.* Possible solutions to the dilemma are: (1) to allow the algorithm to work backwardly to improve on previous decisions or (2) to have the algorithm project forwardly to take future consequences of decisions into account. In adaptive testing, backtracking is impossible; the algorithm is applied in real time and earlier choices can not be undone. Thus, the only possibility left is to have the algorithm project forwardly each time a new item is selected. This is exactly what the new class of algorithms presented in this chapter do.

3. Constrained CAT with Shadow Tests

The basic concept of a shadow test is illustrated in Figure 1. The selection of each new item in the CAT is preceded with the on-line assembly of a shadow test. A shadow test is a full-length test that: (1) meets all the test constraints; (2) contains all items administered previously; and (3) has maximum information at the current ability estimate. The item to be administered is the one with maximum information in the shadow test. The next actions are updating the ability estimate of the examinee, returning the unused items to the pool, and repeating the procedure.

The following pseudo-algorithm gives a more precise summary of the idea:

Step 1: Initialize the ability estimator.

Step 2: Assemble a shadow test that meets the constraints and has maximum information at the current ability estimate.

Step 3: Administered the item in the shadow test with maximum information at the ability estimate.

Step 4: Update the ability estimate

Step 5: Return all unused items to the pool.

Step 6: Adjust the constraints to allow for the attributes of the item already administered.

Step 7: Repeat Steps 2-6 until n items have been administered.

Observe that test length has been fixed in this algorithm. This choice is in agreement with practice in nearly all existing CAT programs. Though a stopping rule based on a predetermined level of accuracy

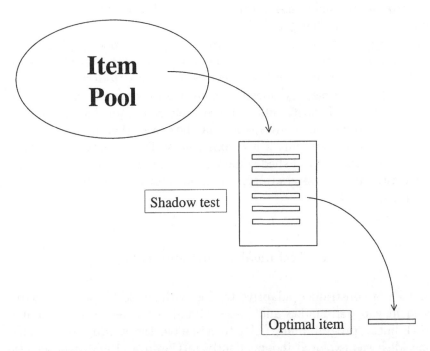

Figure 1. Constrained adaptive testing with shadow tests

for the ability estimator is desirable form a statistical point of view, it seems impossible to guarantee always the same specifications for all examinees for a test with random length.

The ideal in constrained adaptive testing is a test that is feasible (i.e., meets all constraints) and has an optimal value for the objective function in (1) at the true ability of the examinee. Since the true ability is unknown, all one can hope for is an item-selection mechanism with a result approximating this ideal case. The proposed adaptive testing scheme has this feature; it yields *feasible adaptive tests converging to the optimal value for the information function at the true ability of the examinees.*

This claim can be shown to hold as follows. The algorithm realizes all constraints simultaneously for each shadow test. Each next shadow tests contains all items administered previously. Thus, the last shadow test is the actual adaptive test and always meets all constraints. Further, each shadow test is assembled to have a maximum value for the information function in (1) and the item selected from the shadow test has a maximum contribution to this function. For a consistent ability estimator, it follows from Slutsky's theorems (e.g., Ferguson, 1996) that the value for the function in (1) converges to the maximum value possible at the true ability of the examinee. Mild conditions for the ML estimator with maximum-information item selection to be consistent are formulated in Chang and Ying (in press).

This argument assumes an infinitely large item pool with known item parameter values. However, the conclusion is expected to hold closely enough for all practical purposes for any well-designed, finite item pool. Of course, the speed of convergence depends on size and nature of the set of constraints. For a severely constrained adaptive test from a small item pool, convergence may be slower than for a test from a large pool involving only a few constraints. The empirical examples later in this chapter will shed some light on the question how fast the ability estimator converges in typical applications of this procedure for constrained CAT.

4. Technical Implementation

The idea of constrained adaptive testing with shadow tests was introduced in van der Linden and Reese (1998), who used the technique of 0-1 linear programming (LP) to assemble the shadow tests. The same idea was explored independently in Cordova (1998), whose test assembly work was based on the network-flow programming approach introduced in Armstrong and Jones (1992). In principle, any algorithm for automated test assembly that generates an optimal feasible solution and is fast enough for application in real time can be used to implement the above adaptive testing scheme. Even for test assembly heuristics that tend to provide suboptimal solutions, such as the WDM or the normalized weighted absolute deviation heuristic (NWADH) by Luecht (1998; Luecht & Hirsch, 1992), considerable gain over the existing methods of constrained adaptive testing can be expected when implemented in the proposed scheme. A recent review of approaches to automated test assembly is given in van der Linden (1998).

The examples later in this chapter are all based on the technique of 0-1 LP. This technique allows us to deal with virtually any type of constraint that can be met in test assembly and thus offers maximum flexibility when modeling the problem of shadow test assembly. In addition, general software for 0-1 LP is available that can be used to solve such models in real time for the type of CAT pools met in real life.

4.1. BASIC NOTATION AND DEFINITIONS

To maintain generality, a 0-1 LP model for the assembly of shadow tests from an item pool with some of its items organized as sets with a common stimulus is formulated. This testing format has become increasingly popular; several of the item pools used in the empirical examples later in this chapter involved this format. Typically, in testing with set-based items, the numbers of items per stimulus available in the pool are larger than the numbers to be selected in the test.

The following notation will be used throughout the remainder of this chapter:

items in the pool	$: i = 1, ..., I;$
stimuli in the pool	$: s = 1, ..., S;$
set of items in the pool with stimulus s	$: U_s, s = 1, ..., S;$
items in the adaptive test	$: k = 1, ..., n;$
stimuli in the adaptive test	$: l = 1, ..., m.$

Thus, i_k and s_l are the indices of the kth item and lth stimulus in the adaptive test, respectively. Using this notation, $S_{k-1} \equiv \{i_1, ..., i_{k-1}\}$ is defined as the set of the first $k - 1$ items administered in the test. Consequently, $R_k \equiv \{1, ..., I\} \backslash S_{k-1}$ is the set of items remaining in the pool after $k - 1$ items have been selected in the test.

The kth shadow test is denoted as $T_k \equiv \{i_1, ..., i_{k-1}, i'_k, ..., i'_n\}$ whereas $S_l \equiv (s_1, ..., s_l)$ is defined as the set of the first l stimuli in the test. If the constraints on the number of items for lth stimulus in the adaptive test have not yet been satisfied, s_l is called the *active stimulus* and U_{s_l} the *active item set*. If s_l is active, the next item is selected from $U_{s_l} \cap \{i'_k, ..., i'_n\}$. Otherwise, it is selected from $\{i'_k, ..., i'_n\}$. Therefore, the list of *eligible items* in the kth shadow test is defined as

$$A_k \equiv \begin{cases} U_{s_l} \cap \{i'_k, ..., i'_n\} & \text{if the } l\text{th stimulus is active;} \\ \{i'_k, ..., i'_n\} & \text{otherwise.} \end{cases} \tag{3}$$

Let $\widehat{\theta}_{k-1}$ denote the ability estimate updated after the first $k - 1$ items in the adaptive test. It thus holds that the kth item in the adaptive test is

$$i_k \equiv \arg \max_i \{I_i(\widehat{\theta}_{k-1}); i \in A_k\}. \tag{4}$$

When assembling the shadow test, the objective function should be maximized only over the set of items eligible for administration. In particular, if the lth stimulus is active, it may be disadvantageous to maximize the information in the shadow test over items not in U_{s_l} (even though such items are needed to complete the shadow test). To implement this idea for the objective function in the model below, the following set is defined:

$$O_k \equiv \begin{cases} U_{s_l} & \text{if the } l\text{th stimulus is active;} \\ R_k & \text{otherwise.} \end{cases} \tag{5}$$

4.2. 0-1 LP MODEL FOR SHADOW TEST

The model is an adapted version of the one presented in van der Linden and Reese (1998). To formulate its objective function and constraints, 0-1 valued decision variables x_i and z_s are introduced. These variables take the value 1 if item i and stimulus s is selected in the shadow test, respectively, and the value 0 otherwise.

In addition, the following notation is needed to denote the various types of item and stimulus attributes that may play a role in the assemble of the shadow test:

categorical item attribute	: C_e^i, $e = 1, ..., E$;
categorical stimulus attribute	: C_f^s, $f = 1, ..., F$;
quantitative item attribute	: q_i, $i = 1, ..., I$;
quantitative stimulus attribute	: q_s, $s = 1, ..., S$;
sets of mutually exclusive items	: V_g^i, $g = 1, ..., G$;
sets of mutually exclusive stimuli	: V_h^s, $h = 1, ..., H$.

Categorical item attributes, such as item content or format, partition the item pool into classes C_e^i, $e = 1, ..., E$. Note that each attribute involves a different partition. For simplicity, only the case of one attribute is discussed; adding more constraints is straightforward. The model below requires the number of items in the test from class C_e^i to be between n_e^l and n_e^u. Likewise, the set of stimuli in the pool is partitioned by a categorical attribute C_f^s, $f = 1, ..., F$, and the number of stimuli from class C_f^s is required to be between n_f^l and n_f^u. In addition, the items and stimuli are assumed to be described by categorical attributes, such as a word count or an item difficulty parameter. For simplicity, the case of one attribute with value q_i for item i and r_s for stimulus s, respectively, is discussed. The model requires the sum of these values for the items and stimuli in the test to be between in the intervals (q^{il}, q^{iu}) and (q^{sl}, q^{su}), respectively. Note that in the model below in fact average values are constrained since the model also fixes the total numbers of items and stimuli in the test. Finally, the use of logical constraints to model test specifications is illustrated. It is assumed that the item pool has sets of items, V_g^i, $g = 1, ..., G$, and sets of stimuli, V_h^s, $h = 1, ..., H$, that exclude each other in the same test, for instance, because knowing one of them facilitates solving the others.

The model is formulated as follows:

$$\text{maximize} \sum_{i \in O_k} I_i(\widehat{\theta}_{k-1}) x_i \qquad \text{(maximum information)} \qquad (6)$$

subject to

$$\sum_{i=1}^{I} x_i = n, \qquad (\text{\# of items}) \qquad (7)$$

$$\sum_{s=1}^{S} z_s = m, \qquad (\text{\# of stimuli}) \qquad (8)$$

$$\sum_{i \in S_{k-1}} x_i = k - 1, \qquad (\text{items already selected}) \qquad (9)$$

$$\sum_{i \in U_s} x_i {\geq} n_s^l z_s, \qquad s = 1, ..., S, \qquad (\text{\# of items per stimulus}) \qquad (10)$$

$$\sum_{i \in U_s} x_i {\leq} n_s^u z_s, \qquad s = 1, ..., S, \qquad (\text{\# of items per stimulus}) \qquad (11)$$

$$\sum_{i \in C_e^i} x_i {\geq} n_e^l, \qquad e = 1, ..., E, \qquad (\text{categorical attribute}) \qquad (12)$$

$$\sum_{i \in C_e^i} x_i {\leq} n_e^u, \qquad e = 1, ..., E, \qquad (\text{categorical attribute}) \qquad (13)$$

$$\sum_{i=1}^{I} q_i x_i \geq q^{il}, \qquad (\text{quantitative attribute}) \qquad (14)$$

$$\sum_{i=1}^{I} q_i x_i \leq q^{iu}, \qquad (\text{quantitative attribute}) \qquad (15)$$

$$\sum_{i \in C_f^s} z_s \geq n_f^l, \qquad f = 1, ..., F, \qquad (\text{categorical attribute}) \qquad (16)$$

$$\sum_{i \in C_f^s} z_s \leq n_f^u, \qquad f = 1, ..., F, \qquad (\text{categorical attribute}) \qquad (17)$$

$$\sum_{i=1}^{I} q_s z_s \geq q^{sl}, \qquad (\text{quantitative attribute}) \qquad (18)$$

$$\sum_{i=1}^{I} q_s z_s \leq q^{su}, \qquad (\text{quantitative attribute}) \qquad (19)$$

$$\sum_{i \in V_g^i} x_i \leq 1, \qquad g = 1, ..., G, \qquad (\text{mutually exclusive items}) \qquad (20)$$

$$\sum_{s \in V_h^s} z_s \leq 1, \qquad h = 1, ..., H, \qquad (\text{mutually exclusive stimuli}) \qquad (21)$$

$$x_i = 0, 1, \qquad i = 1, ..., I, \qquad (\text{domain of variables}) \qquad (22)$$

$$z_s = 0, 1, \qquad s = 1, ..., S. \qquad \text{(domain of variables)} \qquad (23)$$

The constraints in (9) fix the values of the decision variables of all $k-1$ items already administered to 1. In doing so, the model automatically accounts for the attributes of all these items (Step 6 in the pseudo-algorithm). The constraints in (10)-(11) serve a double goal. On the one hand, they constrain the number of items per stimulus. On the other hand, the variables for the stimuli in the right-hand sides of these constraints keep the selection of the items and stimuli consistent. It is only possible to assign an item from a set to the shadow test if its stimulus is assigned and, reversely, a stimulus cannot be assigned to a shadow test without assigning a permitted number of its items. The other constraints were already explained above.

4.3. NUMERICAL ASPECTS

Commercial software for solving models as in (6)-(23) is amply available. This software can be used to calculate optimal values for the variables x_i, $i = 1, ..., I$, and z_s, $s = 1, ..., S$. Exact solutions to 0-1 LP problems can only be obtained through explicit enumeration, preferably in the form of an implementation of the branch-and-bound (BAB) method. 0-1 LP problems are known to be NP-hard, that is, their solution time is not bounded by a polynomial in the size of the problem. Therefore, special implementations of the method for larger problems are needed. The following ideas have been implemented successfully.

First, note that the constraints in (7)-(23) do not depend on the value of $\widehat{\theta}$. The update of this estimate only affects the objective function in (6). Repeated application of the model for $k = 1, ..., n$ can thus be described as a series of problems in which the space of feasible solutions remains the same but the coefficients in the objective function (i.e., the values for the item information function) change. The changes become small if the ability estimates stabilize. Because solutions can be found much quicker if the algorithm starts from a good initial feasible solution, the obvious choice is to use the $(k-1)$th shadow test as the initial solution for the kth test. This measure has been proven to improve the speed of the solution processes dramatically. The first shadow test need not be calculated in real time. It can be preassembled for the initial estimates of θ before the adaptive test becomes operational.

Second, additional improvements include specifying the order in which the variables are branched on. The variables for the items, x_i, determine the selection of individual items whereas those for the stimuli, z_s, determine the selection of larger sets of items. It always pays off to start branching on the variables that have the largest impact.

Therefore, branching on the stimulus variables should precede branching on the item variables. Also, forcing the slack variables to be integer for the constraints in the model has proven to be an efficient measure; that is, the variables used by the LP program to turn the inequalities in the model into equalities. In the branching order, slack variables for constraints with stimuli (items) should have higher priority then the decision variables for these stimuli (items). For technical details on optimal branching in BAB methods for solving 0-1 LP problems in constrained adaptive testing, see Veldkamp (submitted).

It is a common experience with BAB processes that values close to optimality are found long before the end of the process. If a good upper bound to the value of the objective function is available, it therefore makes sense to stop the process as soon as the objective function approaches the bound satisfactorily closely. Good results have been found for tolerances as small as 1-2% of the value of the upper bound.

All applications later in this chapter used CAT software developed at the University of Twente. To calculate the shadow tests, the software made calls to either the ConTEST package for 0-1 LP based test assembly (Timminga, van der Linden & Schweizer, 1997) or the solver in the CPLEX 6.0 package (ILOG, 1998). In most examples below, the CPU time needed for one cycle of ability estimation and item selection was 2-3 seconds. The only exceptions were for two adaptive tests in Application 3, which took 5-6 and 7-8 seconds per items, respectively and one in Application 4, for which 6-8 seconds per items were needed (all results on a Pentium Pro/166 MHz processor). All these tests had item sets with common stimuli; it is a general experience that solving 0-1 LP problems for tests with items sets tend to take more time (van der Linden, 2000a).

These CPU times are suitable for real-life application of the proposed CAT method. In fact, much larger times would not have involved any problem since it is always possible to calculate ahead. When the examinee works on item $k - 1$, the computer can already calculate two solutions for the kth shadow test, one for the update of θ after a correct and the other after an incorrect response to item $k - 1$.

5. Four Applications to Adaptive Testing Problems

The principle underlying the adaptive testing scheme in this chapter is that every feature an adaptive test is required to have implies a set of constraints to be imposed on the item selection process. The constraints are implemented through the assembly of a shadow test prior to the selection of the item. As shadow tests are full-size linear tests, in principle any feature possible for linear tests can also be realized for an adaptive test. The only thing needed is the formulation of the constraints as a set of linear (in)equalities for insertion into the model in (6)-(23).

The principle is illustrated with four applications each addressing a different aspect of adaptive testing. In the first application, the practicality of the shadow test approach is assessed for a CAT program with an extremely large number of content constraints of varying nature. The second application deals with the problem of differential speededness in CAT. Because each examinee gets a different selection of items, some of them may have trouble completing the test. It is shown how the problem can be solved by inserting a response-time constraint in the shadow test. The question of how to deal with item-exposure control in constrained adaptive testing with shadow tests is addressed in the third example. The method of control is based on Stocking and Lewis' (1998) conditional approach but uses the shadow test to define the list from which the items are sampled for administration. Also, sampling is with replacement. The last example is relevant to the case of a testing program that offers its examinees a choice between an adaptive and a linear version of the same test. It is shown how the observed-scores on the two versions of the test can be automatically equated by inserting a few additional constraints into the model for the shadow test. Each of these applications provides one or more empirical examples using a data set from an existing CAT program.

5.1. CAT with Large Numbers of Nonstatistical Constraints

To check the practicability of a shadow test approach for a CAT program with a large number of nonstatistical constraints, a simulation study was conducted for a pool of 753 items from the Law School Admission Test (LSAT). A 50-item adaptive version of the LSAT was simulated. The current linear version of the LSAT is twice as long; all its specification were therefore reduced to half their size. The specifications dealt with such item and stimulus attributes as item and stimulus content, gender and minority orientation, word counts, and answer key distributions. The set of content attributes defined an elaborate classification system for the items and stimuli for which the test had to meet a large number of specifications. In all, the LP model for the shadow test had 804 variables and 433 constraints.

Three conditions were simulated: (1) unconstrained CAT (adaptive version of the LSAT ignoring all current specifications); (2) constrained CAT with the least severely constrained section of the LSAT first; and

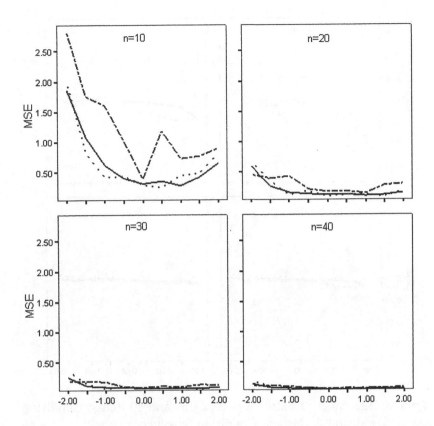

Figure 2. MSE functions after $n = 10, 20, 30,$ and 40 items (Condition 1: dotted; Condition 2: solid; Condition 3: dashed)

(3) constrained CAT with the most severely constrained section first. Mean-squared error (MSE) and bias functions were calculated for each three conditions after $n = 10, 20, ..., 40$ items.

The results are shown in Figure 2 and 3. For all test lengths, the results for the conditions of unconstrained CAT and constrained CAT with the least severely constrained section first were practically indistinguishable. The MSE and bias functions for the condition of constrained CAT with the most severely constrained first were less favorable for the shorter test lengths but matched those of the other two conditions for $n > 20$. The results are discussed in more detail in van der Linden and Reese (1998). The main conclusion from the study is that adding large numbers of constraints to an adaptive test is possible without substantial loss in statistical precision.

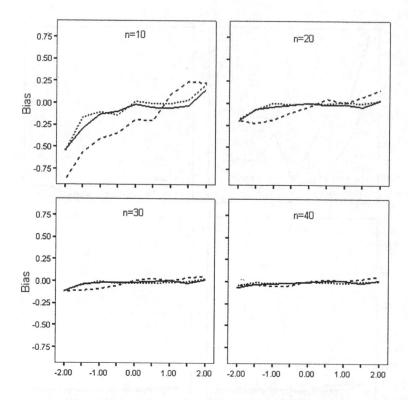

Figure 3. Bias functions after $n = 10, 20, 30,$ and 40 items (Condition 1: dotted; Condition 2: dashed; Condition 3: solid)

5.2. CAT WITH RESPONSE-TIME CONSTRAINTS

Another problem not anticipated before adaptive testing became operational deals with the timing of the test. In the current CAT programs, examinees are typically free to sign up for a time slot in the period of their choice but the length of the slots is constant. Test questions, however, require different amounts of time to complete. Such differences are due to the amount of reading involved in the item or the difficulty of the problem formulated in it. Because each examinee gets an individual selection of items, some examinees may run out of time whereas others are able to finish the test easily in time.

The requirement that adaptive tests should have items that enable each examinee to finish in time is another example of a constraint to be imposed on the assembly of the shadow tests. A constraint to this effect was already proposed in one of the early papers on applying the technique of 0-1 LP to assembling linear tests:

$$\sum_{i=1}^{I} t_i x_i \leq t_{tot}, \qquad (24)$$

where t_i is the amount of time needed for item i, estimated, for instance, as the 95th percentile in the distribution of response times recorded in the pretest of the item, and t_{tot} is the total amount of time available (van der Linden & Timminga, 1989).

A more sophisticated approach is followed in van der Linden, Scrams and Schnipke (1999). Since examinees are also known to show individual variation in response times, they suggest using a lognormal model for the distribution of the response times of an examinee responding to an item with separate parameters for the examinee and the item. They also discuss how the values of the item parameter can be estimated from pretest data and how estimates of the value of the examinee parameter can be updated using the actual response times during the test. A Bayesian procedure is proposed for these updates as well as for obtaining the posterior predictive densities for the response times for the examinee on the remaining items in the pool. These densities are used to establish the constraint that controls the time needed for the test.

Suppose $k - 1$ items have been administered. At this point, both the actual response times for the examinee on the items in S_{k-1} and the predictive densities for the remaining items in R_k are known. Let $t_{ij}^{\alpha_k}$ be the α_kth percentile in the posterior predictive density for item $i \in R_k$. The percentile is formulated to be dependent on k because it makes sense to choose more liberal values of α in the beginning of the test but more conservative ones towards the end.

The following constraint controls the time needed for the test:

$$\sum_{i \in S_{k-1}} t_{ij} x_i + \sum_{i \in R_k} t_{ij}^{\alpha_k} x_i \leq t_{tot}. \qquad (25)$$

Inserting the constraint in the model in (6)-(23) forces the new items in the shadow test to be selected such that the sum of their predicted response times does not exceed the remaining amount of time available.

The procedure was applied to an item pool for the CAT version of the Arithmetic Reasoning Test in the Armed Services Vocational Aptitude Battery (ASVAB). The pool consisted of 186 items calibrated under the model in (1). Response times had been recorded for 38,357 examinees who had taken the test previously. The test had a length of 15 items for which $t_{tot} = 39$ minutes. Percentile α_k was chosen to be the 50th percentile for $k = 1$ and moved up in equal steps to the 95th percentile for the last items.

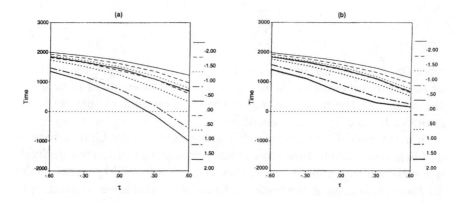

Figure 4. Time left after completion of the test (a) without and (b) with response time constraints

To evaluate the effects of the constraint in (27) on the time needed by the examinees, two different conditions were simulated: (1) the current version of the ASVAB test and (2) a version of the test with the constraint. The results are given in Figure 4. The first panel shows the average time left after completion of the test as a function of the slowness of the examinee, τ, for the condition without the constraint. Different curves are displayed for the different values of θ used in the study. The faster examinees had still left some time after the test but the slower examinees ran out of time. The second panel shows the same information for the condition with the constraint. The left-hand side of the curves remain the same but for $\tau \geq -.30$ the constraints became active. The lower curves run now more horizontally and none of them crosses the dotted line representing the time limit. The effects of the constraints on the bias and MSE in the estimator of θ were also estimated. These effects were of the same small order as for the LSAT in Application 1. For these and other results from this simulation study, see van der Linden, Scrams, and Schnipke (1999).

An interesting phenomenon was discovered during the simulations. In the first panel in Figure 4, the curves are ordered uniformly by the values of the ability parameter examinees, θ. However, the order was counter to the intuitions of the authors who had expected to see the examinees with a low ability and not those with a *high* ability suffer from the time limit. Additional analyses of the data set revealed that there was no correlation between ability and speed (.04). Thus, among the more able examinees there were as many slow as fast examinees. On the other hand, as expected, a strong positive correlation (.65) was found between the item difficulty parameter and the parameter for the amount of time required by the item. Because the test was adaptive, the more able examinees received the more difficult items, particularly towards the end of the test. Consequently, the more able examinees responding more slowly had trouble completing the test.

5.3. CAT WITH ITEM-EXPOSURE CONTROL

Because in CAT items are selected to maximize the information measure in (1), variations in the information in the items can have a large impact on their probability of selection. Items with a maximum contribution at some values of θ are chosen each time the ability estimate is updated to these values. Items that have no maximum contribution for any value of θ are never chosen at all.

This process has two undesired consequences. First, items with a high probability of being selected are frequently exposed and run the danger, therefore, of becoming known to the examinees. Second, items with a low probability of selection represent a loss of resources. They have gone through a long process of reviewing and pretesting and, even if only slightly less then optimal, do not return much of these investments.

To prevent from security problems, CAT programs usually constrain their item selection process to yield item exposure rates not larger than certain target values. As a result, the exposure rates of popular items go down and those of some of the less popular ones go up. A review of the existing methods of item-exposure control is given in Stocking and Lewis (this volume).

Sympson and Hetter introduced the idea of having an additional probability experiment determine if items selected are actually administered or removed from the pool for the examinee. Stocking and Lewis (1998) proposed a conditional version of this experiment. Let $P_i(S \mid \theta)$ denote the probability of selecting item i conditional on the ability θ and A the event of administering the item. It holds that:

$$P_i(A \mid \theta) = P_i(A, S \mid \theta) = P_i(A \mid S, \theta) P_i(S \mid \theta). \tag{26}$$

The exposure rates $P_i(A \mid \theta)$ are controlled by setting the values $P_i(A \mid S, \theta)$ at appropriate levels. However, the probabilities of selecting the items, $P_i(S \mid \theta)$, can only be determined by simulating adaptive test administrations from the item pool. Moreover, the event of selecting item i is dependent on the presence of the other items in the pool. Therefore, the values of the control parameters can be set only through an iterative process with cycles of: (1) simulating the test; (2) estimating the probabilities of selection; and (3) adjusting the values for the control parameters.

In the original version of the Sympson-Hetter method, the probability experiment with the control parameters is repeated until an item is administered. In Stocking and Lewis' modification of the experiment, only one experiment is needed. In both methods, the experiments are based on sampling without replacement; once an item is passed, it is not replaced in the pool for the examinee.

The method of item-exposure control for the adaptive testing scheme in this chapter is the following modification of the Stocking-Lewis experiment: First, for the kth shadow test, the experiment is always run over the list of eligible items, A_k. The list is a ordered by the values of the items for their information functions at $\widehat{\theta}$. The list does not have a fixed length but decreases as the number of items administered increases. Second, the experiment is based on sampling with replacement. All items on the list passed are of better quality than the one actually administered. Not returning such items to the pool would result in a lower optimal value of the objective function for the later shadow tests, or, in the worst case, even in the impossibility to assemble them. On the other hand, the only consequence of replacing the items in the pool is higher probabilities of selection for the best items and lower values for the control parameters to compensate for the increase.

The effects of the proposed method of exposure control on the bias and MSE functions of the ability estimator were assessed for adaptive tests from item pools from three different adaptive tests programs: the Graduate Management Admission Test (GMAT), the Graduate Record Examination (GRE), and the PRAXIS test. Two conditions were simulated: (1) constrained CAT without exposure control and (2) constrained CAT with exposure control. Quantitative details on the models for these tests as well as the target values for the exposure rates used, the number of steps required to find the optimal values for the control parameter, the maximum exposure rate in the item pool realized, and the number of items with exposure rates violating the target are given in Table 1.

Table 1. Results from exposure control study for three different tests

Test	# of Variables	# of Constraints	Target Value	# of Steps	Maximum Value	# of Violations
1	397	78	.29	4	.29	0
2	534	140	.22	5	.26	6
3	900	139	.20	6	.21	2

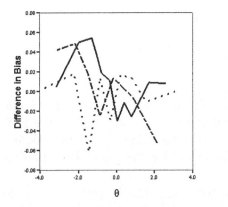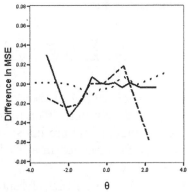

Figure 5. Differences between MSE and bias functions for CAT and without exposure control (Test 1: solid; Test 2: dotted; Test 3: dashed)

For two of these tests a few items in the pool had exposure rates larger than the target; however, both the number and sizes of the violations were low. The number of steps required to find the values for the control parameters were typical (see Stocking & Lewis (1998). In Figure 5, the differences between the bias and MSE functions for the ability estimators in the three tests are plotted. The general conclusion is that these differences are unsystematic and negligibly small. Adding the above method of exposure control to the procedure for constrained adaptive testing with shadow tests did not entail any loss of statistical precision in these examples.

5.4. CAT WITH EQUATED NUMBER-CORRECT SCORES

Testing programs making the transition to an adaptive testing format often want their examinees to have the choice between the former linear version and the new adaptive version of the test. However, this choice is only justified if the scores on both tests are comparable. To achieve comparable scores, the method of equipercentile equating has been applied to equate ability estimates on the adaptive version of the test to the number-right scores on the paper-and-for version (Segall, 1997).

The logic of constrained adaptive testing with shadow tests proposed in this chapter suggests the search for constraints that yield an adaptive test with observed scores automatically equated to those on the linear test. Such constraints are possible using the condition for two tests to have the same conditional number-correct score distributions given θ presented in van der Linden and Luecht (1998). They show that, for any value of θ, the conditional distributions of observed number-correct scores on two test forms with items $i = 1, ..., n$ and $j = 1, ..., n$ are identical if and only if

$$\sum_{i=1}^{n} P_i^r(\theta) = \sum_{j=1}^{n} P_j^r(\theta), \quad r = 1, ..., n. \tag{27}$$

They also show that for $r \to n$ the conditions become quickly negligible and report nearly perfect empirical results for $r = 2$ or 3. Note that the conditions in (27) are linear in the items. Thus, the formulation of constraints for insertion in a 0-1 LP model for the assembly of shadow tests is possible.

Let $j = 1, ..., n$ be the reference test which the adaptive test has to be equated to. The observed number-correct scores on the adaptive test can be equated to those on the reference test by inserting the following set of constraints in the model in (6)-(23):

$$\sum_{i=1}^{n} P_i^r(\widehat{\theta}_{k-1})x_i - \sum_{j=1}^{n} P_j^r(\widehat{\theta}_{k-1}) \leq c, \quad r = 1, ..., R, \leq n \tag{28}$$

$$\sum_{i=1}^{n} P_i^r(\widehat{\theta}_{k-1})x_i - \sum_{j=1}^{n} P_j^r(\widehat{\theta}_{k-1}) \geq -c \quad r = 1, ..., R \leq n \tag{29}$$

where c is a tolerance parameter with an arbitrarily small value and R need not be larger than 3 or 4, say. Note that these constraints thus require the difference between the sums of powers of the response functions at $\widehat{\theta}_{k-1}$ to be in an arbitrarily small interval about zero, $(-c, c)$. They do not require the two sets of response functions to be identical across the whole range of θ. They also require only that sums of powers of their values at $\widehat{\theta}_{k-1}$ be identical, not the powers of the response functions of the individual items. Thus, the algorithm does *not* build adaptive tests that are required to be item-by-item parallel to the alternative test.

To assess the effects of the constraints in (28)-(29) on the observed number-scores in the adaptive test as well as on the statistical properties of its ability estimator, a simulation study was conducted for the same item pool from the LSAT as in Application 1. The conditions compared were: (1) unconstrained CAT and (2) constrained CAT with the above conditions for $R = 1, 2$. In either condition, the true values of the examinees were sampled from $N(0, 1)$. The observed-number correct scores were recorded after $n = 20, ..., 50$ items. As a reference test, a previous form of the LSAT was used.

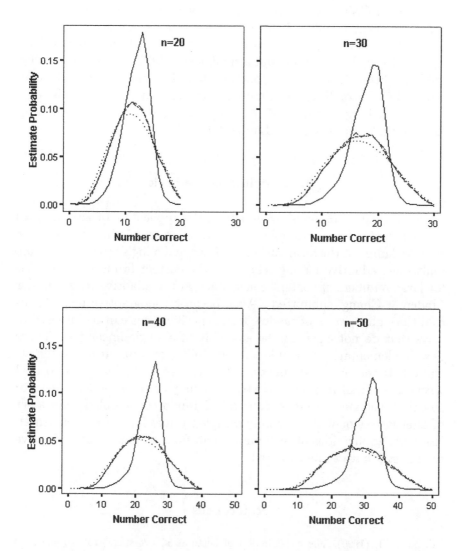

Figure 6. Observed-score distributions for CAT with and without constraints for number-correct score equating (target distribution: dotted; CAT without constraints: solid; CAT with 1 constraint: dashed-dotted; CAT with 2 constraints: dashed)

The results are given in Figure 6. As expected, the observed number-correct distribution for the unconstrained CAT was peaked with a mode slightly larger than $n/2$. After 20 items the observed number–correct distributions for the constrained condition had already moved away from this distribution towards the target distribution on the reference test. After 30 items, the observed number–correct distributions for the constrained CAT and the reference test were indistinguishable for all practical purposes. The choice of value for R did not seem to matter much. The ability estimators in both conditions did not show any differences in the bias function but a small loss in MSE for the constrained CAT condition. The loss was comparable to the one for one of the constrained CAT conditions in Figure 2 (Application 1). A more detailed discussion of the results and other applications of the constraints in (29)-(30) is given in van der Linden (2000b).

6. Concluding Remark

The empirical examples above illustrate the application of several types of constraints. These examples do not exhaust all possibilities. A more recent example is the formulation of Chang and Ying's (1999) α-stratified multistage adaptive testing scheme in the current framework to allow for large numbers of content constraints on the adaptive test (van der Linden & Chang, submitted). Also, because the adaptive test is realized through a *series* of shadow tests, specifications can be imposed in ways that do not exist for the assembly of a single linear test. These new implementations include the possibility of alternating systematically between objective functions for successive shadow tests to deal with cases of multiple-objective test assembly or using models with stochastic constraints (that is, constraints randomly sampled from a set of alternatives or with randomly sampled values for their coefficients). However, applications of such implementations to constrained adaptive testing still have to be explored.

References

Adema, J. J. (1990). The construction of customized two-stage tests. *Journal of Educational Measurement*, 27, 241-253.

Armstrong, R. D., & Jones, D. H. (1992). Polynomial algorithms for item matching. *Applied Psychological Measurement, 16*, 271-288.

Birnbaum, A. (1968). Some latent trait models and their use in inferring an examinee's ability. In F. M. Lord & M. R. Novick, *Statistical theories of mental test scores* (pp. 397-479). Reading, MA: Addison-Wesley.

Chang, H., & Ying, Z. (1999). α-stratified multistage computerized adaptive testing. *Applied Psychological Measurement, 23*, 211-222.

Chang, H.-H., & Ying, Z. (in press). Nonlinear designs for logistic item response models with application in computerized adaptive tests. *The Annals of Statistics*.

Cordova, M. J. (1998). *Optimization methods in computerized adaptive testing*. Unpublished doctoral dissertation, Rutgers University, New Brunswick, NJ.

Ferguson, T. S. (1996). *A course in large-sample theory*. London: Chapman & Hall.

Hetter, R. D. , Sympson, J. B. (1997). Item exposure in CAT-ASVAB. In W. A. Sands, B. K. Waters, & J. R. McBride (Eds.), *Computerized adaptive testing: From inquiry to operation* (pp. 141-144). Washington, DC: American Psychological Association

ILOG, Inc. (1998). *CPLEX 6.0* [Computer Program and Manual]. Incline Village, NV: ILOG.

Kingsbury, G. G. & Zara, A. R. (1991). Procedures for selecting items for computerized adaptive tests. *Applied Measurement in Education*, 2, 359-375.

Lord, F. M. (1980). *Applications of item response theory to practical testing problems*. Hillsdale, NJ: Erlbaum.

Luecht, R.D. (1988). Computer-assisted test assembly using optimization heuristics. *Applied Psychological Measurement*, *22*, 224-236.

Luecht, R.M. & Hirsch, T.M. (1992). Computerized test construction using average growth approximation of target information functions. *Applied Psychological Measurement*, *16*, 41-52.

Luecht, R. M. & Nungester, R. J. (1998). Some practical examples of computer-adaptive sequential testing. *Journal of Educational Measurement*, *35*, 229-249.

Segall, D. O. (1997). Equating the CAT-ASVAB. In W. A. Sands, B. K. Waters, & J. R. McBride (Eds.), *Computerized adaptive testing: From inquiry to operation* (pp.81-198). Washington, DC: American Psychological Association.

Stocking, M. L. & Lewis, C. (1998). Controlling item exposure conditional on ability in computerized adaptive testing. *Journal of Educational and Behavioral Statistics*, 23, 57-75.

Swanson, L. & Stocking, M. L. (1993). A model and heuristic for solving very large item selection problems. *Applied Psychological Measurement*, 17, 151-166.

Sympson, J. B. & Hetter, R. D. (1985, October). Controlling item-exposure rates in computerized adaptive testing. *Proceedings of the 27th annual meeting of the Military Testing Association* (pp. 973-977). San Diego, CA: Navy Personnel Research and Development Center.

Timminga, E. van der Linden, W. J. , & Schweizer, D. A. (1997). *ConTEST 2.0: A decision support system for item banking and optimal test assembly* [Computer program and manual]. Groningen, The Netherlands: iec ProGAMMA

van der Linden, W. J. (1998a). Optimal assembly of psychological and educational tests. *Applied Psychological Measurement*, 22, 195-211.

van der Linden, W.J. (1998b). Bayesian item selection criteria for adaptive testing. *Psychometrika*, 62, 201-216.

van der Linden, W.J. (1999). A procedure for empirical initialization of the trait estimator in adaptive testing. *Applied Psychological Measurement*, 23, 21-29.

van der Linden, W.J. (2000a). Optimal assembly of tests with item sets. *Applied Psychological Measurement*, 24. (In press)

van der Linden, W. J. (2000b). Adaptive testing with equated number-correct scoring. *Applied Psychological Measurement*, 24. (In press).

van der Linden, W. J. (2000c). On complexity in computer-based testing. In G. N. Mills, M. Potenza, J. J. Fremer & W. Ward (Eds.). *Computer-based testing: Building the foundation for future assessments*. Mahwah, NJ: Lawrence Erlbaum Associates.

van der Linden, W. J. & Adema, J. J. (1998). Simultaneous assembly of multiple test forms. *Journal of Educational Measurement*, *35*, 185-198 [Erratum in Vol. 36, 90-91].

van der Linden, W. J. & Chang, H.-H. (submitted) Alpha-stratified multistage adaptive testing with large numbers of content constraints.

van der Linden, W. J. &, Luecht, R. M. (1998). Observed equating as a test assembly problem. *Psychometrika*, 62, 401-418.

van der Linden, W. J. & Reese, L. M. (1998). A model for optimal constrained adaptive testing. *Applied Psychological Measurement,* 22, 259-270.

van der Linden, W. J., Scrams, D. J., & Schnipke, D. L. (1999). Using response-time constraints to control for differential speededness in computerized adaptive testing. *Applied Psychological Measurement,* 23, 195-210.

van der Linden, W. J., & Boekkooi-Timminga, E. (1989). A maximin model for test design with practical constraints. *Psychometrika,* 54, 237-247.

Veldkamp, B.P. (submitted). *Modifications of the branch-and-bound algorithm for application in constrained adaptive testing.*

Wainer, H., & Kiely, G. L. (1987). Item clusters in computerized adaptive testing: A case for testlets. *Journal of Educational Measurement,* 24, 185-201.

Chapter 3
Principles of Multidimensional Adaptive Testing

Daniel O. Segall*
Defense Manpower Data Center
The United States Department of Defense, USA

1. Introduction

Tests used to measure individual differences are often designed to provide comprehensive information along several dimensions of knowledge, skill, or ability. For example, college entrance exams routinely provide separate scores on math and verbal dimensions. Some colleges may elect to base qualification on a compensatory model, where an applicant's total score (math plus verbal) must exceed some specified cut-off. In this instance, the individual math and verbal scores may provide useful feedback to students and schools about strengths and weaknesses in aptitudes and curriculum. In other instances, colleges may elect to base qualification on a multiple-hurdle model, where the applicant's scores on selected components must exceed separate cut-offs defined along each dimension. For example, a college may elect to have one qualification standard for math knowledge and another standard for verbal proficiency. Applicants may be required to meet one or the other, or both standards to qualify for entrance. In all these instances, it is useful and important for the individual component-scores to possess adequate psychometric properties, including sufficient precision and validity.

When the dimensions measured by a test or battery are correlated, responses to items measuring one dimension provide clues about the examinee's standing along other dimensions. An examinee exhibiting a high-level vocabulary proficiency is likely (although not assured) to exhibit a similar high-level of reading comprehension, and visa-versa. Knowledge of the magnitude of the association between the dimensions in the population of interest, in addition to the individual's performance levels can add a unique source of information, and if used properly can lead to a more precise estimate of proficiencies. This cross-information is ignored by conventional scoring methods and by unidimensional item selection and scoring methods used in computerized adaptive testing

---

* The views expressed are those of the author and not necessarily those of the Department of Defense or the United States Government.

W.J. van der Linden and C.A.W. Glas (eds.),
Computerized Adaptive Testing: Theory and Practice, 53–73.
© 2000 *Kluwer Academic Publishers. Printed in the Netherlands.*

(CAT). The challenge discussed in this chapter is to increase the efficiency of adaptive item selection and scoring algorithms by extending unidimensional methods to the simultaneous measurement of multiple dimensions.

The cross-information gathered from items of correlated dimensions can be effectively modeled by multidimensional item response theory. In the case of computerized adaptive testing, this information can aid measurement in two ways. First, it can aid in the selection of items, leading to the choice of more informative items. Second, it can aid in the estimation of ability, leading to test scores with added precision. In order to realize these benefits, two generalizations of unidimensional adaptive testing are necessary, one for item selection, and another for scoring. The benefit of this multidimensional generalization is increased measurement efficiency—manifested by either greater precision or reduced test-lengths.

2. Literature Review

Bloxom and Vale (1987) were the first to formally consider the extension of unidimensional adaptive testing methods to multiple dimensions. They noted that the direct multivariate generalization of unidimensional IRT scoring procedures could easily exceed the computational power of personal computers of the time (mid 1980's). To avoid intensive calculations associated with iterative algorithms and numerical integration, they proposed an efficient scoring procedure based on a multivariate extension of Owen's (1975) sequential updating procedure. Through a series of normal approximations, the multivariate extension provides closed form expressions for point estimates of ability. The issue of efficient item selection was not explicitly addressed by Bloxom and Vale.

Tam (1992) developed an iterative maximum likelihood (ML) ability estimation procedure for the two-dimensional normal ogive model. Tam evaluated this procedure along with several others using such criteria as precision, test information, and computation time. Like Bloxom and Vale, the problem of item-selection was not specifically addressed. Tam's item selection procedures assumed ideal item pools where the difficulty of the item was matched to the current ability level of the examinee.

Segall (1996) extended previous work (Bloxom & Vale, 1987; Tam, 1992) by providing a theory-based procedure for item selection that incorporates prior knowledge of the joint distribution of ability. Segall also presented maximum likelihood and Bayesian procedures for item

selection and scoring of multidimensional adaptive tests for the general
H-dimensional model. By this point in time (mid 1990's), the power of
personal computers could support the computations associated with iterative numerical procedures, making approximate-scoring methods of
the type suggested by Bloxom and Vale (1987) less desirable. The benefits of the Bayesian approach were evaluated from simulations based
on a large-scale high-stakes test: the Computerized Adaptive Testing
version of the Armed Services Vocational Aptitude Battery (Segall &
Moreno, 1999). Segall demonstrated that for realistic item pools, multidimensional adaptive testing can provide equal or higher precision with
about one-third fewer items than required by unidimensional adaptive
testing.

Luecht (1996) examined the benefits of applying multidimensional
item selection and scoring techniques in a licensing/certification context, where mandatory complex content constraints were imposed. He
compared the reliability of the ML multidimensional approach (Segall,
1996) to a unidimensional CAT. Results demonstrated that a shorter
multidimensional CAT with content constraints could achieve about
the same subscore reliability as its longer unidimensional counterpart.
Estimated savings in test lengths were consistent with Segall's findings,
ranging from about 25% to 40%.

van der Linden (1999) presents a multidimensional adaptive testing method intended to optimize the precision of a composite measure,
where the composite of interest is a linear function of latent abilities.
The composite weights associated with each dimension are specified
a priori, based on external developer-defined criteria. Ability estimation proceeds according to ML, and item selection is based on a minimum error variance criterion. The error (or sampling) variance of the
composite measure is obtained from a linear combination of elements
from the inverse Fisher-information matrix. van der Linden demon-
strates that for a two dimensional item-pool, a 50 item adaptive test
provides nearly uniform measurement precision across the ability space.
For shorter tests (of 10 and 30 items), the ML estimates tended to be
considerably biased and inefficient.

As pointed out by van der Linden (1999), a Bayesian procedure such
as the one proposed by Segall (1996) can lead to inferences that are
more informative than those based on ML approaches. This is true
when the dimensions underlying the item-responses are correlated, and
the joint distribution of latent ability is known or estimable with a fair
degree of accuracy. The greatest benefit of a Bayesian approach is likely
to occur for short to moderate test-lengths, or when test-length is short
relative to the number of dimensions. Added information is provided by
the prior distribution which incorporates known dependencies among

the ability variables. The remainder of this chapter provides an explication of Segall's (1996) Bayesian methodology for item selection and scoring of multidimensional adaptive tests.

3. Multidimensional Item Selection and Scoring

Two unidimensional item-selection and scoring approaches, based on maximum likelihood and Bayesian estimation, have direct multidimensional counterparts. Associated with each of the two approaches are adaptive item selection rules, and methods for estimating ability and quantifying the level of uncertainty in the ability estimates. However, the extension based on maximum-likelihood theory has serious deficiencies. Although successful unidimensional applications of maximum-likelihood procedures exist, at least two drawbacks are exacerbated in multidimensional CAT. First, towards the beginning of the adaptive test, item selection is hampered by non-informative likelihood functions which possess indeterminate or poorly defined maxima. Consequently, some *adhockery* is needed to bolster item selection procedures in the absence of sufficient data. Second, the ML item selection approach does not consider prior knowledge about the joint distribution of ability. These shortcomings are remedied by the Bayesian methodology.

The objective of the multidimensional adaptive testing algorithms is the efficient estimation of the H-dimensional vector of ability values $\boldsymbol{\theta} = \{\theta_1, \theta_2, ..., \theta_H\}$. The development of these algorithms is based on five principles and their associated steps drawn primarily from Bayesian theory. The first section below describes the specification of the *Prior Density* function which characterizes all usable information about the latent ability parameters before the data (item-responses) are observed. The second section describes the *Likelihood Function* which provides a mathematical description of the process giving rise to the observed item-responses in terms of the unknown ability parameters. The next section outlines the specification of the *Posterior Distribution*, which summarizes the current state of knowledge (arising from both the observed responses and the prior information). The fourth section casts the issue of *Item Selection* in terms of a Bayes decision problem for choosing among optimal experiments, and derives an expression for item-specific information measures to be used for choosing among candidate items. The final section derives specific *Posterior Inference* statements from the posterior distribution, which consist of point estimates of ability.

3.1. Prior Density

Here we consider a two-stage process which leads to an individual's item responses. First, an individual is sampled from a population (with a known distribution). That is, a value of $\boldsymbol{\theta}$ is sampled from a population with distribution $f(\boldsymbol{\theta})$. Second, this individual (with fixed $\boldsymbol{\theta}$) is administered multiple test items resulting in a set of binary (correct/incorrect) responses. Under this model, the ability parameters $\boldsymbol{\theta}$ are treated as random variables with distribution $f(\boldsymbol{\theta})$. We shall consider the case in which f is multivariate normal

$$f(\boldsymbol{\theta}) = (2\pi)^{-H/2} |\Phi|^{-1/2} \exp\left[-\frac{1}{2} (\boldsymbol{\theta} - \boldsymbol{\mu})' \Phi^{-1} (\boldsymbol{\theta} - \boldsymbol{\mu})\right], \qquad (1)$$

with mean vector $\boldsymbol{\mu} = \{\mu_1, \mu_2, ..., \mu_H\}$ and $H \times H$ covariance-matrix Φ. We further assume that $\boldsymbol{\mu}$ and Φ are known. The prior $f(\boldsymbol{\theta})$ encapsulates all usable knowledge about $\boldsymbol{\theta}$ before the item responses have been collected.

Figure 1 illustrates the joint density function for a bivariate normal prior with centroid $\mu_1 = \mu_2 = 0$, and covariance terms $\phi_{11} = \phi_{22} = 1$, and $\phi_{12} = 0.6$. Also displayed are the marginal distributions for each dimension which are also normally distributed with means and variances equal to their corresponding values in the bivariate distribution (i.e. 0 and 1, respectively). From Figure 1 it is evident that prior information about θ_1 comes from two sources. First, the range of probable values is confined primarily to the interval $(-2, +2)$, as indicated by the marginal distribution. Second, small values of θ_1 tend to be associated with small values of θ_2, and a similar association is observed for moderate and large values of θ_1 and θ_2. Thus a second source of information about θ_1 comes from its association with θ_2. In the general H-dimensional case, prior information derived from correlated dimensions leads to additional precision for the estimation of individual ability parameters.

3.2. Likelihood Function

The modeled data consist of a vector of scored responses from an individual examinee $\mathbf{u}_n = \{u_{i_1}, u_{i_2}, ..., u_{i_n}\}$ to n adaptively administered items. The set of administered items is denoted by $S_n = \{i_1, i_2, ..., i_n\}$, whose elements uniquely identify the items which are indexed in the pool according to $i = 1, 2, ..., I$. For example, if the first item administered was the 50th item in the pool, then $i_1 = 50$; if the second item administered was the 24th item in the pool, then $i_2 = 24$; and so forth. In this case $S_2 = \{50, 24\}$. If Item 50 was answered correctly, and Item

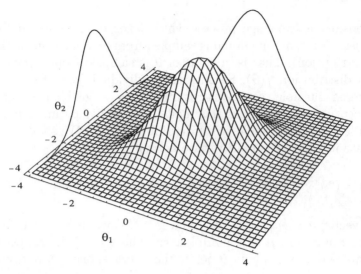

Figure 1. Bivariate normal prior distribution

24 answered incorrectly then $\mathbf{u}_2 = \{1, 0\}$. For notational simplicity, we shall assume that the non-subscripted item response vector $\mathbf{u}$ contains n-elements (i.e. $\mathbf{u} \equiv \mathbf{u}_n$).

Furthermore, an examinee can be characterized by their standing on H traits denoted by the vector $\boldsymbol{\theta} = \{\theta_1, \theta_2, ..., \theta_H\}$, where each trait affects performance on one or more test items. The item response function for item i is given by

$$p_i(\boldsymbol{\theta}) \equiv \mathrm{Prob}\,(U_i = 1|\boldsymbol{\theta}) = c_i + \frac{1 - c_i}{1 + \exp\left[-D\mathbf{a}_i'\,(\boldsymbol{\theta} - b_i\mathbf{1})\right]}\,, \qquad (2)$$

where U_i is the binary random variable containing the response to item i ($U_i = 1$, if item i is answered correctly; and $U_i = 0$ otherwise), c_i is the probability that a person with infinitely low ability will answer item i correctly, b_i is the difficulty parameter of item i, $\mathbf{1}$ is a $H \times 1$ vector of 1's, D is the constant 1.7, and $\mathbf{a}_i'$ is a $1 \times H$ vector of discrimination parameters for item i.

The form of the item response function (2) is a generalization of the three-parameter logistic model proposed by Birnbaum (1968), with the addition of a linear-compensatory rule for multiple latent traits. For a one-dimensional model ($H = 1$), this function reduces to the standard three-parameter logistic model. Also note that the model possesses a single difficulty parameter b_i. Separate difficulty parameters for each dimension are indeterminate and thus cannot be estimated from observed response data.

Table 1. Example item parameters and responses.

Item	a_1	a_2	b	c	u	Item	a_1	a_2	b	c	u
1	1.0	0.0	0.00	.20	1	5	0.0	1.0	0.55	.20	1
2	0.0	1.0	0.50	.20	0	6	1.0	1.5	0.95	.20	0
3	1.0	1.0	0.75	.30	1	7	2.0	0.0	−1.00	.20	1
4	2.0	0.0	0.60	.25	0	8	0.0	1.7	−0.70	.20	1

Another basic model assumption is that of local or conditional independence (Lord & Novick, 1968). According to this assumption, the joint probability function of a set of n responses $\{u_{i_1}, u_{i_2}, ..., u_{i_n}\}$ for an examinee of ability $\boldsymbol{\theta}$ is equal to the product of the probabilities associated with the individual item responses

$$f(U_{i_1} = u_{i_1}, U_{i_2} = u_{i_2}, ..., U_{i_n} = u_{i_n}|\boldsymbol{\theta}) = \prod_{i \in S_n} p_i(\boldsymbol{\theta})^{u_i} q_i(\boldsymbol{\theta})^{1-u_i} , \quad (3)$$

where the product runs over the set of administered (or selected) items $S_n = \{i_1, i_2, ..., i_n\}$, and $q_i(\boldsymbol{\theta}) = 1 - p_i(\boldsymbol{\theta})$. The ability to express $f(U_{i_1} = u_{i_1}, U_{i_2} = u_{i_2}, ..., U_{i_n} = u_{i_n}|\boldsymbol{\theta})$ as a product of terms which depend on individual item-response functions leads to computational simplifications in item selection and scoring. Without the assumption of local independence, expressions required by ML and Bayes methods would be intractable for all but very short tests.

The likelihood function given by

$$L(\mathbf{u}|\boldsymbol{\theta}) = \prod_{i \in S_n} p_i(\boldsymbol{\theta})^{u_i} q_i(\boldsymbol{\theta})^{1-u_i} \quad (4)$$

is algebraically equivalent to the joint probability function (3). The change in notation however reflects a shift in emphasis from the random variables $\mathbf{u}$ with $\boldsymbol{\theta}$ fixed, to the parameters $\boldsymbol{\theta}$, with $\mathbf{u}$ fixed. Since $\mathbf{u}$ are a set of sampled (observed) values of the item responses, the quantity $L(\mathbf{u}|\boldsymbol{\theta})$ is merely a function of the parameters $\boldsymbol{\theta}$.

Figure 2 illustrates the likelihood function for the pattern of responses to the eight-item test displayed in Table 1. As indicated, the region of highest likelihood occurs near the point $(0.257, 0.516)$. However, this value as a point estimate of ability does not consider information contributed by the prior distribution. Both sources of information are however combined by the *posterior density*.

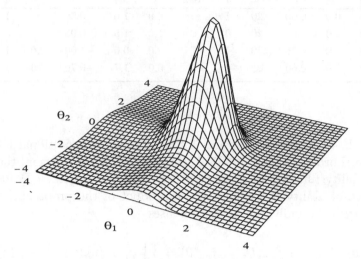

Figure 2. Likelihood function $L(\mathbf{u}|\boldsymbol{\theta})$

3.3. POSTERIOR DENSITY

Given specifications for the prior $f(\boldsymbol{\theta})$ and likelihood $L(\mathbf{u}|\boldsymbol{\theta})$, we are now in a position to make probability statements about $\boldsymbol{\theta}$ given $\mathbf{u}$. These can be made through an application of Bayes' rule which is used to construct the posterior density function

$$f(\boldsymbol{\theta}|\mathbf{u}) = \frac{L(\mathbf{u}|\boldsymbol{\theta})f(\boldsymbol{\theta})}{f(\mathbf{u})} ,\qquad (5)$$

where $f(\boldsymbol{\theta})$ is the multivariate normal density function (1), $L(\mathbf{u}|\boldsymbol{\theta})$ is the likelihood function (4), and $f(\mathbf{u})$ is the marginal probability of $\mathbf{u}$ given by

$$f(\mathbf{u}) = \int_{-\infty}^{\infty} L(\mathbf{u}|\boldsymbol{\theta})f(\boldsymbol{\theta}) \, d\boldsymbol{\theta} .$$

The posterior density $f(\boldsymbol{\theta}|\mathbf{u})$ contains all existing information about $\boldsymbol{\theta}$, and is used as a basis to provide point and interval estimates of ability parameters $\boldsymbol{\theta}$. As implied by the notation, the posterior represents the distribution of $\boldsymbol{\theta}$ for fixed $\mathbf{u}$, where $\mathbf{u}$ is fixed at the observed response values for the n items.

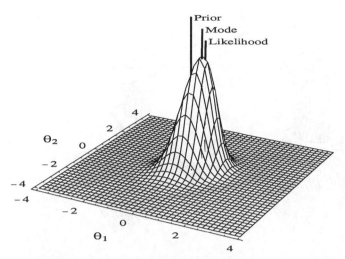

Figure 3. Posterior distribution $f(\boldsymbol{\theta}|\mathbf{u})$.

Figure 3 displays the posterior density function for the pattern of responses displayed in Table 1. According to (5), the height of the posterior density function is proportional to the product of the prior times the likelihood. Since the posterior distribution incorporates the information from the item responses, it is generally less variable than the prior distribution (Figure 1). Note also that the posterior density function forms a compromise between the prior and likelihood (Figure 2) functions. In the example displayed in Figure 3, the mode of the posterior density $(0.233, 0.317)$ forms a compromise between the prior centered at $(0, 0)$ and the data, which suggest that the most likely value of $\boldsymbol{\theta}$ is $(0.257, 0.516)$. In this example, the centroid of the posterior distribution is more heavily influenced by the data than by the prior. In general, the role of the prior diminishes as test-length is increased.

3.4. Item Selection

Item selection in adaptive testing can be framed in terms of a specialized area of Bayesian decision theory, namely the area pertaining to choice of experiments (Bernardo & Smith, 1994, p. 63). Suppose that to assist in the measurement of $\boldsymbol{\theta}$, we can choose among several experiments. In the adaptive testing context, the kth experiment would involve the administration of item i_k from the pool of remaining items, denoted by $R_k = \{1, 2, ..., I\} \setminus S_{k-1}$. Given that $k-1$ items have already been administered, the task is to decide which item is to be administered as the next (kth) item from the set of remaining items R_k.

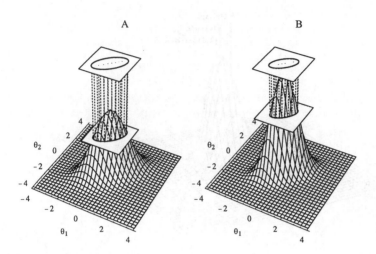

Figure 4. Isodensity contours: (A) $N(\boldsymbol{\mu} = \mathbf{0}; \sigma_1 = \sigma_2 = 1)$;
(B) $N(\boldsymbol{\mu} = \mathbf{0}; \sigma_1 = \sigma_2 = .8)$.

Bayesian decision theory considers the utility of administering each candidate item, and chooses the item with the highest expected utility. The item-specific utility can be expressed as a function of at least two sources: the cost of administering item i, and a loss for making inference $\hat{\boldsymbol{\theta}}$ when the true parameter is $\boldsymbol{\theta}$. When the cost of administration is the same for all items, then it can be shown (O'Hagan, 1994, p. 87) that the following item-selection strategy will maximize utility: *Choose the item that provides the largest decrement in the size of the posterior credibility region.*

This item-selection criterion can be illustrated with an example in which two dimensions are measured, and where the posterior distribution $f(\boldsymbol{\theta}|\mathbf{u}_k)$ is bivariate normal. Figure 4 displays two normal posterior distributions with centroids $(0,0)$, $\rho = .6$, and standard deviations $\sigma_1 = \sigma_2 = 1$ (Figure 4A) and $\sigma_1 = \sigma_2 = .8$ (Figure 4B). Associated with each distribution is an isodensity contour—a cross section of the surface made by a plane parallel to the (θ_1, θ_2)-plane. In general, these contours are elliptical and can be used to define multidimensional credible regions—regions of the posterior distribution containing 50%, 90%, 95% or 99% of the probability under $f(\boldsymbol{\theta}|\mathbf{u}_k)$. The coverage proportions or percentages can be adjusted by raising or lowering the altitude of the intersecting parallel plane, which in turn influences the size of the elliptical region.

The two elliptical regions displayed in Figure 4 each contain about 39% of the probability under their respective densities. Geometrically, this means that the volume of the three-dimensional region between the bivariate normal surface and the (θ_1, θ_2)-plane, bounded laterally by the right elliptic cylinder based on the pictured ellipse is equal to .39 (Tatsuoka, 1971, p. 70). Note however that the size (area) of the elliptical credible region in Figure 4B is considerably smaller than the region in Figure 4A. If these distributions represented the expected posterior outcomes from administering two different items, we would prefer the outcome depicted in Figure 4B—the outcome which provides the smallest credible region.

For a normal posterior distribution, the size (length, area, volume) of the credible region is given by

$$V_i = |\Sigma_i|^{1/2} \times g(H) \times \left[\chi_H^2(p)\right]^{H/2} , \qquad (6)$$

where Σ_i is the posterior covariance matrix based on the administration of item i (for $i \in R_k$), H is the number of dimensions, $g(H)$ is a term which is based on the number of dimensions, and $\chi_H^2(p)$ is the χ^2-value ($df = H$) located at the $p \times 100$ percentile (Anderson, 1984, p. 263). The coverage probability p can be altered by reference to the appropriate percentiles of the χ^2-distribution. When comparisons are made among the credible regions of different candidate items, all terms except the first remain constant in (6). Thus the item with the smallest value of $|\Sigma_i|$ will provide the largest decrement in the size of the posterior credibility region.

Two related issues hamper the direct application of $|\Sigma_i|$ as an item selection criterion. First, the posterior density $f(\theta|\mathbf{u}_k)$ as parameterized by (5) is not normal. Second, the posterior density of interest is based on responses u_{i_k} to candidate items $(i_k \in R_k)$ which have not yet been observed. Both these problems can be solved by approximating the non-normal posterior with a multivariate normal density based on the curvature at the mode. Specifically, the posterior distribution $f(\theta|\mathbf{u}_k)$ (obtained after the administration of item i_k and observation of the associated response u_{i_k}) can be approximated by a normal distribution having mean equal to the posterior mode $\hat{\theta}^{k-1}$, and covariance matrix $\Sigma_{i|S_{k-1}}$ equal to the inverse of the posterior information matrix evaluated at the mode $\hat{\theta}^{k-1}$:

$$\overset{\sim}{\Sigma}_{i|S_{k-1}} = \left[\mathbf{I}_{i|S_{k-1}}\right]^{-1} ,$$

where the information matrix $\mathbf{I}_{i|S_{k-1}}$ is minus the expected Hessian (second derivative matrix) of the log posterior

$$\mathbf{I}_{i|S_{k-1}} = -\mathrm{E}\left[\frac{\partial^2}{\partial\boldsymbol{\theta}\partial\boldsymbol{\theta}'}\ln f(\boldsymbol{\theta}|\mathbf{u}_k)\right],\tag{7}$$

and where the expectation is over the random item response variables $\mathbf{u}_k$. As the conditional notation "$i|S_{k-1}$" implies, the posterior covariance and information matrices associated with the ith item depend on both: (a) the characteristics of the candidate item i itself, and (b) the characteristics of the administered items S_{k-1}.

Note that in (7), the last element of $\mathbf{u}_k$, namely u_{i_k} has not yet been observed. However, by taking the expectation of the matrix of second derivatives, the item response terms $\mathbf{u}_k$ are replaced by their expected values $p_i(\boldsymbol{\theta})$, and the covariance matrix $\Sigma_{i|S_{k-1}}$ can be calculated prior to the administration of the candidate item i. The required posterior mode $\hat{\boldsymbol{\theta}}^{k-1}$ is calculated from the $k-1$ administered items. The information matrix $\mathbf{I}_{i|S_{k-1}}$ is calculated from the item parameters of the $k-1$ administered items, and from the parameters of the ith candidate item. These calculations are detailed in Appendix A.

One additional simplification can be made by noting that the determinant of the inverse of $\mathbf{I}_{i|S_{k-1}}$ is equal to the reciprocal of the determinant (Searle, 1982, p. 130). With this simplification the item selection criterion becomes

$$\left|\widetilde{\Sigma_{i|S_{k-1}}}\right| = \left|\left[\mathbf{I}_{i|S_{k-1}}\right]^{-1}\right|$$
$$= \left|\mathbf{I}_{i|S_{k-1}}\right|^{-1}.\tag{8}$$

Then from inspection of (8) we see that the candidate item which maximizes the determinant of the posterior information matrix $\mathbf{I}_{i|S_{k-1}}$ will provide the largest decrement in the size of the posterior credibility region.

The suitability of the item-selection criterion depends in part on how well the non-normal posterior can be approximated by a normal distribution. Figure 5 displays the normal approximation to the posterior distribution based on the eight sets of item responses and parameters provided in Table 1. The centroid of the distribution was set equal to the mode of the posterior $(0.233, 0.317)$. The covariance matrix Σ was computed from the inverse of the information matrix (7), which provides

$$\Sigma = \begin{pmatrix} .343 & .092 \\ .092 & .367 \end{pmatrix}.$$

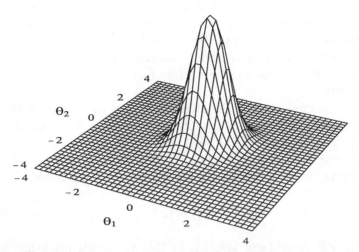

Figure 5. Normal approximation to posterior distribution.

A comparison of Figures 3 and 5 suggests that with these data, the normal approximation provides a close representation of the non-normal posterior. In general, close agreement of the sort displayed here provides support for the use of $\left|\mathbf{I}_{i|S_{k-1}}\right|$ as an inverse-indicator of the credibility region volume.

Selection of the kth adaptively administered item involves evaluation of the determinant of the posterior information matrix for candidate item i, denoted by $\left|\mathbf{I}_{i|S_{k-1}}\right|$. This quantity is computed for each of the unadministered (remaining) items contained in the pool, $i \in R_k$. The candidate item with the largest criterion value will be selected for administration. Computational details are provided in Appendix A.3.

3.5. POSTERIOR INFERENCE

Because the information in the form given by (5) is not readily usable, various numerical summaries of the posterior distribution are used. In item response theory, the posterior distribution is typically characterized by summary measures of central tendency and dispersion. Point estimates of ability are typically defined as the mean or mode of the posterior distribution. In many instances (for tests of moderate to long lengths) these will be nearly identical. However the mode of the posterior distribution (modal estimate) is better suited than the mean for applications involving higher dimensionality, since far fewer calculations are required. In addition to providing a score (posterior summary-measure of ability), the mode is also required for item selection purposes as described in the previous section. Accordingly, it is computed after each item-response to aid in the selection of the next item, and can be computed at the end of the test to provide an overall or final point estimate of ability. Below we drop the subscripts k in $\mathbf{u}_k$ with the understanding that modal estimates can be computed for any set or super-set of responses by straight forward application of the following formulas.

The modal estimates of $\boldsymbol{\theta}$, denoted by $\hat{\boldsymbol{\theta}}$, are those values that correspond to the maximum of the posterior density function: $\hat{\boldsymbol{\theta}} = \max_\theta f(\boldsymbol{\theta}|\mathbf{u})$. The estimate $\hat{\boldsymbol{\theta}}$ can be found by taking the H partial derivatives of the log-posterior density function, setting these equal to zero, and solving the H simultaneous non-linear equations for $\boldsymbol{\theta}$

$$\frac{\partial}{\partial \boldsymbol{\theta}} \ln f(\boldsymbol{\theta}|\mathbf{u}) = \mathbf{0} . \tag{9}$$

Since there is no closed-form solution to (9), an iterative method is required. Suppose we let $\boldsymbol{\theta}^{(m)}$ denote the mth approximation to the value of $\boldsymbol{\theta}$ that maximizes $\ln f(\boldsymbol{\theta}|\mathbf{u})$, then a better approximation is generally given by

$$\boldsymbol{\theta}^{(m+1)} = \boldsymbol{\theta}^{(m)} - \boldsymbol{\delta}^{(m)} , \tag{10}$$

where $\boldsymbol{\delta}^{(m)}$ is the $H \times 1$ vector

$$\boldsymbol{\delta}^{(m)} = \left[\mathbf{M}(\boldsymbol{\theta}^{(m)})\right]^{-1} \times \frac{\partial}{\partial \boldsymbol{\theta}} \ln f(\boldsymbol{\theta}^{(m)}|\mathbf{u}) . \tag{11}$$

The matrix $\mathbf{M}(\boldsymbol{\theta}^{(m)})$ is either the matrix of second partial derivatives $\mathbf{J}(\boldsymbol{\theta})$ (Newton-Raphson method), or the negative posterior information matrix $-\mathbf{I}(\boldsymbol{\theta})$ (Fisher method of scoring)—evaluated at $\boldsymbol{\theta} = \boldsymbol{\theta}^{(m)}$. Modal estimates can be obtained through successive approximations using (10) and (11). Additional approximations are obtained until the elements of $\boldsymbol{\theta}^{(m)}$ change very little from one iteration to the next. Explicit expressions for the required derivatives and information matrix are provided in Appendix A.

4. Example

This section provides a detailed example of multidimensional item selection and scoring calculations based on the methodology presented in the previous section. The calculations presented below are for a single examinee administered a fixed-length four-item adaptive test, where items are selected from a pool of eight items spanning two latent dimensions. Note that all computational formulae presented below can be applied to higher dimensionality ($H > 2$) problems without modification. The matrix notation used enables the calculations to be presented in a way that is independent of the number of dimensions H. Item parameters and associated responses are displayed in Table 1. The prior distribution of ability is assumed to be multivariate normal with unit variances, zero means, and correlated dimensions ($\phi_{12} = .6$). The basic steps consisting of initialization, provisional ability estimation, item selection, and scoring are detailed below.

4.1. INITIALIZATION

First, the provisional ability estimate $\hat{\theta}^k$ (where $k = 0$) is set equal to the mean of the prior distribution of ability. In this example the mean of the prior is $\mu = (0, 0)$. The inverse of the prior covariance matrix is also calculated, since it is used in all subsequent item-selection calculations:

$$\Phi^{-1} = \begin{pmatrix} 1 & .6 \\ .6 & 1 \end{pmatrix}^{-1} = \begin{pmatrix} 1.563 & -0.938 \\ -0.938 & 1.563 \end{pmatrix}$$

4.2. ITEM SELECTION

Item selection proceeds by computing the determinant of the posterior information matrix $|\mathbf{I}_{i|S_{k-1}}|$ for each candidate item ($i \in R_k$), where the information matrix is evaluated at the provisional ability estimate $\hat{\theta}^{k-1}$. From (14) (Appendix A), we see that the posterior information matrix consists of summands arising from three sources

$$\mathbf{I}_{i|S_{k-1}} = \overbrace{\Phi^{-1}}^{\text{Prior}} + \overbrace{\mathbf{W}_{S_{k-1}}}^{\text{Administered Items}} + \overbrace{\mathbf{W}_i}^{\text{Candidate Item}} . \tag{12}$$

The first source is the inverse prior covariance matrix (initialized in the first step). The second source consists of summed $\mathbf{W}$-matrices associated with previously administered items

$$\mathbf{W}_{S_{k-1}} = \sum_{j \in S_{k-1}} \mathbf{W}_j ,$$

where $\mathbf{W}_j$ for item j is defined by (13), and the sum $\sum_{j \in S_{k-1}}$ runs over those items already selected. The final term consists of the $\mathbf{W}$-matrix for the candidate item i, also defined by (13).

Table 2 displays values required to select the first item. These include the $\mathbf{W}_i = \{w_{i(11)}, w_{i(12)} = w_{i(21)}, w_{i(22)}\}$ and posterior information

Table 2. Item selection calculations: First item.

Item i	$w_{i(11)}$	$w_{i(12)}$	$w_{i(22)}$	$I_{i(11)}$	$I_{i(12)}$	$I_{i(22)}$	$|\mathbf{I}_i|$
1	0.482	0.000	0.000	2.044	−0.938	1.563	2.315
2	0.000	0.000	0.330	1.563	−0.938	1.893	2.079
3	0.028	0.028	0.028	1.591	−0.909	1.591	1.703
4	0.302	0.000	0.000	1.865	−0.938	1.563	2.034
5	0.000	0.000	0.310	1.563	−0.938	1.873	2.047
6	0.003	0.005	0.007	1.566	−0.933	1.570	1.588
7	0.287	0.000	0.000	1.850	−0.938	1.563	2.011
8	0.000	0.000	0.672	1.563	−0.938	2.234	2.612

$\mathbf{I}_i = \{I_{i(11)}, I_{i(12)} = I_{i(21)}, I_{i(22)}\}$ matrices and their determinants for the eight candidate items. Since no items have been administered prior to the first item, the posterior information matrix consists of terms from two (rather than three) sources:

$$\mathbf{I}_i \;=\; \overbrace{\Phi^{-1}}^{\text{Prior}} \;+\; \overbrace{\mathbf{W}_i}^{\text{Candidate Item}}$$

From inspection of the last column in Table 2, Item 8 is selected for administration, since it has the largest criterion value: $|\mathbf{I}_8| = 2.612$.

4.3. PROVISIONAL ABILITY ESTIMATION

Once the kth selected item has been administered and scored, the provisional ability estimate $\hat{\boldsymbol{\theta}}^{k-1}$ is updated using the full set of k observed-responses and administered-items to produce $\hat{\boldsymbol{\theta}}^k$. Unfortunately, there is no guarantee that the required iterative numerical procedures (Newton-Raphson or Fisher-Scoring algorithms) will converge if the starting value for the ability parameter $\boldsymbol{\theta}^{(1)}$ in (10) and (11) is far from the maximum. However, satisfactory convergence behavior is generally obtained by setting the starting value $\boldsymbol{\theta}^{(1)}$ equal to the posterior mode obtained from the previous calculations (i.e. $\boldsymbol{\theta}^{(1)} = \hat{\boldsymbol{\theta}}^{k-1}$). The starting value $\boldsymbol{\theta}^{(1)}$ for the first provisional update is set equal to the mean of the prior. Typically if one method fails (Newton-Raphson or Fisher-Scoring) the other of the two methods will converge to the true maximum. In practice it is useful to program both methods, using one as a backup in case the other fails to converge. Using the Newton-Raphson algorithm based on (10) and (11), a correct response to Item 8 results in the posterior mode estimate $\hat{\boldsymbol{\theta}}^1 = (0.102, 0.170)$.

Table 3. Item selection and scoring summary.

Sequence k	Item i	u	Posterior Mode	
			$\hat{\theta}_1$	$\hat{\theta}_2$
0	—	—	0.000	0.000
1	8	1	0.102	0.170
2	1	1	0.457	0.343
3	4	0	0.103	0.171
4	2	0	0.034	−0.075

4.4. ITEM SELECTION AND SCORING CYCLE

The previous two steps of item-selection and provisional ability estimation are repeated until the test termination criterion has been satisfied—in this example, until 4 items have been administered. Tables 3 and 4 display key summary calculations used in item selection and scoring.

Table 3 provides a summary of the administered items i, responses u, and modal ability estimates ($\hat{\theta}_1$ and $\hat{\theta}_2$). As indicated, the first item selected was Item 8. A correct response to this item ($u = 1$) resulted in a two-dimensional Bayes mode estimate of $\hat{\theta}_1 = 0.102$ and $\hat{\theta}_2 = 0.170$. The second, third, and fourth items selected were 1, 4, and 2, respectively. Note that a correct response to an item resulted in higher $\hat{\theta}$ values along *both* dimensions. Similarly, an incorrect response also influenced the provisional ability estimates of both dimensions—resulting in lower $\hat{\theta}$ scores. The final ability estimate after providing an incorrect response to the fourth item was $\hat{\boldsymbol{\theta}}^4 = (0.034, -0.075)$.

Table 4 provides the item selection criteria based on the $|\mathbf{I}_{i|S_{k-1}}|$ indices. The last row displays the provisional ability estimate used in the evaluation of the posterior information matrix $\mathbf{I}_{i|S_{k-1}}$. As indicated the first item selected was Item 8, which had the maximum value of the criterion $|\mathbf{I}_8| = 2.612$. The second item selected was Item 1, which had the largest criterion value among the remaining candidate items, and so forth.

Table 5 displays calculations associated with the selection of the last (fourth) item. The elements of the $\mathbf{W}_i$ matrices evaluated at the provisional ability estimate $\hat{\boldsymbol{\theta}}^3 = (0.103, 0.171)$ are displayed in Columns 2–4. Columns 5–7 display elements of the posterior information matrices $\mathbf{I}_{i|S_3}$ for candidate items (those not previously administered). These matrices are computed from (14), which after the third administered

Table 4. Item selection indices $\left|\mathbf{I}_{i|S_{k-1}}\right|$

	Adaptive Test Sequence k			
Item i	1	2	3	4
1	2.315	$\Rightarrow 3.265$	—	—
2	2.079	2.893	3.782	$\Rightarrow 5.247$
3	1.703	2.591	3.920	4.617
4	2.034	3.247	$\Rightarrow 5.572$	—
5	2.047	2.864	3.754	5.201
6	1.588	2.355	3.390	4.345
7	2.011	2.692	2.992	4.658
8	$\Rightarrow 2.612$	—	—	—
$\hat{\mu}^{k-1}$	$(0.000, 0.000)$	$(0.102, 0.170)$	$(0.457, 0.343)$	$(0.103, 0.171)$

Table 5. Item selection calculations: Fourth item.

Item i	$w_{i(11)}$	$w_{i(12)}$	$w_{i(22)}$	$I_{i(11)|S_3}$	$I_{i(12)|S_3}$	$I_{i(22)|S_3}$	$\left|\mathbf{I}_{i|S_3}\right|$
1	0.491	0.000	0.000	—	—	—	—
2	0.000	0.000	0.396	2.538	−0.938	2.413	5.247
3	0.058	0.058	0.058	2.597	−0.879	2.075	4.617
4	0.485	0.000	0.000	—	—	—	—
5	0.000	0.000	0.378	2.538	−0.938	2.395	5.201
6	0.010	0.015	0.022	2.548	−0.923	2.039	4.345
7	0.206	0.000	0.000	2.745	−0.938	2.017	4.658
8	0.000	0.000	0.455	—	—	—	—

item take the form

$$\mathbf{I}_{i|S_3} \;=\; \overbrace{\Phi^{-1}}^{\text{Prior}} \;+\; \overbrace{\mathbf{W}_1 + \mathbf{W}_4 + \mathbf{W}_8}^{\text{Administered Items}} \;+\; \overbrace{\mathbf{W}_i}^{\text{Candidate Item}} .$$

The item selection criteria computed from the determinant of the posterior information matrices are displayed in the last column of Table 5. The maximum value is associated with Item 2, which was administered as the fourth item in the adaptive sequence.

5. Discussion

The multidimensional item-selection and scoring methods presented here provide an opportunity for increased measurement efficiency over unidimensional adaptive testing methods. However, before these benefits can be fully realized, several practical issues including item parameter specification and item-exposure must be addressed. Segall (1996) provides a discussion of a straight forward approach for item-parameter specification based on unidimensional 3PL estimates. Also discussed is an approach to exposure control that places a ceiling on the administration rates of the pool's most informative items, while sacrificing only small to moderate amounts of precision.

By applying Bayesian principles to multidimensional IRT, item-selection and scoring algorithms can be specified which enhance the precision of adaptive test scores. This increase in precision or efficiency can be potentially large for test-scores obtained from batteries that measure several highly correlated dimensions. However, the magnitude of the efficiency-gain over unidimensional methods is likely to be test- or battery-specific. For specific applications, efficiency gains can be investigated through a direct comparison of unidimensional and multidimensional approaches. To this end, this chapter presents the underlying theoretical and computational bases for the multidimensional approach—increasing the accessibility of this new methodology to interested researchers and practitioners.

Appendix

A. Computational Formulas

A.1. FIRST PARTIAL DERIVATIVES

$$\frac{\partial}{\partial \boldsymbol{\theta}} \ln f(\boldsymbol{\theta}|\mathbf{u}) = D \sum_{i \in S} v_i \mathbf{a}_i - \Phi^{-1} (\boldsymbol{\theta} - \boldsymbol{\mu}) \ ,$$

where the sum runs over items contained in S, and

$$v_i = \frac{[p_i(\boldsymbol{\theta}) - c_i] \, [u_i - p_i(\boldsymbol{\theta})]}{(1 - c_i) \, p_i(\boldsymbol{\theta})} \ .$$

A.2. SECOND PARTIAL DERIVATIVES

$$\mathbf{J}_S(\boldsymbol{\theta}) \equiv \frac{\partial^2}{\partial \boldsymbol{\theta} \partial \boldsymbol{\theta}'} \ln f(\boldsymbol{\theta}|\mathbf{u}) = D^2 \sum_{i \in S} \mathbf{a}_i \mathbf{a}_i' w_i - \Phi^{-1} \ ,$$

where

$$w_i = \frac{q_i(\boldsymbol{\theta}) \left[p_i(\boldsymbol{\theta}) - c_i\right] \left[c_i u_i - p_i^2(\boldsymbol{\theta})\right]}{p_i^2(\boldsymbol{\theta}) \left(1 - c_i\right)^2} .$$

A.3. POSTERIOR INFORMATION MATRIX

The information matrix for a set of items S is given by

$$\mathbf{I}_S = -\mathrm{E}\left[\frac{\partial^2}{\partial\boldsymbol{\theta}\partial\boldsymbol{\theta}'} \ln f(\boldsymbol{\theta}|\mathbf{u})\right] = \Phi^{-1} + \sum_{i\in S}\mathbf{W}_i \, ,$$

where

$$\mathbf{W}_i = D^2\mathbf{a}_i\mathbf{a}_i' w_i^* \tag{13}$$

and where

$$w_i^* = \frac{q_i(\boldsymbol{\theta})}{p_i(\boldsymbol{\theta})} \times \left[\frac{p_i(\boldsymbol{\theta}) - c_i}{1 - c_i}\right]^2 .$$

The posterior information matrix associated with candidate item i

$$\mathbf{I}_{i|S_{k-1}} = \Phi^{-1} + \mathbf{W}_i + \sum_{j\in S_{k-1}}\mathbf{W}_j \tag{14}$$

is formed from $\mathbf{W}$-terms associated with previously administered items S_{k-1}, and from a $\mathbf{W}$-term associated with candidate item i.

References

Anderson, T. W. (1984). *An introduction to multivariate statistical analysis.* New York: John Wiley & Sons.

Bernardo, J. M., & Smith, A. F. M. (1994). *Bayesian theory.* New York: John Wiley & Sons.

Birnbaum, A. (1968). Some latent trait models and their use in inferring an examinee's ability. In F. M. Lord & M. R. Novick (Eds.), *Statistical theories of mental test scores* (pp. 395–479). Reading, MA: Addison-Wesley.

Bloxom, B. M., & Vale, C. D. (1987, June). *Multidimensional adaptive testing: A procedure for sequential estimation of the posterior centroid and dispersion of theta.* Paper presented at the meeting of the Psychometric Society, Montreal.

Lord, F. M., & Novick, M. R. (1968). *Statistical theories of mental test scores.* Reading, MA: Addison-Wesley.

Luecht, R. M. (1996). Multidimensional computerized adaptive testing in a certification or licensure context. *Applied Psychological Measurement, 20,* 389–404.

O'Hagan, A. (1994). *Kendall's advanced theory of statistics: Bayesian inference* (Vol. 2B). London: Edward Arnold.

Owen, R. J. (1975). A Bayesian sequential procedure for quantal response in the context of adaptive mental testing. *Journal of the American Statistical Association,* 70, 351–356.

Searle, S. R. (1982). *Matrix algebra useful for statistics.* New York: John Wiley & Sons.

Segall, D. O. (1996). Multidimensional adaptive testing. *Psychometrika, 61,* 331–354.

Segall, D. O., & Moreno, K. E. (1999). Development of the computerized adaptive testing version of the Armed Services Vocational Aptitude Battery. In F. Drasgow & J. B. Olson-Buchanan (Eds.), *Innovations in computerized assessment.* Hillsdale, NJ: Lawrence Erlbaum Associates.

Tam, S. S. (1992). *A comparison of methods for adaptive estimation of a multidimensional trait.* Unpublished doctoral dissertation, Columbia University, New York City, NY.

Tatsuoka, M. M. (1971). *Multivariate analysis: Techniques for educational and psychological research.* New York: John Wiley & Sons.

van der Linden, W. J. (1999). Multidimensional adaptive testing with a minimum error-variance criterion. *Journal of Educational and Behavioral Statistics, 24,* 398–412.

Chapter 4
The GRE Computer Adaptive Test: Operational Issues

Craig N. Mills & Manfred Steffen
American Institute of Certified Public Accountants, USA
Educational Testing Service, USA

1. Introduction

In 1993, the Graduate Record Examinations (GRE) Program introduced the first high-stakes, large-scale computer adaptive admissions
test. The introduction of the GRE General Test Computer Adaptive
Test (CAT) marked the culmination of several years of research and
development. As with the introduction of any new technology, the end
of the research and development activities was only the first step in the
computerization of the test. In the intervening years, the GRE Program
gained a wealth of operational experience with the CAT and identified
a number of issues that required additional investigation. The purposes
of this chapter are to briefly describe the GRE program, explain the
factors associated with several of the operational issues, present the
results of investigations into those issues, and identify both solutions
and areas in which additional research is warranted. The chapter is
divided into several sections. In the first section, a brief overview of
the GRE Program is presented. The development of the GRE CAT is
discussed in the second section. The next three sections are devoted
to an explanation of some of the operational issues faced by the GRE
Program (item overlap, maintaining item pool quality over time, and
scoring incomplete tests) and summaries of the activities conducted by
the GRE Program to investigate those issues. The final section includes
concluding remarks and suggestions for additional research.

2. Description of the Graduate Record Examinations Program

The GRE Program provides assessments to assist graduate programs
in the United States in their selection of graduate students. The program consists of the GRE General Test, a broad measure of developed
abilities, and a series of Subject Tests that measure achievement in specific areas of study (e.g. Engineering and Psychology). The GRE was

W.J. van der Linden and C.A.W. Glas (eds.),
Computerized Adaptive Testing: Theory and Practice, 75–99.
© *2000 Kluwer Academic Publishers. Printed in the Netherlands.*

ceded to Educational Testing Service (ETS) by the Carnegie Foundation at the time ETS was formed and has remained the property of ETS since that time. However, policy oversight is provided by the GRE Board, which consists predominantly of deans of graduate schools. Two associations, the Association of Graduate Schools and the Council of Graduate Schools, each appoint one Board member every year. The GRE Board elects two other at large Board members annually. Board members serve four-year terms.

The purpose of the GRE Program is to assist graduate schools in the identification and selection of graduate students. Services also exist to help prospective students identify graduate programs of interest to them. In this sense, the GRE Program is a "full service" testing program. In addition to the various GRE tests, the Program produces study materials, publications designed to help graduate students identify programs, and lists of prospective graduate students for graduate schools based on the schools' recruitment criteria. The Program also sponsors Forums on Graduate Education, which provide an opportunity for graduate school recruiters and prospective students to interact. The GRE Board also supports an active program of research on psychometric issues related to the tests and on issues in graduate recruitment, admissions, and retention.

The best known component of the GRE Program is the GRE General Test. In 1992 a computer version of the General Test was introduced. The following year, it was replaced with a computer adaptive test. To date, the Subject Tests have not been offered on computer. Therefore, the remainder of this chapter discusses only the General Test.

The General Test includes measures of analytical, quantitative, and verbal reasoning. Approximately 350,000 individuals take the test each year. Most examinees are U.S. citizens, but about 20 percent are citizens of other countries. Examinees are required to sit for all three measures in one administration. Scores on each measure are reported on scales that range from 200 to 800 in increments of ten points. The GRE General Test has been shown to be a useful predictor of graduate student performance (first year graduate grades) when used in conjunction with other indicators of graduate success such as undergraduate grades.

The GRE General Test is composed of four- and five-option multiple-choice questions. Four types of items, reading comprehension, sentence completion, analogies, and antonyms comprise the verbal measure. The quantitative measure contains discrete, problem solving, and quantitative comparison items. Two types of items, analytical reasoning and logical reasoning are included in the analytical measure.

3. Development of the GRE CAT

Development of the GRE CAT began in the fall of 1988. An important factor in the development of the GRE CAT was the program's policy decision to introduce the CAT while continuing to offer the traditional paper-and-pencil version of the GRE General Test. This policy drove a number of decisions about the design of the CAT version of the test since the new, adaptive test had to produce scores that were comparable to the paper-and-pencil test. Two possible sources of non-comparability between the CAT and the traditional test were identified: the mode of delivery (computer or paper) and the testing model (adaptive or traditional/linear).

Two studies were conducted to establish comparability. The first, which compared the traditional paper-and-pencil test with the same traditional test administered via computer (Schaeffer, Reese, Steffen, McKinley, & Mills, 1993), was supportive of claims of comparability across modes of assessment. The second study assessed the comparability of traditional and adaptive tests when both were administered via computer (Schaeffer, Steffen, Golub-Smith, Mills, & Durso, 1995). Again, the results were supportive of claims of comparability although an equating study was required to adjust for differences in the traditional and adaptive analytical reasoning measure. Based on the results of these two studies, the GRE CAT was introduced operationally in November 1993.

There are a number of differences between the structure of the GRE CAT and its paper-and-pencil counterpart. Since the CAT contains only about half the number of items as appear on the traditional form of the assessment, test specifications had to be modified. To modify the test specifications, test development staff created target specifications at several different test lengths based on a logical analysis of the existing specifications. Simulation studies were then conducted to determine which test lengths were most likely to yield appropriate scores (Mills, 1999).

Introduction of the GRE CAT also required a number of decisions about test administration, such as whether or not to allow examinees to return to previously answered questions, how to select the first item to be administered, and so on. These issues are explained in Mills and Stocking (1996).

In the five years following the introduction of the GRE CAT, over 500,000 individuals took the adaptive version of the test. Over 75 unique CAT item pools were developed and used operationally. Thus, the GRE Program has a wealth of data about the operational issues associated with adaptive testing. While the vast majority of GRE CATs

have been administered without difficulty, a number of operational issues have arisen and have been investigated using the accumulated experience and data of the program. In the next sections of the chapter,
we will address three such issues.

4. Item Overlap

The popular descriptions of adaptive testing describe such tests as ones
that are created while the examinee is taking the test with the result
that every test-taker receives a unique test. However, whenever two
examinees are administered a CAT drawn from the same finite item
pool (or from different pools with some items in common), there is a
non-zero probability that the examinees' tests will contain one or more
items in common. Since the GRE CAT is administered on an almost
daily basis, this item overlap between examinees could pose a security
risk.

The extent to which item overlap is, in fact, a security risk is a matter
of some debate. Several factors have to be considered when evaluating
these security risks for two individual test-takers. First, what is the
likelihood that test takers will be administered common test questions?
Second, how many common questions are likely to be administered to a
given pair of test takers? Third, what is the likelihood that the two test
takers will be in communication with one another? Finally, in the event
that two test takers are in communication with one another, what is
the likely effect of their communication on test scores?

Steffen and Mills (1999) have reported the result of several analyses
of item overlap using operational data from the GRE Program. They
conducted separate analyses for Low, Medium, and High ability examinees on each of the three measures of the GRE General Test. For
purposes of their analyses, low ability examinees were defined as those
with a test score between 300 and 400; medium ability was defined
as a score between 500 and 600 and scores between 700 and 800 were
designated as high ability.

In the Steffen and Mills (1999) study, several overlap conditions were
evaluated. These were item overlap between examinees of (1) similar
ability within a single item pool, (2) different ability within a single
pool, (3) similar ability across multiple pools[1] and (4) different ability
across multiple pools.

---

[1] The GRE Program employs a pool rotation strategy based on the rolling development of multiple item pools and the frequent rotation of pools in the field. No
new item pool may be created that has more than 20 percent of its items in common
with any previously developed pool.

As expected, examinees with similar abilities who took tests derived from the same item pool had the highest degree of overlap. On the analytical measure, less than seven percent of the examinees had no items in common across ability levels. Less than one percent of examinees had no items in common for the other measures of the General Test. Across all measures and ability levels between 48 and 64 percent of examinees were administered one to five common items. Six to ten common items were administered to 32 to 45 percent of examinees.

The use of different, but overlapping item pools resulted in dramatic improvements. For examinees of similar ability, the percent of examinees receiving no items in common was over 80 percent in seven of nine comparisons. In eight of nine comparisons over 90 percent of examinees with different ability had no items in common.

While these results are encouraging, another threat to test security could arise if several test takers collude to memorize and record test questions with the intent of making the collection of items available to others. If a group of examinees could reproduce a substantial portion of an item pool and another examinee were to study items prior to taking the test, it could result in a meaningful elevation of the examinee's score.

Understanding the impact of collusion, in a CAT environment, is quite complex. In a linear examination, collusion necessarily results in scores that are greater than or equal to the scores that would have been produced on any collection of items encountered. However, in a CAT environment, tests are adaptively constructed on the fly. Following the administration of each item, an examinee's ability estimate is recomputed and the next item to be administered is selected based, in part, on this updated ability estimate. As the test progresses, the adaptive algorithm will adjust for correct answers that are not consistent with the rest of the examinee's responses. It is possible, therefore, for an examinee who answers an item correctly through collusion to receive a lower score than would have resulted normally. Of course, as the number of items about which examinees have prior information increases, the ability of the algorithm to successfully recover from "unexpected correct responses" is impeded.

Steffen and Mills (1999) reviewed item overlap in the group context as well by way of a simulation experiment. Adaptive tests were generated for 15 high ability simulees, denoted "sources." A source was assumed to be taking the test for purposes of memorizing questions and creating a bank of items that others may study.

Steffen and Mills (1999) then generated 2000 low, medium, and high ability "recipient" simulees. A recipient simulee was presumed to have access to the bank of items developed from the source simulees.

Table 1. Results of collusion under conditions of perfect and imperfect memorization of items.

Pool	Collusion	Ability	Analytical		Quantitative		Verbal	
			Gain	Imp	Gain	Imp	Gain	Imp
Same	Perfect	Low	20.0	435	16.1	444	12.0	355
		Medium	14.7	247	12.8	251	9.0	243
		High	8.8	54	6.5	54	3.9	53
Different	Perfect	Low	0.6	9	1.1	18	0.9	11
		Medium	1.6	23	1.4	23	1.4	23
		High	1.8	16	0.5	5	0.7	12
Same	Imperfect	Low	3.5	83	3.0	97	2.2	48
		Medium	6.3	112	5.9	135	5.8	160
		High	6.6	50	4.8	51	3.3	52
Different	Imperfect	Low	>0.0	>0	0.2	4	0.4	4
		Medium	0.1	1	0.3	6	0.5	6
		High	0.9	8	0.3	3	0.7	12

Six adaptive tests were then simulated for each recipient examinee. Three tests were administered from the pool administered to the source simulees and three were administered from a different, overlapping pool. Within each of these conditions, tests were simulated in which there was (a) no cheating–item performance was based solely on ability, (b) perfect cheating–recipients were credited with a correct response for all items administered to the source simulees, and (c) imperfect cheating–recipients were credited with a correct response for items seen by five or more source simulees.

Among the indicators of the potential impact of collusion were the number of items answered correctly as a result of collusion (Gain), and the impact of the gain in number of items answered correctly on the reported score (Imp). Table 1 summarizes these statistics for the simulations described above. It includes the average results for three ability levels for the simulations in which collusion was modeled for all items administered to sources.

When the number of compromised items available within a pool is relatively large and memorization is effective, the impact of collusion is quite dramatic. The "Gain" columns in Table 1 indicates that, for the Low ability Analytical simulees, the expectation is that 20.0 items would be answered correctly that would otherwise have been missed. Able examinees also benefit from the collusion, at least in terms of number of items answered correctly. For the Analytical measure, High ability examinees are expected to answer 8.8 items correctly that they would otherwise have missed.

The "Imp" column in Table 1 summarizes the mean score increase as a result of collusion. On average, Low ability Analytical examinees are expected to increase their reported score by 435 points as a result of collusion. The mean reported score after collusion is 780. The impact on High ability examinees is less severe (only 54 reported score points). However, since the scale is bounded at 800, there is a ceiling effect.

The impact of collusion is markedly dampened when items are selected from different (but overlapping) pools. Most notably, the advantage for low ability students who engage in collusion is virtually eliminated. Even for examinees of moderate ability, the score gain when different pools are administered is less than the conditional standard error of measurement on the tests. Low ability Analytical simulees benefit by only 9 points. The bias introduced by collusion, when testing with different pools, is approximately .20 conditional standard errors of measurement.

The lower portion of Table 1 presents results when collusion is less than 100 percent effective. The results of testing with a different pool are nearly identical for the baseline and collusion conditions. The tests delivered are virtually identical under both conditions and the impact on reported scores is negligible.

Table 2 summarizes the score changes for low, medium, and high ability simulees respectively under conditions of perfect and partial memorization when tests are administered to recipient simulees from the same pool as was administered to source examinees. The table includes the percent of simulees receiving different numbers of items administered to source examinees and the mean and standard deviation of the resulting score changes. The general pattern of results is similar in all three cases. When memorization on the part of both recipient

and source examinees is perfect, the impact is substantial both in terms of the percentage of simulees obtaining score improvements and the magnitude of the score increase. As noted previously, high ability simulees show less improvement than low or medium ability simulees due to a ceiling effect. This general pattern holds as well when memorization is less effective. In this case, both the percentage of students benefiting from collusion and the magnitude of the score improvement is lessened.

Table 2. Distribution of score impact by number of items delivered.

Memorization/ Ability	Impact	Number of Compromised Items Delivered						
		00	01-05	06-10	11-15	16-20	21-25	26-28
Perfect	Pct	0.0	0.2	0.1	0.1	0.0	7.1	92.6
Low	Mean	0.0	25.0	75.0	250.0	0.0	447.2	445.2
	S.D.	0.0	25.0	25.0	0.0	0.0	56.9	55.5
Partial	Pct	66.1	10.7	0.2	0.2	19.6	3.4	0.0
Low	Mean	0.0	6.1	116.7	300.0	415.5	416.2	0.0
	S.D.	0.0	28.1	85.0	35.4	58.9	50.3	0.0
Perfect	Pct	0.0	0.0	0.1	0.0	0.0	7.9	92.1
Medium	Mean	0.0	0.0	50.0	0.0	0.0	256.3	251.9
	S.D.	0.0	0.0	0.0	0.0	0.0	55.0	53.4
Partial	Pct	27.4	14.6	0.3	2.9	45.7	9.2	0.0
Medium	Mean	0.0	10.3	83.3	174.1	231.5	241.0	0.0
	S.D.	0.0	21.5	47.1	45.7	49.4	49.6	0.0
Perfect	Pct	0.0	0.0	0.0	0.0	0.0	6.4	93.6
High	Mean	0.0	0.0	0.0	0.0	0.0	48.4	53.0
	S.D.	0.0	0.0	0.0	0.0	0.0	42.4	43.0
Partial	Pct	0.0	0.9	1.0	9.0	76.7	12.6	0.0
High	Mean	0.0	5.6	44.7	54.7	49.4	49.4	0.0
	S.D.	0.0	15.7	35.9	36.0	41.8	42.2	0.0

Almost 93 percent of low ability simulees benefited substantially from collusion when their test was drawn from the same pool and memorization was perfect. Between 26 and 28 of the items administered in this simulation had been administered to at least one source examinee with the result that scores increased an average of 445 points. An additional 7 percent of the simulees received between 21 and 25 compromised items in this simulation and had a similar score gain. The situation improves when memorization is less than perfect. Approximately two-thirds of the low ability simulees receive no compromised items in this case and only three percent receive over 20 compromised items. However, even when memorization is imperfect, almost a quarter of the low ability examinees could expect score gains of over 400 points.

The results for simulees of medium ability are similar when memorization is perfect. Over 90 percent of the simulees receive between 26 and 28 compromised items with an average score gain of over 250 points. Contrary to the case with low ability simulees, however, collusion continues to have a substantial impact even when memorization is less effective. Over fifty percent of the simulees had a gain of over 200 points in this simulation.

The results for high ability examinees also show extensive overlap when high ability examinees are considered. As before, over 90 percent of the simulees have between 26 and 28 compromised items administered. However, the impact is only slightly over 50 points due to the previously mentioned ceiling effects. For high ability simulees, 50 point score gains continue to be seen for over 90 percent of the group when memorization is imperfect.

We did not produce tables depicting the distribution of score gains when recipient examinees are administered tests from different pools. Rotating pools effectively reduces the impact of collusion to very small levels.

These results are very salient and might cause tremendous concern, particularly if a single pool was used for an extended period of time with a large volume of examinees. The results when examinees test with a different, but overlapping, pool are also compelling and much less troubling. One conclusion that could be drawn is that rotating pools effectively eliminates the impact of collusion. Based on these data alone, such a conclusion is not unreasonable, particularly since the assumption of perfect memorization by recipient examinees is unlikely to be correct. However, it is important to keep in mind that the collusion here is focused on a single "source" pool. Since subsequent pools are constructed to have minimal overlap with any single source pool, the results are not overly surprising. If an attack were undertaken on a series of "source" pools, the results would likely be much more dramatic even when testing with different pools. This would be due to the fact that a large proportion of the items within any single pool have appeared in at least one other pool. Of course, in that case, the memorization task for recipient examinees would become quite large and one could posit that memorization would be less effective.

5. Maintaining Pool Quality Over Time

The GRE Program offers the computer adaptive GRE General Test on a frequent basis, typically six days a week, three weeks per month. For security purposes, many different item pools are distributed to the field and used operationally. Arbitrary rules have been developed regarding the frequency of item pool development. These rules stipulate that recently used items are not eligible for inclusion in a new pool, that items exceeding certain usage rates in the past cannot be included in new pools and that items exceeding a maximum usage level are retired from use in the operational program[2]. Way, Steffen, and Anderson (1998) have described the GRE pool development process in some detail.

If new items are not added to the inventory of items available for pool construction (i.e. the inventory is static), the quality of successive item pools[3] will decline over time. This decline in quality could be gradual if the initial item pool is very large or it could be precipitous if the pool is small. However, even in a large, but finite inventory, pool quality will decline as items reach their maximum number of administrations and are retired. If item pools are generated sequentially, one would expect a steady decline in the quality of pools with the development of each successive pool.

One way to retard the decline in pool quality would be to develop multiple pools simultaneously. This practice, if implemented, would create the maximum possible number of item pools simultaneously with all pools have roughly the same quality. Another method for maintaining the quality of item pools is to generate new test questions such that the number of newly available items matches or exceeds the number of items that become unavailable for one reason or another. In order for this method to work, the newly available items must not only become available in sufficient quantity, they must also replace the content and statistical characteristics of the removed items. Thus, development of new item pools needs to take into consideration the length of time pools are used in the field, the rate at which items are withheld from administration, the rate at which items are retired altogether, and the rate at which new items become available.

The GRE currently obtains new items via pretesting in both its paper-based and computer-adaptive tests. Therefore, there is a small,

---

[2] These items are typically disclosed to the public and incorporated into test preparation materials.

[3] Item pool quality could be assessed by measures such as precision of the ability estimates derived from the pool and number of times test content specifications cannot be met.

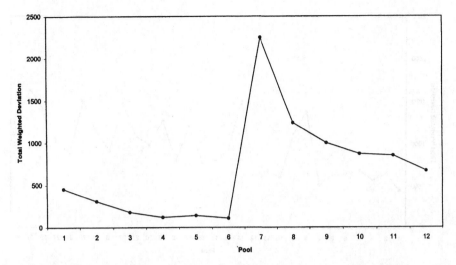

Figure 1. Total weighted deviations for the development of 12 item pools from a finite pool in one year excluding items administered to 200 examinees in the prior two month period.

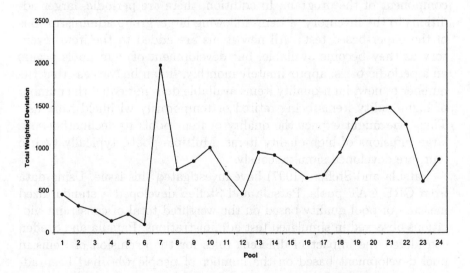

Figure 2. Total weighted deviations for the development of 24 item pools from a finite pool in one year excluding items administered to 200 examinees in the prior two month period.

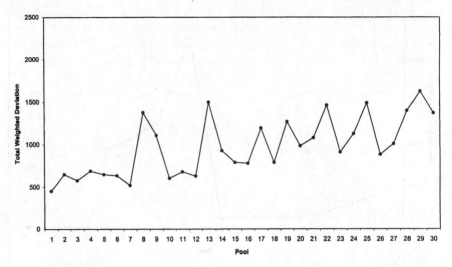

Figure 3. Total weighted deviations for the development of 30 item pools from a finite pool in one year excluding items administered to 500 examinees in the prior two month period.

but steady influx of new items through the computer-based testing component of the program. In addition, there are periodic, larger additions to the inventory of items following large, group administrations of the paper-based test[4]. All new items are added to the item inventory as they become available, but development of item pools occurs on a periodic basis, approximately monthly. It can be the case that the number of new, high-quality items available does not equal the number of high-quality items being retired or temporarily withheld from use. Thus, one might expect the quality of item pools to decline between large infusions of high-quality items. Multiple pools, typically two to four, are developed simultaneously.

Patsula and Steffen (1997) have investigated this issue. Using data from GRE CAT pools, Patsula and Steffen developed a standardized measure of pool quality based on the weighted total of constraint violations observed in simulated test administrations. Patsula and Steffen considered the impact of different rules for the exclusion of items in pool development based on the number of people who had been administered specific items in a defined period of time and the number of pools developed within a defined time period on the quality of the

---

[4] The GRE Program is reducing its use of paper-based testing. As a result, the number of new items and the frequency with which they become available through this mode of testing is decreasing. More items are becoming available through computer-based pretesting, but the number that become available in any particular (short) period of time remains relatively small.

resultant item pools. Sampling from a fixed inventory of items, Patsula and Steffen evaluated selection rules that embargoed items from appearance in new pools if they had been administered to 100, 200, 300, 400, or 500 test takers. They also considered the impact of developing 6, 12, 18, 24, and 30 pools from the inventory for administration in, say, a one year period.. Figures 1, 2, and 3, taken from Patsula and Steffen (1997) demonstrate the effect of changes in these variables. Figure 1 depicts the total weighted deviations when 6 item pools are developed and items are excluded from reuse if they have been administered to 200 or more people in previous pools. Figure 2 retains the 200 administration maximum, but calculates the deviations that occur when 24 pools are developed. Comparison of these figures indicates that pool quality fluctuates on a cyclical basis. These cycles are dependent primarily on the return of high quality items to the inventory following an embargo on their use due to the number of prior exposures. It is also clear that pool quality will deteriorate more quickly when more pools are developed and used in a given period of time than when fewer pools are used. That is, in both Figure 1 and Figure 2, pool quality declines after about three pools. However, since pools are used for about two months in the situation depicted in Figure 1 and they are used for only two weeks in Figure 2, the decline in quality occurs earlier in the testing year in the latter scenario. Thus, while the security concerns described in the previous section argue for the development of many pools and frequent rotation of those pools, such a tactic will introduce greater variation in quality among pools. Figure 3 shows a similar pattern, however, the extent of variation is lessened when the threshold for the exclusion of items is raised. As a result, for a given collection of test items, a testing program will need to balance the needs of security against the consistency of pool quality.

6. Scoring Incomplete CATs

A practical decision that was required at the time the GRE CAT was introduced was how to handle incomplete tests. Although time limits were established with the expectation that virtually all examinees would complete the test, policies were required to govern the generation of test scores for those who did not answer all items. Since the GRE Program could not locate empirical guidance on this issue, logical analysis was used to establish a policy.

Ultimately, GRE decided to implement a policy that came to be known as the "80% Rule" since the primary feature was to report scores for any examinee who completed 80 percent or more of the items on the test. Thus, the scores were based only on the items answered. However, in an attempt to thwart inappropriate test-taking strategies, the policy did provide for a "No Score" in the event an examinee failed to complete 80% of the test. The selection of 80% as the minimum number of items to be answered to receive a score was based on the following considerations:

(1) although individual test results might be somewhat less stable, the overall reliability of the test would not be adversely affected if a small number of scores were reported based on tests that were between 80 and 100 percent complete,

(2) most critical content constraints could be satisfied for most examinees in the first 80 percent of the items, and

(3) initial projections were that only a small percentage of examinees would receive "No Score"

The announcement of the 80% rule immediately generated questions from test takers and coaching schools about test taking strategies. Test takers desired advice concerning whether it was to their advantage to work quickly and attempt to finish the test or to work at a more deliberate pace in order to increase their odds of answering items correctly, but not completing the test. These questions were impossible to answer at the time since individuals would differ in their success on the questions they answered and since it was not possible to determine the effect of working at different speeds for individual examinees.

Initially, the GRE Program described the 80% rule as follows:

> Since the computer uses your answers and the test design specifications to select your next question, you must answer all questions to get a balanced test. For that reason, you cannot leave a section before time has expired unless you've answered all the questions. There should be plenty of time for you to finish each section. However, a few examinees may not be able to finish. In these cases, a score will be generated if at least 80 percent of the questions have been answered (Educational Testing Service, 1993).

As volumes increased, there were concerns that this language might not be sufficiently clear. Therefore, in the fall of 1995 the GRE Program included a more extensive discussion of the 80% rule in a publication describing the GRE General Test. The 1995 instructions are reproduced below:

Maximize your score through effective time management.

— Answer as many questions as you can in each section. If time expires and you have already answered the minimum number of questions required for a section, you *will* receive a score for the section.

— Once you have answered the minimum number of questions required to earn a score for that section, your score may increase or decrease if you continue. There is no single strategy that is best for everyone. At this point in your test, use the following guidelines in making your decision to continue:

  • If you feel confident that you know the correct answer to a question, it is likely that answering it will increase your score.

  • If you only have enough time to carefully consider one question but you are not confident you can answer the question on your screen correctly, do not attempt to answer any more questions.

  • If you have time to carefully consider several more questions but are not confident you can answer the question on your screen correctly, it might be to your advantage to answer that question by guessing. You will then have the opportunity to answer additional questions that you might find easier. Several different question types are included in each section, and you might feel more confident with some types than with others.

  • Randomly responding to questions at the end of your test will likely decrease your score. Leaving the last few questions unanswered will not decrease your score (Educational Testing Service, 1995).

Over the next several months, changes in test taker behavior were noted on the tests. Furthermore, as data accumulated, it became clear that there were differences among examinee scores for individuals who completed the test and those who appeared to be using the 80% rule to spend more time on each item, but to answer fewer items.

Table 3 depicts completion rates for three months prior to and three months following the change in test completion guidelines. Included in Table 3 are the average scores and the mean score differences for individuals who answered all items on the test and those who answered enough items to obtain a score, but who did not complete the test. The dates shown in Table 3 cover the months of June, July, August,

Table 3. Percent of examinees answering different numbers of items and score differences on the GRE CAT under the 80% rule under original and revised test taking guidelines.

									Mean Score Comp. Tests	Mean Score Incomp. Tests
		Percent of Examinees Answering Different Numbers of Items								

Analyt.

Month	28	29	30	31	32	33	34	35	Comp. Tests	Incomp. Tests
6/95	1	6	9	5	4	4	5	67	497	573
7/95	1	6	6	7	4	4	5	67	509	593
8/95	1	8	10	5	4	4	4	64	518	606
10/95	9	7	10	5	4	3	6	59	537	614
11/95	11	8	10	5	4	4	6	52	542	617
12/95	15	12	8	5	4	4	6	45	530	624

Quant.

Month	23	24	25	26	27	28	Comp. Tests	Incomp. Tests
6/95	0	2	3	2	5	87	499	613
7/95	0	2	3	3	4	88	500	596
8/95	0	3	3	2	5	87	512	598
10/95	3	4	4	6	6	78	528	612
11/95	4	4	4	6	6	76	538	611
12/95	5	5	4	6	6	74	531	619

Verbal

Month	24	25	26	27	28	29	30	Comp. Tests	Incomp. Tests
6/95	0	1	1	2	5	5	90	471	483
7/95	0	1	5	3	3	4	86	469	472
8/95	0	1	2	2	5	4	84	481	468
10/95	2	6	3	4	4	4	78	490	484
11/95	3	5	4	3	4	4	77	506	505
12/95	3	6	4	4	4	4	75	511	504

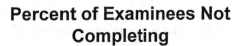

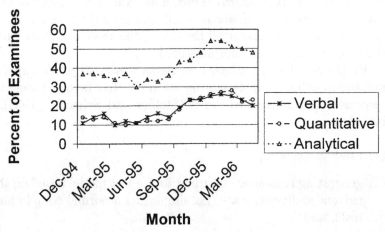

Figure 4. Percent of examinees not completing each measure of the GRE general test CAT before and after changes in Test Taking Guidelines (October 1995).

October, November, and December of 1995. The new guidelines took effect in October 1995. Non-completion rates actually began to increase in September 1995. These data are not included in the table since we believe the change in test taking behavior was a result of people reading the October 1995 publication and applying the new guidelines in September in the erroneous belief that the new guidelines applied at that time. Figure 4 depicts non-completion rates for the three General Test measures for the period of December 1994 through April 1996 inclusive and provides a further demonstration of the impact of the change in test taking guidelines. Table 3 and Figure 4 clearly demonstrate that the change in test taking instructions had a dramatic effect on test completion rates. Table 3 also shows that for the analytical measure, which is speeded and for the quantitative measure which might be slightly speeded, there are clear score differences between individuals who answer all questions and those who do not. For the Analytical Measure, the monthly average score difference ranges from 59 to 94 points. Average Quantitative Measure differences range from 66 to 114 points. The data above clearly indicate that the 80% scoring rule was inadequate and that a scoring method was needed that would both encourage examinees to complete the test and provide a fair estimate of ability for those who were truly unable to complete the test in the time allotted. One proposed option was proportional scoring (Lewis, 1996). In proportional scoring, an adjustment is applied before calculating the final scaled score to reflect the proportion of the test completed (number of items answered/test length). Thus, if an examinee completes the entire test, the proportion answered is 1.0 and no adjustment results. If an examinee does not answer all the items, the score is adjusted to reflect the proportion of items not answered.

Proportional scoring based on items answered was compared to a number of other possible scoring methods that involved generation of responses to not reached items (Steffen and Schaeffer, 1996). These included

(1) random guessing,

(2) generating responses according to the test model based on the most current ability estimate (i.e. simulating a normal completion of the test), and

(3) generating incorrect options for all not reached items.

Figure 5 contains a comparison of these methods of treating incomplete tests. It shows the impact of the various options for individuals who answered 19 items on the 35-item Analytical Measure. The vertical axis is the score that would have been assigned after the 19th item if a score had been generated. The horizontal axis is the score after application of a scoring adjustment. For completeness Figure 5 contains the adjustment that would have resulted if all subsequent items were answered correctly in addition to the options listed previously. Inspection of Figure 5 shows that proportional scoring results in a lower final score for examinees at all ability levels than all other methods except for some low ability examinees when the scoring method is to mark all unanswered items incorrect. Of particular importance, proportional scoring provided a lower score than guessing. This result was obtained for analyses conducted at various test lengths. Although a very small number of cases were noted in which proportional scoring was not lower than guessing, this result was considered to be sufficiently unlikely that it was reasonable to institute proportional scoring and to advise examinees that they should guess to complete their test if they were unable to complete the test in the allotted time. That is, proportional scoring appeared to meet the criteria of fairness and, more importantly, to meet the goal of encouraging examinees to work at a pace that would ensure they completed the test. Proportional scoring was implemented for the GRE Program in the fall of 1996. Test taking guidelines were revised to say "Answer as many questions as you can in each section Random guessing through the test could reduce your score. On the other hand, if you are running out of time and there are still unanswered questions, it is to your advantage to guess at those last questions rather than to leave them unanswered." (Educational Testing Service, 1996).

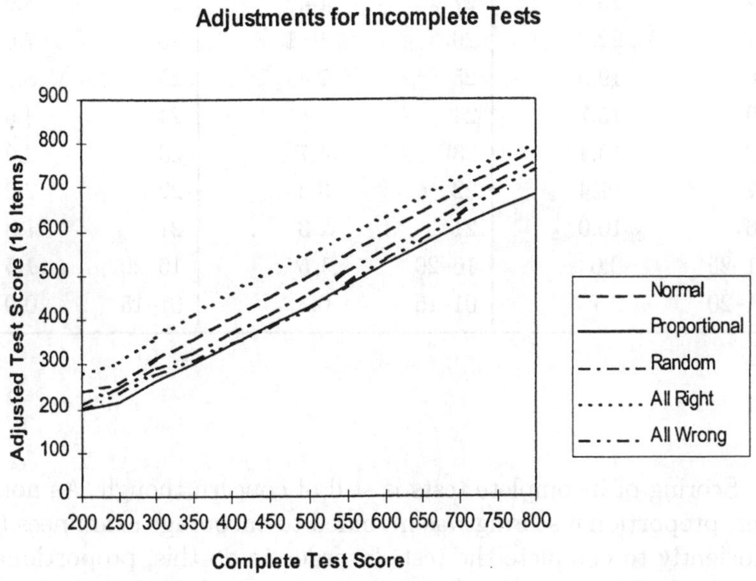

Adjustments for Incomplete Tests

Figure 5. Comparison of the impact of several incomplete test treatments when 19 of 35 items have been answered.

The implementation of proportional scoring had an effect on test-taking behavior as can be seen in Table 4. Table 4 reports the cumulative percentage of examinees answering various numbers of items on the analytical, quantitative, and verbal measures of the General Test under the proportional scoring rule. For all three measures, substantial more examinees are completing the test than was the case under the 80% scoring rule. Less than ten percent of the examinees fail to complete three or more items on the quantitative and verbal measures. Non-completion remains higher on the analytical measure due to speededness, however the non-completion rates are much lower than was previously the case.

Table 4. Cumulative percent of examinees answering different numbers of items in four recent CAT pools under proportional scoring (Educational Testing Service, 1998).

Analytical		Quantitative		Verbal	
Items Answered	Cumulative Percent	Items Answered	Cumulative Percent	Items Answered	Cumulative Percent
35	100.0			30	100.0
34	34.9			29	14.2
33	29.3	28	100.0	28	10.9
32	25.3	27	14.2	27	8.9
31	22.6	26	10.1	26	7.0
30	19.9	25	7.4	25	5.5
29	15.7	24	5.8	24	4.0
28	13.4	23	4.7	23	2.9
27	11.4	22	3.3	22	2.2
26	10.0	21	2.3	21	1.8
21–25	8.6	16–20	1.5	16–20	1.5
01–20	2.1	01–15	0.3	01–15	0.0

Scoring of incomplete tests is still of concern though. As noted earlier, proportional scoring was intended to encourage examinees to work efficiently to complete the test. In order to do this, proportional scoring resulted in a penalty for non-completion of the test that was more severe than random guessing. However, proportional scoring does not always provide the most severe penalty. In some cases, guessing provides a larger score reduction than does proportional scoring. The GRE Program continues to monitor this issue. In order to provide the most accurate advice to examinees as possible, the program reworded its test taking advice again in 1997 telling examinees "If you are stuck on a difficult question, or if you are running out of time and there are still unanswered questions, eliminate as many answer choices as possible and then select and confirm the answer you think is best." (Educational Testing Service, 1997, p. 49).

To investigate the extent to which proportional scoring effects scores, we conducted an analysis using simulated data from the quantitative measure. Table 5 shows the results for low, middle, and high ability examinees who fail to complete the quantitative measure. The number of unanswered items ranges from one to ten. The average score for

Table 5. Comparison of average scores resulting from four methods for scoring incomplete adaptive tests for low, medium, and high ability simulees on the GRE Quantitative Measure.

	# Unanswered	Completion Method			
		Normal	Guessing	Incorrect	Proportional
Low	10	331	319	275	246
Ability	9	331	320	279	255
	8	331	324	289	263
	7	331	324	295	273
	6	331	324	300	282
	5	330	325	304	291
	4	330	324	308	301
	3	329	326	312	310
	2	329	328	320	320
	1	329	330	327	329
Middle	10	534	479	419	397
Ability	9	534	484	432	413
	8	534	493	450	430
	7	534	500	466	446
	6	534	506	479	461
	5	534	509	485	479
	4	534	514	497	492
	3	533	517	504	505
	2	533	523	514	520
	1	533	529	525	533
High	10	749	679	621	568
Ability	9	749	689	644	588
	8	749	689	644	588
	7	749	707	682	628
	6	749	713	691	648
	5	749	720	704	668
	4	749	728	716	688
	3	749	734	726	718
	2	749	739	735	729
	1	749	745	743	749

Table 6. Distribution of score differences between proportional scoring and guessing for low, medium, and high ability simulees on the GRE Quantitative Measure.

	# Unanswered	Differences (Proportional Scoring–Guessing)							
		≤ -60	-50	-40	-30	-20	-10	0	≥ 10
Low	10	72.3	10.1	5.3	3.7	2.9	1.6	3.0	1.3
Ability	9	65.3	12.5	7.3	5.0	3.6	1.9	3.1	1.5
	8	57.7	15.7	11.0	6.2	3.4	2.1	2.7	1.4
	6	27.3	15.3	18.6	17.2	10.7	4.5	4.2	2.4
	5	17.1	11.4	16.3	21.2	16.8	8.6	5.1	3.6
	4	7.1	7.1	11.8	18.2	22.2	17.4	9.8	6.6
	3	3.6	2.9	6.5	14.9	19.9	26.5	14.3	11.6
	2	1.8	1.4	2.9	5.6	14.3	35.9	22.8	15.4
	1	1.2	0.5	1.4	2.4	4.9	13.9	54.7	21.1
Middle	10	81.7	4.9	4.5	2.4	1.9	1.2	0.7	2.9
Ability	9	73.1	8.2	6.7	3.3	2.6	2.3	1.1	2.8
	8	64.4	12.5	8.3	4.9	3.2	2.3	1.1	3.4
	7	51.9	15.3	11.3	8.4	5.0	3.0	1.7	1.3
	6	36.5	15.9	15.7	12.2	7.8	5.4	2.7	4.0
	5	17.9	13.5	17.4	16.2	13.8	8.6	5.6	7.2
	4	7.2	8.2	13.6	18.1	18.5	14.0	9.4	10.1
	3	2.5	3.5	7.4	13.6	18.6	20.7	15.4	18.5
	2	0.9	1.7	2.5	6.6	14.7	22.4	21.3	20.1
High	10	94.1	1.4	1.4	1.1	0.5	0.3	0.3	0.9
Ability	9	92.2	2.7	2.0	1.1	0.5	0.5	0.2	0.9
	8	90.7	3.9	2.1	1.2	0.9	0.5	0.2	0.7
	7	85.3	6.9	3.4	2.3	1.0	0.6	0.3	0.4
	6	54.3	12.3	8.2	3.9	2.4	1.1	0.4	0.7
	5	47.6	20.1	13.8	9.1	5.4	1.7	1.3	1.0
	4	18.9	21.2	22.1	19.6	10.0	4.9	2.2	1.2
	3	3.2	6.3	18.3	25.7	23.0	13.2	6.7	3.6
	2	0.5	0.7	3.4	11.1	22.2	29.9	19.7	12.7
	1	0.2	0.1	0.5	1.8	4.4	14.2	32.0	47.0

normal test completion, completion by guessing, completion by marking all subsequent items incorrect, and proportional scoring is shown. In general, as expected, the average score declines as the scoring rule moves from normal completion though, in order, guessing, all incorrect, and proportional scoring. The only exceptions to this pattern are when one item is unanswered for low or high ability examinees or when three or fewer items are left unanswered for medium ability examinees.

To understand better the effect of proportional scoring, Table 6 shows the distribution of differences in assigned scores between proportional scoring and completing the test by guessing. As expected, proportional scoring results in lower scores for the vast majority of examinees. However, it is important to note that, at all three ability levels and for every test length, at least some test takers would receive a higher score under proportional scoring than by guessing to complete the test. For high ability students in particular, it appears that examinees would fair better to leave the last question unanswered than to guess. The results reported in Table 6 suggest that additional research is necessary to understand the effect of proportional scoring and to consider alternate methods for scoring incomplete tests.

One interesting area of research with regard to proportional scoring is its effect on the scores of examinees who complete the test, but do so by guessing rapidly at the last several items on the test. In this case, the score assigned will be based on a test in which the examinee has responded to items in two different ways during the administration of the test, giving careful consideration to some items and little or no consideration to others. In this case, the impact of proportional scoring on the final scaled score is likely to be affected by not only the responses to the last items, but also the functioning of the item selection algorithm (including any considerations of item exposure that influence item selection), the items that have been administered previously, and the consistency of the examinee's responses to the earlier items. This type of test taking behavior could, for example, be contrasted with the effect on scores of working steadily, but somewhat more quickly than is optimal on each item in order to have time to consider all items equally during the test administration. In either case, the percentage of examinees that might be affected to any substantial degree is likely to be small. Nonetheless, if proportional scoring is to be used routinely, more information is needed about its impact on the validity of the final test scores.

7. Conclusions

Computer-based testing is rapidly expanding. The GRE Program, the first large-scale high-stakes admissions testing program to offer its test on computer has over five years of experiewith computer adaptive testing. As a result, several important practical issues have been identified that require analysis and, possibly, the development of additional techniques for application in operational testing environments. This chapter has explained three of those issues:

1. Item overlap between and among examinees,

2. Maintaining the quality of item pools over time, and

3. Scoring incomplete adaptive tests.

For each topic, this chapter has attempted to describe the issue, explain current procedures and the rationales underlying them, and present data relevant to the problem.

References

Educational Testing Service. (1993). *The GRE Computer Adaptive Testing Program (CAT): Integrating Convenience, Assessment, and Technology.* Princeton, NJ: Educational Testing Service.

Educational Testing Service. (1995). *Graduate Record Examinations 1995-96 General Test Descriptive Booklet.* Princeton, NJ: Educational Testing Service. (1995)

Educational Testing Service. (1996). *Graduate Record Examinations 1996-97 Information & Registration Bulletin.* Princeton, NJ: Educational Testing Service.

Educational Testing Service. (1997). *Graduate Record Examinations 1997-98 Information & Registration Bulletin.* Princeton, NJ: Educational Testing Service.

Educational Testing Service. (1998). *Graduate Record Examinations GRE CAT Pools Analysis Report: Analytical, Quantitative, and Verbal Pools 72 through 75.* Unpublished Statistical Report. Princeton, NJ: Educational Testing Service.

Lewis, C. (1996). An idea for scoring an incomplete adaptive test. Personal communication.

Mills, C.N. (1999). Development and introduction of a computer adaptive Graduate Record Examinations General Test. In F. Drasgow & J. B. Olson-Buchanan, (Eds.), *Innovations in computerized assessment* (pp. 135-178). Hillsdale, NJ: Lawrence.Erlbaum.Publishing.

Mills, C. N., & Stocking, M. S. (1996). Practical issues in large-scale computerized adaptive testing. *Applied Measurement in Education, 9,* (287-304).

Patsula, L. N., & Steffen, M. (1997). *Maintaining item and test security in a CAT environemnt: A simulation study.* Paper presented at the annual meeting of the National Council on Measurement in Education, Chicago, IL.

Schaeffer, G. A., Reese, C.M., Steffen, M., McKinley, R.L., & Mills, C.N. (1993). *Field test of a computer-based GRE General Test.* Research Report, RR-93-07. Princeton, NJ: Educational Testing Service.

Steffen, M., & Mills, C. N. (1999). *An investigation of item overlap and security risks in an operational CAT environment.* Unpublished manuscript.

Steffen, M., & Schaeffer, G. A. (1996). *Comparison of scoring models for incomplete adaptive tests.* Presentation to the Graduate Record Examinations Technical Advisory Committee for the GRE General Test.

Stocking, M. L., & Swanson, L. (1993a). A method for severely constrained item selection in adaptive testing. *Applied Psychological Measurement, 17,* (277-292).

Stocking, M. L., & Swanson, L. (1993b). A model and heuristic for solving very large item selection problems. *Applied Psychological Measurement, 17,* (151-166).

Way, W. D., Steffen, M., & Anderson, G. S. (1998). *Developing, maintaining, and renewing the item inventory to support computer-based testing.* Paper presented at the colloquium Computer-Based Testing: Building the Foundation for Future Assessments, Philadelphia, PA.

Chapter 5
MATHCAT: A flexible testing system in mathematics education for adults

Alfred J. Verschoor & Gerard J.J.M. Straetmans
National Institute for Educational Measurement (Cito),
The Netherlands

1. Introduction

Adult basic education in the Netherlands consists of a program of several courses of various levels. One of the courses in the program is mathematics, offered at three different course levels. The majority of the students have a foreign background. Due to a large variation in background, most of their educational histories are unknown or can only be determined unreliably. In the intake procedure of the program, a placement test is used to assign students to a course level. As the students' abilities vary widely, the paper-and-pencil placement test currently used has the two-stage format described in Lord (1971). In the first stage, all examinees take a routing test of 15 items with an average difficulty corresponding to the average proficiency level in the population of students. Depending on their scores on the routing test, the examinees then take one of the three follow-up tests. Each follow up test consists of 10 items. There are several drawbacks to this current testing procedure:

1. Test administration is laborious because of the scoring that has to take place after the routing test;

2. Preventing disclosure of the test items is difficult due to the flexible intake procedure inherent in adult basic education. Disclosed items can easily lead to misclassifications (assignment of prospective students to a course level for which they lack proficiency).

3. Because only one branching decision is made, possible misroutings cannot be corrected (Weiss, 1974) and measurement precision may be low.

A computerized adaptive placement test might offer a solution to the problems mentioned. First, such tests have as many branching decisions as items in the test. Erroneously branching to items that are

W.J. van der Linden and C.A.W. Glas (eds.),
Computerized Adaptive Testing: Theory and Practice, 101–116.
© 2000 *Kluwer Academic Publishers. Printed in the Netherlands.*

too easy (the response was wrong by mistake) or too difficult (the candidate made a good guess) will be corrected later in the test. Second, computerized test administration offers the advantage of immediate test scoring and feedback. Remedial measures can be taken right after the test. Third, preventing disclosure of testing material is less of a problem because, in theory, each examinee takes a different test.

These features of computerized adaptive testing are very interesting, particularly because all colleges offering adult basic education in The Netherlands already have or are in the process of installing well-equipped computer rooms. Besides, the technology of computerized adaptive testing is flexible enough to deliver tests from the same pool serving other purposes than making placement decision, such as testing the students' achievements during the course or grading them at the end of it. A testing system with these additional features would be very helpful in supporting the current movement towards a more flexible school system in adult education.

In this chapter, a description is given of MATHCAT, the adaptive testing system that has replaced the old paper-and-pencil two-stage test. MATHCAT delivers tests serving two different educational purposes. One type of test is a placement test that can be used to assign examinees to courses in arithmetic/mathematics at three possible levels available. The other is an achievement test that can be used to obtain an estimate of a student's level of achievement during these courses. In our description of the system, we will first focus on the item bank. The quality of the item bank has strong implications for the utility of the test scores. Then, the testing algorithms for the placement test and the achievement test are discussed and results from an evaluation of these algorithms are presented. Finally, we will discuss some features of the student and teacher modules in MATHCAT.

2. An Item Bank for Numerical and Mathematical Skills

Adaptive testing requires an item bank, preferably calibrated using an IRT model. The item bank currently used in MATHCAT contains 578 items, of which 476 were calibrated using the following model (Verhelst & Glas, 1995):

$$p_i(\theta) \equiv P(X_i = 1|\theta) \equiv \frac{\exp(a_i(\theta - b_i))}{1 + \exp(a_i(\theta - b_i))}. \qquad (1)$$

The response to item i, X_i, is either correct ($X_i = 1$) or incorrect ($X_i = 0$). The probability of answering an item correctly in (1) is an increasing function of latent proficiency, θ, and depends on two item characteristics: the item difficulty b_i and the discriminatory power of an item a_i. All parameter values were estimated using the OPLM software (Verhelst & Glas, 1995). The software iteratively chooses integer values for the item parameters a_i, computes conditional maximum likelihood estimates (CML) of the item parameters b_i, and tests for model fit, until acceptable estimates of the values of the item parameters a_i and b_i are obtained. The distribution of θ was estimated using the MML method. The estimated mean and standard deviation were $\hat{\mu} = 0.074$ and $\hat{\sigma} = 0.519$. The uncalibrated items are so-called 'seeded items'. They are new items not used in the adaptive test for the purpose of estimating the proficiency of the examinees but only to collect responses for their future calibration.

Cutoff scores on the proficiency scale are used to define the three course levels. The cutoff scores were derived through the following procedure. First, content specialists defined subsets of items by labeling them as Level 1, Level 2, or Level 3 items. Second, the mean difficulty of each subset of items was computed. Third, using the basic equation of the OPLM model in (1), the cutoff scores were defined as the abilities that had a minimum probability of success equal to .7 for all items labeled as Level 1 and 2, respectively. This procedure resulted in $\theta_{12} = -0.544$ (cutoff score between Level 1 and 2) and $\theta_{23} \doteq -0.021$ (cutoff score between Level 2 and 3).

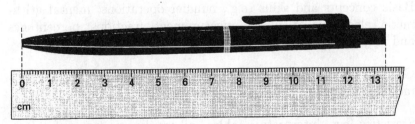

The length of the pen is ...... centimeters and ...... millimeters.

Figure 1. Sample Item #1 (Domain: Basic concepts and skills; Level: 1; Format: Short answer)

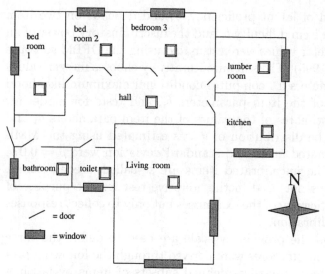

Eric and Fiona have bought a new house. This is the floor plan.
Which rooms have South-facing windows?

Figure 2. Sample Item #2 (Domain: geometry; Level: 2; Format: select
all that apply)

The items in the item bank cover end-of-course objectives in the
following four cognitive domains:

1. Basic concepts and skills (e.g., number operations, mental arithmetic, electronic calculator, measurement, fractions, percentages
   and proportions);

2. Geometry (e.g., orientation in space, reading maps, identifying geo-
   metrical figures);

3. Statistics (e.g., interpreting tables and graphs, measures of centre,
   probability);

4. Algebra (e.g., relations between variables, formulas).

Most items are of the short-answer type. Other item formats frequently used are the multiple choice and the select-all-that-apply formats.

In the Figures 1 through 4, four items from the MATHCAT pool
are shown. These items were selected to represent the three different
course levels, the four domains as well as the dominant item formats.

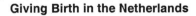

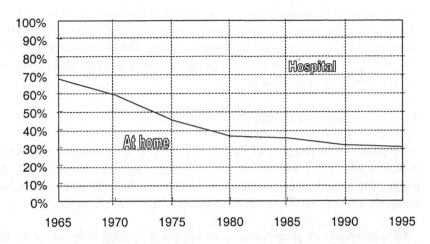

The graph above shows the percentage of women giving birth to a child at home or in hospital. What is the percentage of women giving birth in hospital in 1985? .... percent.

Figure 3. Sample Item #3 (Domain: Statistics; Level: 3; Format: Short answer)

The following is a procedure for converting degrees Fahrenheit (F) into degrees Celsius (C):

1. Take a particular temperature in degrees Fahrenheit.
2. Subtract 32.
3. Multiply the resulting difference by 5.
4. Divide the resulting product by 9

Which formula is a correct representation of the above procedure?
A. $C = F - 32$ x 5 : 9
B. $C = (F - 32)$ x 5 : 9
C. $C = F - (32$ x 5) : 9
D. $C = F - 32$ x (5 : 9)
E. $C = F - (32$ x 5 : 9)

Figure 4. Sample Item #4 (Domain: Algebra; Level: 3; Format: Multiple choice)

3. Item-Selection Algorithm

The item-selection algorithm is the heart of an adaptive test. It determines how the test must be started, continued, and stopped. Different purposes of the test should be supported by different algorithms. The algorithms used for the placement decisions and achievement testing are discussed in this section.

3.1. PLACEMENT TESTING

Purpose   The purpose of the placement test is to assign prospective students of adult basic education to three different course levels. An important practical requirement is that tests for this purpose should be as short as possible, with a maximum length of 25 items.

Administration procedure   In adaptive testing, the choice of the next item is dependent on the current estimate of the examinee's proficiency. However, when testing begins, no previous information about the proficiency level of the examinee is available. This holds particularly in placement testing, where often new students are tested to decide on their optimal level of instruction. In many CAT programs, this problem is resolved by selecting an item optimal at the average proficiency of the examinees in the calibration study.

In MATHCAT, a different strategy is used. The reason has to do with the fact that the kind of examinees being tested are often poorly educated and have bad recollections of attending school. In addition, many of them suffer from test anxiety. To encourage examinees to feel more comfortable, the first two items in the placement test are selected at random from a subset of relatively easy (Level 1) items.

Mental arithmetic is an important topic in adult mathematics education. Its importance is reflected in the relatively large percentage of mental arithmetic items in the item bank (89 of 476 items). Mental arithmetic items should be answered by performing mental calculations and not by using paper-and-pencil. To guarantee that this condition is met, the testing algorithm selects the first four items from a subset of mental arithmetic items. As soon as an examinee has responded to the fourth item in the test, he or she receives the following message from the software: 'From now on you are free to use paper and pencil.'

Wainer (1992) has suggested that using CAT for tests with a cutoff score is not worth the trouble. According to him: "The most practical way to make the test adaptive is to choose an adaptive stopping rule. Thus after each item we make the decision 'pass', 'fail', or 'keep on testing'. If testing is continued, an item is selected whose difficulty matches that of the cut-score as closely as possible" (p.4). The trouble Wainer refers to is the process of estimating the examinee's proficiency each time an item has been responded to. However, modern computers have enough power to perform these calculations very quickly.

Another reason why Wainer's suggestion was not followed was because it would require an item bank with very large numbers of items at or about the cut-off points. The present item bank, however, was designed to provide sufficient numbers of items along the full achievement continuum. This was necessary because of the two different purposes of MATHCAT (i.e., placement and achievement testing).

In the placement test, items are selected using the maximum-information criterion (van der Linden & Pashley, this volume). This criterion selects items with maximum information at the current proficiency estimate for the examinee. The test stops as soon as the examinee can be assigned to a course level with 90 percent certainty. This is the case, if the 90% confidence interval about the examinee's current proficiency estimate no longer covers any of the cutoff scores. This rule is used in combination with the requirement that the test length be between 12 and 25 items.

Figure 5 depicts the process of administering the placement test to a high-proficiency examinee (Straetmans & Eggen, 1998).

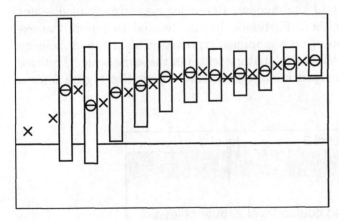

Figure 5. Placement test taken by a high-proficiency student

In the graph, the horizontal axis represents the successive items in the test. On the vertical axis both the difficulty values of the items selected (denoted by crosses) and the proficiency estimates of the examinee (denoted by circles) are projected. The two horizontal lines parallel represent the cutoff scores between Level 1 (easy) and 2 (moderate) and between Levels 2 and 3 (hard). To put the examinee at ease, the first two items were selected at random from a subset of relatively easy items. After the examinee responded to the second item, the proficiency was estimated for the first time. Of course, this estimate could not be very precise. The bar around the estimate represents a 90%-confidence interval for the examinee's proficiency. As both cutoff scores fell in this confidence interval, it was not yet possible to determine which course level this examinee should be assigned to. Therefore, testing had to be continued. From this point, the adaptive test did what it is supposed to do and selected items with difficulty values close to the proficiency estimates. If the examinee gave false answers, the estimate went down; for correct answers it went up. As demonstrated by the size of the 90%-confidence interval, in either case the uncertainty about the examinee's proficiency level was decreased. After twelve items, the test was stopped because the confidence interval no longer covered either of the cutoff scores. Because the lower bound of the confidence interval lay above the higher cutoff score, we could be fairly sure that the examinee had to be assigned to the course at Level 3.

Reporting of results   Immediately after testing has stopped, the student gets information about his or her performance. It was difficult to find a straightforward, yet sufficiently informative way to do this. In adaptive testing, the number-correct score is not a good indicator of the performance of the student. Reporting an estimate of θ is not very informative either. Therefore, it was decided to report the examinees' performances on a graphical representation of the proficiency scale along with a short explanatory text. On the same scale, the three course levels are marked in various shades of grey. See Figure 6 for a sample report.

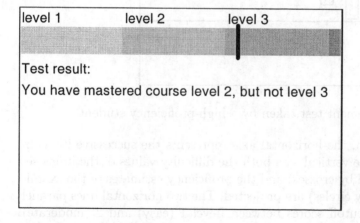

Figure 6. Sample results for the placement test

Evaluation of the placement test   An important criterion for the evaluation of a placement test is the accuracy of its decisions. To determine the accuracy of the MATHCAT system the following simulation studies were performed. Values for the proficiency parameter were drawn from the population distribution. For each value, an adaptive test was simulated using the algorithm above. From each course level, 1000 values were drawn, dividing the proficiency ranges into ten equal-size intervals with 100 draws from each interval. The tests were stopped using the rule above.

Table 1 shows how many simulees were placed at which level on the basis of their test results.

Table 1. Accuracy of placement tests

True Course	Observed	Course	Level	% of Correct
Level	1	2	3	Decisions
1	920	80	0	92.0
2	93	831	76	83.1
3	0	53	947	94.7

The accuracy was largest for proficiency values from Level 3 and smallest for values from Level 2. This result follows from the fact that Level 2 has two adjacent levels, while the others have only one. It never occurred, for instance, that a Level 1 examinee was placed in Level 3, or vice versa. A simulation study for the current paper-and-pencil two-stage placement test resulted in percentages of correct decisions equal to 88.5% (Level 1), 81.6% (Level 2), and 91.1% (Level 3). Thus the new adaptive test was more accurate.

Table 2. Length of placement tests

Course Level	Average Test Length (SD)	Percentage of Minimum Test Lengths	Percentage of Maximum Test Lengths
1	17(5.9)	49	32
2	20(5.6)	19	48
3	16 (5.4)	51	22

Table 2 shows the average test lengths (with standard deviations) as well as the percentages of tests that had minimum (12 items) and maximum length (25 items). Compared to the current two-stage test (fixed length of 25 items), the average length of the adaptive tests was considerably shorter.

3.2. ACHIEVEMENT TESTING

Purpose   The second purpose of MATHCAT was to test the achievements of the examinees placed in a course at a certain level. The two questions that should be answered by the test results are: (1) To which extent are the objectives of the course level met? (2) What are the strong and weak points in the achievements of the student?

Identification of strong and weak points in the achievements implies that the relevant content domains must be represented in the test by a reasonable number of items. A most-informative test does not necessarily represent the domains to be reported on well, while good representation is necessary to focus on the strong and weak points of the student. Thus, the item-selection algorithm for the achievement test has to deal with several content constraints on the test. For a review of techniques to impose such constraints, see van der Linden (this volume).

Item-selection procedure   The procedure used to implement the content constraints on the item selection process consists of three phases.

The goal of the first phase is to provide an initial estimate of the proficiency of the student. This information is then used to determine which test content specification should be used. At first, the use of the results from the placement test administered before the course was considered for this purpose but two reasons complicated this approach. First, the students do not always take the placement test. Second, these test results might have become obsolete, particularly if much time has elapsed between the administration of the placement test and the subsequent administration of the achievement test. In the first phase, ten items are administered; the first four are mental arithmetic items, the remaining six are drawn from Domain 1. (An overview of the content specifications for the achievement test is given in Table 3).

Dependent on the proficiency estimate, the items administered in the second phase have to obey one from three different sets of content specifications. If $\widehat{\theta} \leqslant -0.544$, the specifications belonging to the objectives of Level 1 are chosen. If $\widehat{\theta} \geqslant -0.021$, the specifications for Level 3 are chosen. For the intermediate values, the content specifications for Level 2 are chosen.

In the second phase, 20-30 items are administered. The items for Level 1 are mainly from Domain 1. For this level the other domains are covered only marginally. Also, within Domain 1, most items are taken from Subdomains 1.1 and 1.4. As an additional constraint, it holds that in the first and second phase together at least ten items should be taken from these two subdomains. The other ten items are selected freely from any domain. The items for Level 2 are approximately from the same domain as those for Level 1. However, one exception is that ten items should be selected from Subdomain 1.3. For Level 3, items are selected predominantly from Domains 2, 3 and 4.

Table 3. Numbers of items from (sub)domains in achievement test

			Level		
Domain		Subdomain	1	2	3
1.	Basic concepts and skills				$\geqslant 10$
		1.1 Number operations, mental arithmetic	$\geqslant 10$	$\geqslant 10$	
		1.2 Electronic calculator			
		1.3 Fractions, proportions and percentages		$\geqslant 10$	
		1.4 Measurement	$\geqslant 10$	$\geqslant 10$	
2.	Geometry				$\geqslant 8$
3.	Statistics				$\geqslant 8$
4.	Algebra				$\geqslant 8$
	Total		30	35	40

In the third phase of the test, the five seeded items referred to earlier are administered. The responses on these items are not used for proficiency estimation but for calibration purposes. To prevent from examinees taking seeded items that are too easy or too difficult, the items are assigned according to the following rule:

$\theta \leqslant -0.544$: random selection of items that measure the achievements in the domains belonging to Level 1 and 2.

$-0.544 \leqslant \theta \leqslant -0.021$: random selection of items that measure the achievements in the domains belonging to all three levels.

$\theta \geqslant -0.021$: random selection of items that measure the achievements in the domains belonging to Level 2 and 3.

Reporting test results Immediately after finishing the test, the results are presented. Again, the results are reported graphically along with a short explanatory text. Not only the general proficiency level is depicted but also proficiency estimates for relevant domains or subdomains. If an estimate for a domain is significantly lower than the overall estimate, a warning is given that there might be a deficiency in the student's knowledge. An example is given in Figure 7.

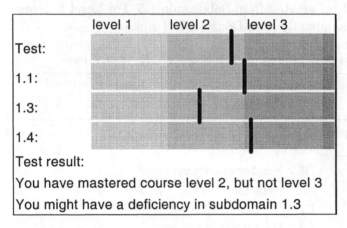

Figure 7. Sample results for the achievement test

Evaluation of the achievement test To assess the accuracy of the test, a simulation study was performed. The accuracy of the first phase in the test was evaluated by the percentage of correct branching decisions made for a typical population of examinees. The accuracy of the whole test was evaluated by the standard error of the proficiency estimate based on the combination of responses from the first and second phases.

Evaluation of first phase To this end, 1000 proficiency values from each course level were drawn. That is, 1000 values were drawn from the interval [-1, -0.54]; 1000 values from the interval [-.54, 0.002]; and 1000 values from the interval [-.02, .46]. The values were drawn to have a uniform distribution over these intervals. Table 4 shows how many examinees were branched towards the different levels.

Table 4. Accuracy of phase 1 of achievement tests

True Course Level	Observed	Course	Level	% of Correct
	1	2	3	Decisions
1	848	152	0	84.8
2	122	760	118	76.0
3	0	122	878	87.8

The accuracy was largest for proficiency values from Level 3 and smallest for values from Level 2. The latter result follows from the fact that Level 2 has two adjacent levels, while the others have only one. Examinees were never placed more than one levels from their true level.

Evaluation of final proficiency estimate   To assess the accuracy of the final proficiency estimate, 5000 proficiency values were drawn from the population distribution, which was normal with mean 0.074 and standard deviation 0.519. For each value the adaptive test was simulated and the test lengths and final proficiency estimates were recorded, Table 5 gives some statistics on the distribution of estimates and test lengths.

Table 5. Some statistics on achievement tests

	Mean	SD
Estimated Proficiency	0.071	0.530
SEM	0.088	0.022
Test Length	37.147	3.692

The statistics were also calculated for each course level separately. The results are shown in Table 6.

Table 6. Statistics on achievement tests per course level

Level	Mean Estimated Proficiency	SEM	Sample Size
1	-.761	.101	675
2	-.239	.082	1480
3	.430	.088	2842

Finally, a number of statistics specific for each (sub)domain were calculated. These results are given in Table 7. The deficiency warnings in the last column of Table 7 refer to erroneous warnings; no true deficiencies were simulated.

Table 7. Statistics on achievement tests per (sub)domain

	Mean Estimated Proficiency	SEM	Sample Size	# of Deficiency Warnings
Subdomain 1.1	-.405	.170	2155	126 (5.8%)
Subdomain 1.3	-.252	.145	1480	52 (3.5%)
Subdomain 1.4	-.393	.164	2155	71 (3.3%)
Domain 1	.434	.181	2842	175 (6.2%)
Domain 2	.417	.189	2842	139 (4.9%)
Domain 3	.423	.230	2842	219 (7.7%)
Domain 4	.420	.186	2842	140 (4.9%)

4. The MATHCAT Software

The MATHCAT software consists both of a student and a teacher module. The student module administers the test and presents the test results to the student. The teacher module can be used to perform the following tasks:

1. Adding and removing students;
2. Planning a test for a student (a student is only allowed to take a test that the teacher has planned for him/her);
3. Viewing the most recent test results for all students (group report);
4. Viewing all available test results for a selected student (individual report).

Group report   In Figure 8, an example of a group report is given:

Student ID	Student Name	Placement Test	Achievement Test
1	E. Long	79 (4/1/99)	86 (10/3/99)
2	R. Smith	91 (4/2/99)	
3	S. Baker	103 (4/2/99)	XXX

Figure 8. Example of a group report

To report more realistic numbers, the estimated proficiencies in the reports are transformed by $\widehat{\theta} = 28.68\widehat{\theta} + 96.44$. Thus, the cutoff score between Level 1 and 2 in the reports is at 82 and the one between Level 2 and 3 is at 96. If no test result is shown, the student has not yet taken the test. If the test result is reported as 'XXX', the student is currently taking the test.

Individual report   An example of an overview of all test results by a selected student is given in Figure 9. The overview not only depicts the transformed proficiency estimates for the student but also the level at which these estimates were classified. Next to the overall test results, the relevant profile scores are shown, together with their course levels, as well as diagnostic warnings.

Name:	E. Long		
Date:	4/1/99	6/2/99	10/3/99
Placement Test:	79 - Level 1		
Achievement Test:		83 - Level 2	86 - Level 2
Subdomain 1.1:		87 - Level 2	*88* - Level 2
Subdomain 1.3:		**75** - Level 1	85 - Level 2
Subdomain 1.4:		86 - Level 2	*86* - Level 2
Domain 1:			
Domain 2:			
Domain 3:			
Domain 4:			

Figure 9. Example of individual report

The report provides diagnostic warnings of two different types. The first type is a detected deficiency. In Figure 9, this type of warning is printed in boldface: the score of **75** for Subdomain 1.3 taken at 6/2/99 is suspect. The second type is an absence of progress issued when the estimated proficiency is not substantially higher than the previous estimate. In Figure 9, this type of warning is printed in italic. For example, the score of *88* for Subdomain 1.1 taken on 10/3/99 is not substantially higher than the previous result of *87* on 6/2/99.

5. Conclusions

Since January 1999 the MATHCAT testing system has been available to Dutch colleges for basic adult education. The use of MATHCAT has several advantages: greater accuracy, shorter test lengths, and greater ease of use. Decisions based on these tests are slightly more accurate than for the previous two-stage paper-and-pencil placement test (89.9% vs. 87.3% of correct placements). At the same time, however, the placement tests are considerably shorter. The software has been proven simple to use in practice. All correction work, previously done by hand, is now done by the software. Unlike the previous two-stage test, no manual scoring after a first subtest is necessary. In sum, the main advantage of the system is that test administration has become less time consuming, both for the students and the teachers. Thus teachers can now spend more time on their core activity: teaching.

References

Lord, F.M. (1971). A theoretical study of two-stage testing. *Psychometrika, 36,* 227-242.

Straetmans, G.J.J.M., & Eggen, T.J.H.M. (1998). Computerized adaptive testing: what it is and how it works. *Educational Technology, 38,* 45-52.

Verhelst, N.D., & Glas, C.A.W. (1995). The generalized one parameter model: OPLM. In: G.H.Fischer & I.W.Molenaar (Eds.). *Rasch models: their foundations, recent developments and applications* (pp. 215-237). New York: Springer.

Wainer, H. (1992). *Some practical considerations when converting a linearly administered test to an adaptive format.* (Research Report No. 92-13). Princeton, NJ: Educational Testing Service.

Weiss, D. J. (1974). *Strategies of adaptive ability measurement.* (Research Report No. 74-5). Minneapolis: University of Minnesota, Dep. of Psychology, Psychometric Methods Program.

Chapter 6
Computer-Adaptive Sequential Testing

Richard M. Luecht & Ronald J. Nungester
University of North Carolina at Greensboro, USA
National Board of Medical Examiners, USA

1. Introduction

This chapter describes a framework for the large scale production and administration of computerized tests called *computer-adaptive sequential testing* or CAST (Luecht, Nungester & Hadadi, 1996; Luecht, 1997; Luecht & Nungester, 1998). CAST integrates test design, test assembly, test administration, and data management components in a comprehensive manner intended to support the mass production of secure, high quality, parallel test forms over time. The framework is a modular approach to testing which makes use of modern psychometric and computer technologies.

CAST was originally conceived as a test design methodology for developing computer-adaptive[1] versions of the United States Medical Licensing Examination™ (USMLE™) Steps (Federation of State Medical Boards and National Board of Medical Examiners, 1999). The USMLE Steps are high stakes examinations used to evaluate the medical knowledge of candidate physicians as part of the medical licensure process in the U.S.. Although the Steps are primarily mastery tests, used for making pass/fail licensure decisions, total test scores and discipline-based subscores are also reported. Therefore, an adaptive-testing component is also attractive to economize on test length, while maximizing score precision.

The CAST framework has been successfully used in two empirical computerized field trials for the USMLE Step 1 and 2 examinations (NBME, 1996, 1997; Case, Luecht and Swanson, 1998; Luecht, Nungester, Swanson and Hadadi, 1998; Swanson, Luecht, Gessaroli, and Nungester, 1998). In addition, CAST has undergone rather extensive simulation research with positive outcomes (Luecht, Nungester and Hadadi, 1996; Luecht and Nungester, 1998).

In terms of USMLE, what CAST offers is a comprehensive means to develop mastery tests, with adaptive capabilities. CAST also helps

---

[1] The computerized versions of the USMLE™ Steps were introduced in 1999 as non-adaptive tests. CAST versions are slated for implementation in the near future.

W.J. van der Linden and C.A.W. Glas (eds.),
Computerized Adaptive Testing: Theory and Practice, 117–128.
© 2000 *Kluwer Academic Publishers. Printed in the Netherlands.*

guarantee that every test form can satisfy the extensive and complex content requirements needed to represent the spectrum of medical knowledge covered by each of the USMLE Steps. Within that more general context, the CAST framework was recognized as a useful way to design many different types of high stakes computerized tests, where the quality assurance of individual test forms is essential.

The basic test design model for CAST is a multistage test, with adaptive testing capabilities at the level of subtests or testlets (e.g. Adema, 1990; Sheehan and Lewis, 1992; Luecht, Nungester and Hadadi, 1996). That basic design can be engineered to produce a specific type of adaptive test and then be used as a template for creating future test forms. As part of the template, CAST incorporates explicit targeting of measurement precision where it is most needed and simultaneously contends with potentially complex content requirements, item exposure issues, and other constraints on test construction. The CAST test forms can also be preconstructed, allowing test developers to assure quality before releasing the tests. This quality assurance capability is viewed as essential for testing programs like the USMLE Steps, where content validity and test form quality are mandatory for EVERY test form.

2. The Basic CAST Framework

On the surface, the CAST framework presents as a multistage test, with adaptive capabilities. In that respect, CAST is hardly a new concept (see, for example, Cronbach and Gleser, 1965; Lord, 1971, 1980). However, CAST formalizes a set of statistical targets and all of the other test specifications into a template that can be used in conjunction with automated test assembly to manufacture in large scale adaptive test forms with the desired parallel statistical and content characteristics. CAST also allows the test forms to be created before test administration takes place. This capability provides some important advantages for interventions in terms of quality assurance and heading off security risks due to overexposed materials.

Figure 1 displays a simple CAST configuration. A single instance of this configuration is referred to as a *panel*. When all of the statistical targets and categorical content constraints are added to the panel, it becomes a template for test assembly. Each instance of the CAST panel can be treated as an independent test administration entity that can be uniquely assigned to an examinee an used for tracking responses conducting data parity checks, etc..

The panel shown in Figure 1 is made up of seven *modules*, labeled 1M, 2E, 2M, 2H, 3E, 3M, and 3H (top to bottom and left to right).

The abscissa shows both the relative item difficulty scale (Easy to Difficult) and the examinee proficiency scale (Low to High proficiency). Easier modules are therefore targeted for lower proficiency examinees and more difficult modules are targeted for higher proficiency examinees.

Modules are the building blocks of CAST. A module is a test assembly unit denoting a group of items assigned to a particular CAST stage within a particular panel instance. Modules can range in size from several items to more than 100 items. Each of the seven modules in Figure 1 represents a specific set of items. The letter designations denote the average difficulty of the modules. "E" denotes easy, "M" denotes moderate difficulty and "H" denotes hard. It may be helpful to think of modules as subtests or testlets, however, those terms tend to carry a variety of ambiguous connotations in the testing literature. As building blocks, CAST modules may have their own specifications and can be split or joined, as needed, for test construction.

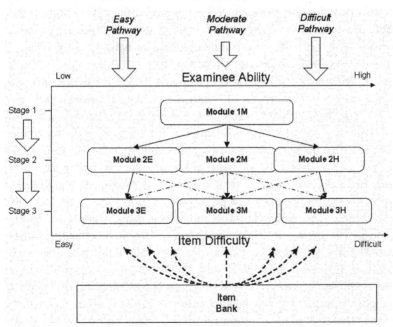

Figure 1. A three stage CAST configuration.

In a CAST panel, each module is uniquely assigned to a particular testing *stage*. The minimum number of stages is two; there is no upper bound. Figure 1 depicts the design for a three stage CAST panel. In general, more stages implies more potential flexibility in adapting the test to individual examinees. However, adding more stages also dramatically increases the complexity of the test assembly problem, without necessarily adding much to the measurement precision of the final test forms for moderate to long tests (Luecht, Nungester and Hadadi, 1996; Luecht and Nungester, 1998) .

All of the modules assigned to the same stage have an equivalent number of items, however, the module size can vary across stages. For example, referring to Figure 1, Modules 2E, 2M, and 2H would have the same number of items. Similarly, Modules 3E, 3M, and 3H would each have an equal number of items. However, the module size at Stage 1 could be different than at Stage 2 or at Stage 3. Varying the number of items by stage directly affects the total usage of items for the panel as well as the adaptive properties for the various combinations of modules over stages.

For a given panel, items can be shared by the modules within stages but not across stages. The modules assigned to a particular stage are mutually-exclusive in terms of their selection during the examination (i.e. an examinee could see module 2E or 2M at Stage 2, not both). Some items could be therefore shared among Modules 2E, 2M, and 2H, if the item inventories were sparse in crucial content areas. Similarly, Modules 3E, 3M, and 3H could share items, as needed. However, no items could be shared across the three stages.

An important concept in CAST is the *pathway*. Pathways are the allowable sequences of modules an examinee can see, based upon pre-defined routing rules that govern how an examinee progresses from module-to-module, across the CAST stages. The adaptive routing rules and/or sequential decision rules are implemented between the stages and produce the pathways. Pathways can vary from allowing all possible routes to using only a very restricted set of routes, as a matter of testing policy. For example, if the pass/fail cut point on a mastery test was located near the middle of the moderate difficulty region of the score scale in Figure 1, one pathway might route all borderline failing examinees to only the diagnostic modules (e.g. following the pathway of Modules 1M+2E+3E in Figure 1) and preclude those examinees from taking any type of "recovery" routes such as 1M+2E+3M. This type of strategic policy might be invoked to prevent unscrupulous examinees from taking the test just to memorize items near the pass/fail point of the scale. CAST allows that type of strategy to be implemented by simply shutting down the most risky pathways.

As more pathways are allowed, CAST begins to represent a more traditional computer-adaptive test with many possible routes. However, by restricting the number of possible pathways, test developers can exercise greater control over the quality of the individual test forms seen by particular examinees. For example, the primary (i.e. most likely) pathways can each have explicit statistical targets associated with them. Referring once again to Figure 1, there might be one target for the easiest pathway, 1M+2E+3E, another target for the pathway likely to be taken by examinees of moderate proficiency (1M+2M+3M) and a third target for the most difficult pathway (1M+2H+3H). The arrows at the top of the figure denote those three primary pathways.

The statistical targets are used to control average item difficulty and score precision for examinees likely to follow that pathway. This approach of using targets allows test developers to generate tests having parallel statistical characteristics over time for examinees of similar proficiency.

The targets for the pathways can be designed using IRT test information functions to indicate the desired statistical measurement at various regions of the score scale[2]. A target test information function (TIF) denotes the amount of measurement precision needed within various regions of the latent proficiency scale (Birnbaum, 1968). The use of test information targets is fairly common in the literature on automated test assembly procedures (e.g. van der Linden, 1998; van der Linden and Luecht, 1995; Luecht, 1992, 1998). Of course, we need to determine the amount of estimation error variance we are willing to tolerate in various regions of the proficiency score scale. We can then determine the target TIF(s) necessary to achieve that magnitude of precision. The targeted amount of information should represent a reasonable or average amount, given the overall characteristics of the item bank (Luecht, 1992; Luecht and Nungester, 1998). Using average information targets is a robust way to ensure that the testing program can support the construction of many parallel test forms over time.

The target TIFs become part of the template specifications for building subsequent instances of the CAST panel. Each CAST panel will also have content and other categorical or quantitative specifications. These specifications are specified as constraints during test assembly. *Note that CAST specifications may exist at the module level or at the test level or at both levels.* Allowing specifications to exist at any level is an important feature of CAST. For programs with complex content structures at the test level-like the USMLE[TM] Steps-it may be impossible to satisfy the overall test specifications by only specifying module-level constraints. Instead, it is necessary to consider how the particular com-

---

[2] Module level targets can be derived, if test assembly is done at the module level. Deriving targets for [primary] pathways provides more comprehensive control over the aggregate test characteristics for the majority of test forms available from a particular panel instance.

binations of modules, administered across the CAST stages, will meet the overall specifications.

Once all of the specifications are established for the panel, automated test assembly software (e.g. Luecht, 1998; Luecht & Nungester, 1998) can be used to select items from the active item bank and to assign them to the particular modules in each instance of each panel. Multiple panel instances can be constructed. In our example, the ATA software would need to create multiple instances of the seven modules shown in Figure 1: 1M, 2E, 3E, 2M, 3M, 3H, and 3H, subject to meeting all of the content requirements and the statistical targets for the three primary pathways.

3. CAST Design Strategies for the Specifications

Assembling multiple instances of CAST panels is a systematic process and requires some clever uses of automated test assembly (ATA) heuristics and algorithms (Luecht, Nungester and Hadadi, 1996; Luecht, 1998; Luecht & Nungester, 1998). For example, it may be necessary to approach test construction as a multiple step assembly problem, where the first stage modules are initially constructed and incorporated as preselected items in the construction of modules for subsequent stages.

Modules can also be combined for purposes of test assembly and later split to form the final modules assigned to the CAST stages. For example, combining modules 2E+3E, 2M+3M, and 2H+3H would effectively reduce the CAST ATA problem posed by the design configuration in Figure 1 to a two-stage problem. Once constructed, those larger modules could be split to produce the smaller Stage 2 and 3 modules (perhaps using primary content specification as the partitioning strata). The capabilities of the ATA software will largely dictate how the test assembly should proceed.

In general, a CAST panel can be designed using one of three strategies: (1) a bottom-up strategy, (2) a top-down strategy, or (3) a mixture strategy. The choice of strategy needs to be decided on the basis of the importance of content specificity and the item writing and production capabilities that will replenish the item bank to provide consistent tests over time.

3.1. BOTTOM-UP STRATEGY

The bottom-up strategy requires module level specifications for the statistical targets and also for all content and other categorical test features. If comprehensive module-level specifications can be developed, multiple instances of each module can be built. Adema (1990) advocated this approach and formalized a linear programming solution. Similarly, Sheehan and Lewis (1992) provided a design framework for building testlets of equivalent difficulty and with fixed content requirements, for use in computer-mastery testing.

Unfortunately, not all testing programs have content blue prints and test specifications that can be adequately partitioned to the module level. Nonetheless, if the bottom-up strategy can be employed, it has the advantage of true "modularity". That is, the modules can be mixed-and-matched to create multiple permutations of the panel instances. This provides a type of "plug in" capability which has positive implications for examination security as well as test parallelism.

3.2. TOP-DOWN STRATEGY

The top-down strategy requires only test-level specifications for the statistical targets and for all of the categorical content and other categorical constraints. The primary pathways are viewed as the main test assembly units. That is, the modules must be combined in prescribed ways - usually along primary pathways within each panel - to produce the desired test level properties. Modules are not necessarily exchangeable across panels under the top-down strategy. This top-down approach allows comprehensive and complex content constraints to be employed but also requires introducing practical limitations on the number of pathways that can be considered. For example, in the three stage CAST panel displayed in Figure 1, we could simultaneously build three test forms corresponding to each of the three main pathways: (1) 1M+2E+3E, (2) 1M+2M+3M, and (3) 1M+2H+3H (where 1M represents the same items used for each pathway). Adding additional pathways for auxiliary or secondary routes would quickly add to the complexity of the test assembly problem and could reduce the overall quality of the solutions.

3.3. MIXTURE STRATEGY

The mixture strategy allows some of the specifications to exist at the module/stage level and others to be aggregated at the test level. For example, we previously described the three primary pathways indicated in Figure 1 as the "easy" pathway, the "moderate difficulty" pathway, and the "hard" pathway. The secondary pathways, while important, are not likely routes for the majority of examinees and can be ignored for purposes of initial test construction. Those auxiliary pathways can be evaluated as part of the final quality assurance of each panel, prior to activation.

We could therefore specify three target TIFs: one for the 1M+2E+3E combination of modules, a second for the 1M+2M+3M combination of modules, and a third for the 1M+2H+3H combination of modules. The automated test assembly software would build the panels by selecting items from the bank to match each of the three TIFs as closely as possible.

The test content specifications, along with other relevant categorical, qualitative item features considered in test assembly (e.g. item types, etc.), would need to be allocated to each the CAST stages, where possible, or treated in aggregate across all of the stages. The aggregated test level specifications would specify the content and other categorical constraints for the total test length along the three primary pathways. For example, we might specify the minimum and maximum items counts in various content areas for a test composed of modules 1M, 2M, and 3M. The same specifications would hold for the other pathways.

The test assembly optimization model solves a set of simultaneous objective functions, subject to meeting the aggregated, total test and module level item content constraints across and within the CAST stages (Adema, 1990; Luecht, Nungester and Hadadi, 1996; Luecht and Nungester, 1998). In the context of the three primary pathways in Figure 1, this would effectively amount to setting up the same content specifications for three tests of unequal difficulty. We want to guarantee that those three pathways (i.e. 1M+2E+3E or 1M+2M+3M and 1M+2H+3H) will independently achieve the aggregate statistical target TIFs as well as meet all of the principal test level content and categorical constraints.

In addition, constraints could be specified for each of the three stages (i.e. we could set up module level constraints). Luecht (1998) provided the details of a heuristic that sequentially solves this type of problem. That same heuristic was used for the CAST simulations reported by Luecht and his colleagues. It has also been effectively used for building experimental CAST forms used in USMLE computerized testing field studies (e.g. Case, et al, 1998).

There is no single way to approach the design of CAST panels. Luecht and Nungester (1998) discuss some additional CAST design strategies, including letting the computer apply a "partitioning" algorithm to optimally allocate the total test content requirements to each stage so that the marginal sums of items meet exactly the total test content specifications across stages.

4. Building CAST Panels

CAST assembly occurs at the panel level. For most large scale testing programs, simultaneous construction of multiple panel instances is preferable. That is, we might build any where from 10 to 100 (or more) simultaneous panel instances for subsequent activation, using a single item bank.

Technically speaking, building CAST panels can be viewed as a multiple objective function, constrained optimization problem. Carrying out the actual item selection - especially for multiple instances of panels - is not feasible without the use of automated test assembly (ATA) computer software. ATA implements optimization algorithms and/or heuristics to select items from a bank which satisfy specific goals. Most ATA goals can be formally stated in terms of a mathematical optimization model comprised of an objective function to be minimized or maximized, subject to various constraints on the solution (e.g. van der Linden, 1998; Luecht, 1998). For example, in the present context, the goal is to select items for that modules that, in aggregate, match each of the target IRT test information functions, subject to test level and module level requirements and restrictions on the frequencies of items in various content areas, as well as other categorical features. Word count restrictions and other quantitative variables can also be incorporated into either the constraints or into the objective functions.

Once a mathematical optimization model is developed (i.e. the objective functions and systems of constraints are formally specified) it can be solved using a computer program. Most of the available ATA computer programs use linear programming algorithms, network flow algorithms, or other heuristics to select the items from an input bank which locally or globally satisfy the stated goals. There are commercially-available computer software packages for doing ATA. A number of CAST applications have been successfully carried out using ATA software written by the first author.

An extensive description of ATA algorithms and heuristics is beyond the scope of this chapter. It is worth noting, however, that CAST presents unique challenges for ATA. First, there is the potential need to simultaneously optimize more than one objective function in the presence of multiple and different target information functions. Second, there is the possibility of having different content specifications and constraint systems for various modules or even for different pathways. Third, there is the obvious the need to simultaneously build multiple instances of each module, pathway, or panel (i.e. different versions of the modules and panels with item overlap carefully controlled within and across panels by the test assembly software). Finally, the ATA procedures must have a dealing with the multiple relationships among the modules across the stages and across panel instances.

5. Test Security Under CAST

Test security is one of the paramount problems facing organizations moving to computer-based testing (Stocking and Swanson, 1998). Specifically, item exposures due to memorized item can unfairly benefit examinees who can get access to the exposed materials. CAST allows security measures to be implemented in several ways. First, items can be randomly presented within modules. Second, item and module level exposure controls can be implemented as part of the ATA item selection process to reduce item overlap across panels (at least under a heuristic solution - see Stocking and Swanson, 1998 or Luecht, 1998). Third, empirical item overlap can be explicitly constrained as part of the ATA process, when building multiple instances of the panels. Fourth, panels can be activated on the basis of having minimal [pair-wise] item overlap for a particular period of time. Finally, explicit pathways can be periodically evaluated for potential over-exposure and the entire panel deactivated when a certain threshold is reached.

In short, CAST allows virtually all of the same security measures as any computer-adaptive test. But, CAST adds the capability to empirically evaluate security risks for panels, before and after they are activated. Furthermore, because the panel instances can be established as legitimate data base entities, it should be possible to rather easily de-activate specific panels deemed at risk.

6. Conclusions

CAST is a highly-structured test development framework for manufacturing computer-adaptive tests. The controls in CAST can be relaxed, as needed, however, they provide some distinct advantages in terms of knowing which pathways examinees can follow. Simply put, CAST sacrifices adaptive flexibility for control. In doing so, CAST also provides the capability to preconstruct all of the test forms. That, in turn, allows test developers to implement various quality assurance procedures that are not possible under the traditional computer-adaptive testing paradigm.

Perhaps most importantly, by creatively using targeted test information and automated test assembly procedures, CAST allows test developers to engineer the exact type of adaptive test they need. Of course, there are practical trade-offs. For example, adding more stages increases the adaptive flexibility of the test, but also requires more items per panel and can complicate the test assembly process. Likewise, adding more modules per stage, some at increasingly more variable levels of difficulty, increases the adaptive nature of the test, but comes at the expense of requiring more expansive item banks to build the more extreme modules and "costs" more items per panel. In general, our experience has been that any statistical gains in precision dissipate quickly enough to suggest that a more controlled panel design - i.e. one containing fewer stages and limited modules per stage - is preferable to a highly adaptive framework. The three stage CAST configuration demonstrated in this chapter is a fairly robust model to consider.

CAST is a practical solution. For high stakes tests like the USMLE Steps, it seems important to exercise the necessary controls to guarantee that every test form is a high quality examination. Unfortunately, computer algorithms, alone, are not yet capable of making those guarantees. So CAST provides for human intervention in the test construction process. In that regard, it follows a long tradition of producing quality tests that may benefit other testing programs.

References

Adema, J. J. (1990). *Models and algorithms for the construction of achievement tests*. Doctoral Dissertation. Enschede, The Netherlands: FEBO.

Birnbaum, A. (1968). Test scores, sufficient statistics, and the information structures of tests. In F. M. Lord and M. Novick (Eds.). *Statistical theories of mental test scores*. Reading, MA: Addison-Wesley.

Case, S. M., Luecht, R. M., & Swanson, D. B. (1998, April). *Design and automated assembly of test forms for the USMLE step 1 field test*. Symposium conducted at the National Council on Measurement in Education Annual Meeting, San Diego.

Cronbach, L. J. & Gleser, G. C. (1965). *Psychological tests and personnel decisions, 2nd Ed.*. Urbana: University of Illinois Press.

Federation of State Medical Boards, National Board of Medical Examiners. (1999). *United States Medical Licensing Examination(tm) Bulletin of Information for Computer-Based Testing (CBT) 1999*. Philadelphia, PA: FSMB and NBME Joint Publication.

Lord, F. M. (1971). The self-scoring flexi-level test. *Journal of Educational Measurement, 8*,147-151.

Lord, F. M. (1980). *Applications of item response theory to practical testing problems*. Hillsdale, NJ: Lawrence Erlbaum Associates.

Luecht, R. M. (1992, April). *Generating target information functions and item specifications in test design*. Paper presented at the Annual Meeting of the National Council on Measurement in Education, San Francisco, CA.

Luecht, R. M. (1998). Computer-assisted test assembly using optimization heuristics. *Applied Psychological Measurement, 22*, 224-236.

Luecht, R. M. & Nungester, R. J. (1998). Some practical examples of computer-adaptive sequential testing. *Journal of Educational Measurement, 35,* 239-249.

Luecht, R. M., Nungester, R. J. & Hadadi, A. (1996, April). *Heuristic-based CAT: balancing item information, content and exposure.* Paper presented at the Annual Meeting of the National Council of Measurement in Education, New York, NY.

Luecht, R. M., Nungester, R. J., Swanson, D. B., & Hadadi, A. (1998, April). *Examinee performance, perceptions, and pacing on a computer adaptive sequential test.* Symposium conducted at the National Council on Measurement in Education Annual Meeting, San Diego.

NBME. Author. (1996) The 1996 Step 2 Field Test Study of a Computerized System for USMLE. *The National Board Examiner, 43(4)* . Philadelphia, PA: National Board of Medical Examiners.

NBME. Author. (1997) Summary of the 1997 USMLE Step 1 Computerized Field Test. *The National Board Examiner, 44(4).* Philadelphia, PA: National Board of Medical Examiners.

Sheehan, K. & Lewis, C. (1992). Computerized mastery testing with nonequivalent testlets. *Applied Psychological Measurement, 16,* 65-76.

Stocking, M. L. & Swanson, L. (1998). Optimal design of item banks for computerized adaptive testing.. *Applied Psychological Measurement, 22,* 271-279.

Swanson, D. B., Luecht, R. M., Gessaroli, M. E., & Nungester, R. J. (1998, April). *Impact of systematic manipulation of subtest difficulty on examinee performance, perceptions, and pacing.* Symposium conducted at the National Council on Measurement in Education Annual Meeting, San Diego.

van der Linden, W. J. (1998). Optimal assembly of psychological and educational tests. *Applied Psychological Measurement, 22,* 195-211.

van der Linden, W. J. & Adema, J.J. (1998). Simultaneous assembly of multiple test forms. *Journal of Educational Measurement, 35,* 185-198.

van der Linden, W. J. & Luecht, R. M. (1995). An optimization model for test assembly to match observed score distributions. In G. Englehard & M. Wilson (Eds.). *Objective Measurement: Theory into Practice (Vol. 3).* Norwood, NJ: Ablex.

Wainer, H. & Kiely, G. L. (1987). Item clusters and computerized adaptive testing: a case for testlets. *Journal of Educational Measurement, 24(3),* 185-201.

Chapter 7
Innovative Item Types for Computerized Testing

Cynthia G. Parshall, Tim Davey, & Peter J. Pashley
IIRP, University of South Florida, USA
ACT, Inc., USA
Law School Admission Council, USA

1. Introduction

Initial applications of a new technology often retain many of the characteristics of the older products or processes that they are displacing. Early automobiles were horse-drawn carriages simply fitted with motors. Television programming remains firmly rooted in the formats pioneered by radio. And air travel initially featured the elegance and service that passengers had come to expect on board trains and ships. The experience so far with computerized testing suggests that it can be added to this list of examples. While the use of computerized tests has grown dramatically in the decade since it became practical, to date most operational computerized tests rely on items developed for conventional, paper-and-pencil administration. However, the computer is capable of much more, offering the promise of new and better forms of assessment that incorporate features and functions not possible with conventional test administration. We call test items that make use of these capabilities innovative item types. Innovative features that can be used by computer-administered items include sound, graphics, animation and video. These can be incorporated into the item stem, response options, or both. Other innovations concern how items function. For example, examinees answering computerized items are not limited to just selecting one of several available response alternatives. They can instead highlight text, click on graphics, drag or move objects around the screen, or re-order a series of statements or pictures. The computer's ability to interact with examinees provides further possibilities. Items are not restricted to merely accepting a response and tests are not restricted to stepping through a fixed set of items in a fixed order. Items can instead be designed to branch through a series of steps or outcomes and tests can branch through a changing series of items, all contingent on an examinee's responses. Finally, researchers are developing algorithms that will allow the computer to score items in which examinees generate, rather than simply select their responses. This could allow

129

W.J. van der Linden and C.A.W. Glas (eds.),
Computerized Adaptive Testing: Theory and Practice, 129–148.
© 2000 *Kluwer Academic Publishers. Printed in the Netherlands.*

complex, performance-based tasks to be graded reliably and at minimal cost. This chapter describes how innovative test items can make use of the computer's capabilities to improvement measurement. Improvement can stem either from innovations that enable us to measure something more than we could before, or to measure it better. Our presentation is organized around a five-branch taxonomy for item innovation. We describe each branch below, illustrating our description with examples of items developed for experimental or operational use.

2. A Taxonomy for Innovative Items

Several ways of categorizing innovative items have already been proposed. Parshall, Stewart, and Ritter (1996) position innovations along a measurement "continuum", in which item formats move from highly constrained (e.g., multiple-choice) to highly open (e.g., essay writing). Koch (1993) categorizes innovative items according to a 4-level hierarchy: 1) traditional items with minor modifications, 2) items that make fuller use of graphics and graphic capabilities, 3) "multidimensional" items (so called because they require that information be visualized or manipulated in several dimensions not because they measure several traits), and 4) situated items, which are items presented in an on-line environment with a high degree of real-world congruence. Luecht and Clauser (1998) discuss item types they term "complex". They note that item complexity may result from features of the stimulus, the nature of the responses elicited, and from the range of permitted examinee actions or activities. We will provide a more comprehensive framework for innovative item types by arranging them along five dimensions. These are: 1) item format, 2) response action, 3) media inclusion, 4) level of interactivity, and 5) scoring algorithm. Item format defines the sort of response collected from the examinee. Two major categories of item formats are selected response and constructed response, with multiple choice being the most common example of the former and written essays an example of the latter. Response action refers to the means by which examinees provide their responses. Keyboard entry, mouse clicks and touch screens are common. Media inclusion covers the use of elements such as sound or video in an item. Level of interactivity describes the extent to which an item type reacts or responds to examinee input. Finally, scoring method addresses how examinee responses are translated into quantitative scores. As will quickly become evident, the five dimensions are far from independent in application. For example, including a video presentation in an item may well also change the ways in which examinees are allowed to interact with and required to respond to an item. Similarly, highly interactive items with complex response actions may require equally complex scoring models. Although we attempt to differentiate the aspects of item innovation in the descriptions below, some of the items we present as examples of one sort of innovation confound matters by being innovative in one or two other ways as well.

2.1. ITEM FORMAT

Conventional, paper-and-pencil tests generally make use of a limited number of item formats. Multiple choice is clearly the most common, but the success of the format has spawned a number of variants. These include multiple response, ordered response, and matching items. Formats not derived from multiple choice include fill-in-the-blank, short answer, and essay. These require examinees to generate or construct rather than select their responses and so are less amenable to machine scoring. All of these item types can be adapted for presentation on computer, along with many more. The most important of these are detailed below.

2.1.1. *Selected Response Items*

The most familiar selected response item format is multiple choice, where examinees choose answers from a list of alternatives. Most often, the item consists of a question, or stem, and a set of anywhere from two to five possible responses. Computerized adaptations of this format can expand on this in order to reduce guessing or afford a more direct form of assessment. For example, items may ask examinees to click on and select the proper sentence from a reading passage, or to select one part of a complex graphic image. Because the number of available options can be much greater than the usual four or five, and can vary from item to item, the possibility of guessing correctly is substantially reduced. However, the effect of items like these on either examinees or item statistics, has not been systematically investigated. Multiple-response items ask examinees to select more than one option, with the number to be selected either specified or left open. O'Neill and Folk (1996) describe an application that uses the former method (e.g., "select two of the following alternatives"). The test presentation software in their application was constrained so that examinees were not able to select either more or fewer elements that the directions specified. Koch (1993) and Parshall, Stewart, and Ritter (1996) describe applications of the latter, more open-ended method (e.g., "click on all correct responses"). It seems plausible that these two approaches could result in cognitively different tasks for the examinees, and that one method might yield better or different information from the other. The inclusion of partial-credit scoring might also be a useful addition. Validity studies are needed to determine examinee reactions to the level of instruction and constraint, impact of alternate scoring approaches, effects on the item difficulty and omit rates, and the value of the resulting information. Another selected response item type might be termed figural response. In figural response items, examinees respond by selecting a part of a figure or graphic. For example, Parshall, Stewart, and Ritter (1996) display an item that contains a portion of a database table. Examinees respond by identifying the column, row, or field in the database that correctly answers a question. O'Neill and Folk (1996) describe math item types in which examinees select a point on a figural histogram, scale, or dial. Ordered response items present examinees with a list of elements that they are asked to order or sequence in accord with some rule. O'Neill and Folk (1996) discuss a verbal version of this item format, in which examinees indicate the correct order or timeline for a series of events. They also describe a quantitative version in which examinees order numerical elements from smallest to largest. Davey, Godwin, and Mittelholtz (1997) describe still another variant of selected response items. Their test of writing skills is designed to simulate the editing stage of the writing process. Examinees are confronted with a passage that contains various grammatical and stylistic errors, but no indication is given as to the location of these errors. Examinees read the passage and use a cursor to point to sections that they feel should be corrected or changed. They are then presented with a list of alternative ways of rewriting the suspect section. Examinees can select one of the alternatives or choose to leave the section as written. If an alternative is chosen, the replacement text is copied into the passage so that the changes can be reviewed in their proper context. The reasoning is that the error-identification portion of this task adds to the cognitive skills assessed by the items, even though conventional multiple choice items are used.

Bennett and Sebrechts (1997) present a multiple choice variant that assigns the same set of response options to a series of questions. Their application, intended to measure more cognitively complex skills, asked examinees not to solve math problems, but rather to sort or categorize the problems into groups. Specifically, examinees were provided with four prototypes and a series of "target" problems. The task is to match each target to one of the prototypes by correctly identifying their underlying structural characteristics.

There are other selected response item formats that could be easily adapted to computer administration. One example is an item type where the examinee is asked to rank a set of response options in order of their degree of correctness. Some form of partial credit, as opposed to dichotomous scoring would probably be appropriate for these items. Another example is an item to which an examinee is able to respond multiple times, possibly with feedback regarding the correctness of each response. The final item score for the examinee could be based on percent correct out of number-attempted, or percent-attempted-until-correct. The goal of each of the selected response item formats described above is to improve measurement in some sense. Some of the formats may tap slightly different cognitive constructs than do traditional multiple choice items. For example, the ordering and multiple-response types add a level of complexity to the task of responding. In contrast, figural response items eliminate a level of abstraction by allowing examinees to indicate their responses more directly. Several formats are intended to reduce the effect of guessing by expanding the range of possible responses. For example, an examinee has a one in four chance of answering a four-option multiple choice item correctly, but only a one in twenty-four chance with a four-option ordering item.

2.1.2. *Constructed Response Items*

A wide range of constructed response items has also been considered, varying from fairly simple formats that are easy to score, to far more complex formats that require the use of elaborate scoring algorithms. The simplest examples are items that require the examinee to type a numerical answer to a quantitative question (e.g., O'Neill & Folk, 1996), or a very short response to a verbal question. These responses are scored, usually dichotomously, by comparing each response to a list of acceptable answers that may include alternate mathematical formulations or acceptable misspellings. Another constructed response item format extends the selected figural response type described above. The constructed version of this type allows examinees to mark on, assemble, or interact with a figure on the screen. Martinez (1993) developed figural response biology items to which examinees respond by copying and moving graphical objects onto a figure. In this particular application, the items were scored using partial-credit models. Another example of this type of item is provided in French and Godwin (1996)'s investigation of science lab items. The electronic circuitry items in this study required examinees to use graphics representing resistors, a battery, and a glow lamp to assemble either a parallel or a sequential circuit. Bennett, Morley, and Quardt (1998) introduce three mathematics constructed response item formats. These items were developed to broaden the range of mathematical problem solving measured by computerized tests, specifically by assessing problem areas that are less tightly structured. The first item type, mathematical expressions, has examinees respond by typing a formula. Because these responses are in the form of symbolic expressions, a correct answer may take any number of forms. Despite this, scoring algorithms have been able to successfully score the great majority of responses. Their second item type, generating expressions, confronts examinees with loosely structured problems and asks them to provide an instance that meets the problem constraints. Typically, responses will be in the form of values or expressions. The third item type developed, graphical modeling, has examinees respond by plotting points on a grid. Breland (1998) investigated a constructed response text-editing task. This application was similar to the selected response editing task described by Davey, Godwin, and Mittelholz (1997) in that examinees are presented with a text passage that contains errors. However, in Breland's case sections containing errors are enclosed by brackets; examinees select the bracketed text and retype it in order to correct it. Automated scoring is provided by matching the examinee's response with a table of correct solutions. The few constructed response item formats described above just scratch the surface of what is possible. Simple, discrete, quick tasks can be added to testing programs fairly easily, and when well designed, can provide a number of measurement benefits. Further, the addition of partial credit or other alternative scoring model ought to enable us to collect more and better information about examinees' proficiency. Some additional, more complex constructed response item types will be addressed further in the sections on "Interactivity" and "Scoring" below.

2.2. RESPONSE ACTION

"Response action" refers to the physical action that an examinee makes to respond to an item. While the item format defines what we ask the examinee to tell us, the response action defines how we ask them to do so. In paper-and-pencil items, the action called for most often is bubbling in an oval associated with a multiple choice response option. Computerized tests most often use the keyboard and the mouse in lieu of a #2 pencil as the preferred input devices. Examinees respond through the keyboard by typing numbers, characters or more extended text. "Command" keys such as <Enter> may also be used. Multiple choice options may be chosen by typing a letter or clicking a box attached to a chosen response. Examinees may also be asked to click on a graphic, on part of a graphic, or on part of a text passage. They may need to drag icons to create or complete an image, or to drag text, numbers, or icons to indicate a correct sequence of events. As an alternative to dragging, some computerized exams ask examinees to click on an image, and then click a second time to mark the new location where the image should be placed. Examinees may need to use the mouse for purposes other than to respond to questions. These include getting on-line help or hints, accessing computerized calculators or reference materials, playing audio or video prompts, or running a simulation. The mouse may also be used to access pull-down menus (Baker & O'Neil, 1995), select numbers and mathematical symbols from an on-screen "palette" (Bennett, Steffen, Singley, Morley, & Jacquemin, 1997), or to identify a specific frame in an on-line video or animation sequence (Bennett, Goodman, Hessinger, Ligget, Marshall, Kahn, & Zack, 1997).

How we ask examinees to respond or interact with computerized tests raises a number of issues. Most of these revolve around the simplicity or complexity of the software's user interface. Do examinees have the necessary computer skills or experience to read, interact with and respond to items? Are directions or tutorials both clear enough and detailed enough to impart the required skills? Is the interface simple enough to be easily learned yet complex enough to provide examinees the power to do all that is needed to efficiently process and respond to items? A superficial consideration of such issues might advocate computerized tests that use only the least common denominator of input devices and response actions. However, there may be good measurement reasons for requiring less common devices and actions. Buxton (1987, p. 367) considered the effect of various input devices on the user interface and concluded that "different devices have different properties, and lend themselves to different things". For particular applications, use of input devices such as touch screens, light pens, joysticks, or trackballs may benefit measurement. For example, very young examinees may be assessed with less error using touch screens or light pens. Particular skills, such as those that are highly movement-oriented, may be better measured using joysticks or trackballs. The input devices listed above are not in the least exotic. All are currently available, relatively prevalent, and fairly cheap. Furthermore, the near future will allow us to call on far more advanced devices. Speech recognition software and microphones will let us collect, and possibly score, spoken responses to oral questions (Stone, 1998). Haptic devices (Gruber, 1998) that use force feedback to simulate touch in a 3-D environment, could greatly increase the real-world congruence of measurement in some arenas.

The choice of input devices, along with the design of the software interface, shapes and defines the response actions available to an examinee. Bennett and Bejar (1999) have discussed the concept of "task constraints" which are the factors in the structure of a test that focus and limit examinee responses. Clearly, standardized paper-and-pencil testing is highly constrained. These constraints affect not only the ways

in which examinees can respond, but also the kinds of questions that
can be asked and possibly even the ways in which examinees can think
about these questions. With computerized tests, the task constraints
are most evident in the software interface. It is here that the dimensions of item format, response action, media, interactivity, and scoring
algorithm all converge and collectively act on examinees. It has frequently been pointed out that user interface concerns deserve serious
attention. Vicino and Moreno (1997, p. 158), stated that "reactions
to computer tasks are largely dependent on the software interface".
Bugbee and Bernt (1990, p. 97) addressed the idea of task constraints
directly by noting that "computer technology imposes unique factors
that may effect testing." Booth argued more strongly that "CBT item
presentation involves a complex fusion of cognitive, psychological, and
perceptual dynamics... We have to look beyond surface characteristics
of the medium: CBT item presentation represents a synergic dynamic
where the overall perceptual, cognitive and psychological, impact of the
medium is greater than the sum of its technological parts." (1991, p
282).

2.3. MEDIA INCLUSION

Many innovative items are entirely text-based, providing innovation
through item format, interactivity, or automated scoring. However, a
major advantage of administering tests via computer is the opportunity
to include non-text media in the items. The use of these media can
reduce dependence on reading skills, as well as enhance the validity
and task congruence of a test.

2.3.1. *Graphics*
Graphics are the most common type of non-text media included in
computerized tests. Of course, paper- and-pencil tests can also include
graphics, but lack the computer's facility for interaction. On computer,
examinees can be allowed to rotate, resize, and zoom in or out of a
scaled image. Graphics can be used as part or all of either questions or
response options. Examinees can respond by selecting one of a set of
figures, by clicking directly on some part of a graphic, or by dragging
icons to assemble a meaningful image. In the graphical modeling items
presented by Bennett, Morley, and Quardt (1998), examinees respond
by plotting points on a set of axes or grid, and then use either curve
or line tools to connect the points. In French and Godwin's (1996) figural response items, examinees manipulate graphic images to assemble
electronic circuits on screen.

2.3.2. Audio

To date, audio has been investigated and developed mainly with computerized tests of music and language skills, content areas that have often included listening skills when assessed conventionally. Paper-and-pencil tests in these fields use tape decks to play the item prompts, making them inconvenient for both proctors and examinees. Converting these tests to computer therefore provides immediate practical benefits. Computerized tests that use audio clips are administered on computers with headphones. Sound quality is improved and examinees, rather than proctors, can be allowed to control when, how often and at what volume audio clips are played (Godwin, 1999; Nissan, 1999; Parshall, Balizet, & Treder, 1998; Perlman, Berger, & Tyler, 1993). At the present, audio clips are most commonly attached only to the item stem or passage. The most typical application presents the examinee with a spoken passage or "lecture", after which a series of traditional multiple choice questions are asked. However, Parshall, Stewart, and Ritter (1996) describe an item that incorporates audio throughout. This item begins by presenting the first part of a spoken sentence as the stem. The response alternatives attached to this stem each speak a possible completion of the sentence. As stated, language and music are the traditional areas in which listening skills have been assessed. Accordingly, most of the development of audio computerized tests has occurred in these areas. Audio clips have been used with recent tests of English proficiency of non-native speakers (e.g., ACT, Inc., 1999; ETS, 1998; Godwin, 1999; Nissan, 1999), and with tests of music listening skills (Perlman, Berger, & Tyler, 1993; Vispoel & Coffman, 1992; Vispoel, Wang, & Bleiler, 1997).

Audio may also find applications outside of language and music (Parshall, 1999). As Vispoel, Wang, and Bleiler point out, "a substantial amount of general life experiences and academic activities involves the processing of information that comes to us through listening" (1997, p. 59). Bennett, Goodman, et al. (1997) discuss a sample application of audio as a form of "historical document" for examinees to interpret and analyze. In one example, actual historical radio spots are included in an item. (Short videos are also provided for the same application, as another example of non-text, historical documents.) Another possible application of audio is a listening comprehension test being investigated by the Law School Admissions Council for possible inclusion in their exam program (ACT, Inc., 1998). Listening comprehension has always been an important skill for their examinee population, but testing it in a conventional setting has never been practical. The development of new assessments of listening skills might also be important because the visual and audio channels of communication tap different cognitive processes. For example, there is evidence that while audio information places greater demands on short-term memory, multiple streams of information can also be processed concurrently more easily and accurately when communicated aurally (Fitch & Kramer, 1994). Audio need not replace text as the means of conveying information to examinees. It can instead be used to supplement or reinforce what the examinee reads or sees. An example of this is found in a prototype computer- adaptive test developed for the National Assessment of Educational Programs, or NAEP (Williams, Sweeney, & Bethke, 1999, April). This test does not directly assess listening skills in a content area. Rather, items and instructions are read aloud as well as being displayed on screen. Audio is simply available as a second channel of communication. A final potential use of audio in computerized tests is to incorporate sound into the user interface itself (Parshall, 1999). Non-speech forms of audio could be used for this application. For example, a simulation exam could have specific sounds mapped to each computer feature or action available to the examinee. Utilizing sound in this manner may provide an interface that is easier to learn and more intuitive to use (Gaver, 1989).

2.3.3. *Video*

Just as some conventional tests have long incorporated audio, so have others used video. Video discs and cassettes, despite their inconvenience, have long been used in tests of business and interpersonal interactions, medical diagnosis and treatment, and aircraft operations (ACT, 1995; Bosman, Hoogenboom, & Walpot, 1994; Shea, Norcini, Baranowski, Langdon, & Popp, 1992; Taggart, 1995). These tests have all of the practical limitations of audio cassette based tests, and then some. For example, an examinee's location in the room may provide an inappropriate advantage (i.e., a better view of the screen or better sound quality). Furthermore, item prompts are typically played in an "assembly-line" fashion, without regard to an individual examinee's readiness to continue. Computer-delivered tests provide more consistent audio/visual quality and return the item-level timing, and initiative, to the examinee. Bennett, Goodman, et al. (1997) demonstrated the use of video in items assessing historical events in a realistic context. They also developed physical education items that include video clips of students performing athletic skills. Olson-Buchanan, Drasgow, Moberg, Mead, Keenan, and Donovan (1998) developed and validated a video-based test of conflict resolution skills. Their test presents scenes of conflict in the workplace. After viewing a scene, examinees are asked which of several actions might best resolve the conflict presented. The examinee chooses an action, after which a brief additional video scene is played wherein the action moves forward and the conflict progresses.

2.3.4. *Animation*

Paper is inherently a static medium, incapable of imparting more than a sense of movement to text or graphics. In contrast, computers can use motion and animation to improve assessment in areas of dynamic processes. Although little work has been done in this area, a few examples of the possibilities can be found in the domain of computerized science lab experiments. Shavelson, Baxter, and Pine (1992) made use of simple animated objects. One application had students conduct simulated experiments by putting sow bug "icons" into an on-screen dish and manipulating certain environmental conditions. The bugs moved about the screen, either randomly or systematically, in response to these conditions. Other examples include French and Godwin's (1996) study, which included animation for science assessment. In their test, an item stem showed the effects of lightning striking a building. Bennett, Goodman, et al. (1997), extended the use of animation by displaying a series of static maps in quick succession, to demonstrate changes in nation boundaries over time. Examinees responded by identifying the particular static map that answered a question. Bennett, Goodman, et al. (1997) also noted that in many science and health related fields, professionals need to be able to read and listen to a variety of electronic instruments. They therefore developed an item that includes a static electrocardiogram strip, an animated heart monitor trace, and an audio file of the related heart sound. Finally, Martinez (1991) suggested figural response items that include stimuli such as time series processes (e.g., cell division). Animation has several advantages over video clips in certain applications, despite the relative popularity of the latter. Animation uses far less computer memory than video to store or display, and may be less expensive to produce. More substantively, animation is likely to be simpler. In the same way that a line drawing can emphasize critical features better than a photograph, animation can focus the examinee on essential aspects of the movement more specifically than video is likely to do. For other applications, the "real-world" detail and complexity inherent in video may be essential.

2.4. INTERACTIVITY

The majority of innovative items are still discrete, single step items. The examinee takes an action (e.g., makes a selection), and the item is complete. A few item types use a limited kind of feedback, increasing item-examinee interaction slightly. With these, the examinee acts and the computer responds with some sort of reaction. For example, when examinees click on a histogram, scale, or dial they may see the bar or gauge move to reflect this selection (O'Neil & Folk, 1996). Examinees who manipulate the environmental conditions of the simulated sow bugs see the animated effects of those conditions on the bugs (Shavelson, Baxter, & Pine, 1992). Examinees who edit text may see the new text embedded within the original passage, letting them re-read the passage with the change reflected (Breland, 1998; Davey, Godwin, & Mittelholz, 1997). Examinees who specify a sequence for a set of elements may see the elements rearranged in the new order (O'Neil & Folk, 1996). Finally, examinees who indicate the points on a grid where a line or curve should be plotted, may see that line or curve displayed (Bennett, Morley, & Quardt, 1998). Examinees can use any of these forms of feedback to help them decide whether their initial selection had the desired effect. However, they are not directly informed about the correctness of a response, although that information might be appropriate in instructional applications. Instead, this limited form of feedback simply allows examinees to consider the item and their response in the proper context. More extensive interactivity has so far rarely been attempted either operationally or in research settings. A notable exception is the video-based test of conflict skills described above. Drasgow, Olson-Buchanan, and Moberg (1999) and Olson-Buchanan, Drasgow, Moberg, Mead, Keenan, and Donovan (1998), refer to the test delivery model they used as *interactive video assessment.* However, even here the interactive scenarios are allowed to branch only once. The initial video scene is followed by a multiple choice question the response to which determines the concluding scene that is shown. Despite this limitation, it is likely that interactivity, along with the use of video prompts, greatly increases the realism of the testing process.

Another sort of interactivity has been investigated with items measuring knowledge of elementary science (G. G. Kingsbury, personal communication, May 7, 1998). These items require examinees to work through multiple steps or stages and so are termed sequential process items. They begin by presenting examinees with the first element in a number or pattern series. The next and subsequent elements in the series are then added to the display automatically at some suitable time interval. The examinee's task is to discover the rule by which the sequence is building. The item ends either when the examinee responds or when the sequence is complete. Scoring could consider both whether the response was correct and the number of elements displayed at the time the response was offered. Godwin (personal communication, October 14, 1998) proposed a more elaborate sort of interaction for a test of student research skills. Here, examinees are presented with a list of possible research topics and asked to select one they find interesting. The computer then branches to a list of potential key words for use in a library search. The examinee identifies those key words considered most likely to lead to useful information. This action, in turn, results in a set of book or periodical titles that may or may not be germane, depending on the examinee's choice of search criteria. The process could continue through several stages, as examinees refine the search and collect appropriate information. The final result might be computerized "note cards" in which the examinee has summarized the most salient information available, or an outline the examinee has assembled for a potential research paper.

There are few examples of operational tests with high degrees of interactivity. One such operational exam is the credentialing test created by the National Council of Architectural Registration Boards (NCARB). This test, as described by Braun (1994) includes a performance or simulation component, during which examinees operate with several "vignettes". Each of the vignettes presents an architectural problem or assignment. The examinee must use computerized drawing tools to design a solution to the problem within specified criteria. These problems have clear similarities to the constructed figural response item type, although the tasks, and the procedures for scoring responses, are far more complex. Another example of interactivity is offered by Clauser, Margolis, Clyman, and Ross (1997), who describe a computerized performance test intended to measure the patient management skills of physicians. The test is comprised of several cases, each of which presents the examinee with a brief description of a patient who is complaining of some malady. The examinee can order medical tests or procedures, interpret the results of those procedures, diagnose the condition, and monitor changes in status over time and in response to actions taken. All of the steps taken by the examinee are recorded, along with the simulation-based time in which each action occurred. Current research is directed towards developing an optimal approach to automating the scoring of the resulting list of transactions.

While there are many instructional applications that utilize the computer's ability to interact with users (and, of course, most computer games are highly interactive), far less work has been conducted for testing purposes. This limited development is probably related to the challenges inherent in developing interactive assessments. Several of these difficulties are noted by Drasgow, Olson-Buchanan, and Moberg (1999). Although the application they have developed consists of only two stages of interaction, they point out that even this limited level of interactivity results in numerous test paths that are taken by few if any examinees. Test developers are therefore required to produce much more content than will be seen by any given examinee. God-win (personal communication) noted a second potential difficulty with development of branching item types. Without care, test developers can produce interactive tests in which a great number of entire paths, and all of their associated options, are incorrect. Examinees may drift further and further away from correct choices, or from providing any useful assessment information. A final challenge of highly interactive assessments is the subject of our next section, namely how these as-sessments are scored. The scoring models used by complex interactive items must consider many factors, including the optimal weight of var-ious components of the task, and the dependency of the examinees' various responses to the task.

2.5. SCORING ALGORITHM

Many of the important practical benefits of computerized testing re-quire that scoring of items and tests be automated. Tests can be adapted to examinees only when item responses are instantly scored by the com-puter. And score reports can be issued immediately after testing only when test scores are determined by the computer. For the majority of the item types discussed up to this point, the computer can easily provides immediate scoring. However in some instances the computer is used solely to administer the items and collect the responses, with scoring done later by human raters (Baker & O'Neil, 1995; O'Neill & Folk, 1996; Shavelson, Baxter, & Pine, 1992). Ideally, even com-plex constructed response items could be delivered on computer, and scored automatically and immediately. Examples of automatic scoring for some moderately complex tasks include the constructed response editing task described in Breland (1998). In this task, examinees re-placed incorrect, marked text by typing their own text. The edited text was scored by matching it to "templates" provided by the test devel-opers. Future versions of this task may use natural language processing to satisfy more complex scoring needs. Davey, Godwin, and Mittel-holtz (1997) incorporated some innovations into the scoring algorithm of their on-line editing task, in which examinees must find and replace text that contains errors. Their scoring scheme recognized that not all errors are equal. The worst sort of error would be to replace correct text with an incorrect alternative. Missing or ignoring glaring problems in the text is only slightly better. An examinee who correctly identifies an error but is unable to correct it should receive more credit than an examinee who did not even spot the problem. Correct answers do not count equally either. Leaving errorless text untouched, while correct, is not as valued as identifying problems and fixing them. Results of an investigation of a complex, constructed response mathematical reasoning item type were reported by Bennett, Steffen, Singley, Morley, and Jacquemin (1997). Examinees in this study responded to items by providing complex mathematical expressions. The computer scored these answers by evaluating the expressions and comparing the results to a key. For example, the expressions $x^{1/2}$, $x^{2/4}$, and $\sqrt{x}$ would evaluate as equivalent. Automated scoring was able to successfully score over 99% of examinee responses. At the present time, only dichotomous scoring is supported, but the authors suggest some possible future approaches that could provide partial-credit scoring as well.

The majority of the examples discussed up to this point are small, discrete tasks, and the result of the automated scoring process is usually a dichotomous item score. As applications become more complex, more complex scoring programs are needed. One such is example is the architectural performance test described earlier. Its scoring involves extracting salient features from examinee responses and comparing these to a complex set of rules (Bejar, 1991). Responses are first processed so that the solutions are represented as lists of design elements. The critical features of an examinee solution are then extracted at an elemental level (e.g., a line that comprises one wall, in a rectangle that represents a bedroom, in the plan of a house). Scoring considers whether an examinee's design satisfies task requirements, both of form and function. The rules that define a correct solution are obtained from experts who score an initial set of responses. These rules may include both a set of criteria to be satisfied, and a means of weighting the relative importance of those criteria. Several automated scoring approaches have been investigated for the exam of physicians' patient management skills (Clauser, Margolis, Clyman, & Ross, 1997; Clauser, Ross, et al., 1997). In Clauser, et al. (1997), a comparison of two methods was conducted. Both a rules-based method and a regression-based approach were evaluated, with scores assigned by the two methods compared to scores given by a panel of experts. Although both methods showed reasonably high correlations with expert ratings, the regression method was found slightly superior to the rules based approach. However, a rules-based approach may be easier to explain and justify both to examinees and to credentialing boards.

Several methods for automating full essay scoring have also been investigated (Burstein, Kukich, Wolff, Lu, & Chodorow, 1998; Page & Petersen, 1995). These methods use different analytical approaches to isolate critical features of the essay responses. For example, the Project Essay Grade, or PEG method (Page & Petersen, 1995) measures a large set of descriptive features of an essay (e.g., overall length), and correlates those variables to raters' scores of the essays. The e-rater system

(Burstein, Kukich, Wolff, Lu, & Chodorow, 1998) utilizes advanced computational linguistics techniques to extract the essay features used by human raters in their assignment of scores (e.g., rhetorical structure). In both of these cases, the essay scoring computer program is first "trained" on a specific essay prompt, by using a small set of essay responses, human scores, and statistical regression techniques to determine the optimal relative weights of essay response features. The weighted set of essay response characteristics is then used by the computer program to score additional essay responses. Both of these methods have shown very promising results, and can produce essay scores that correlate more highly with a single human rater than a second rater typically does. Currently, a number of standardized testing programs with performance tasks or essay prompts use multiple raters to score each examinee response. One approach to implementing computerized, automated scoring would be to use a single human rater, along with the computer program. This could yield scores of comparable reliability, in less time and at a lower cost than the use of multiple human raters, although it would not provide the immediate scoring available for other computerized tasks. Further, research has not established whether these methods can adequately resist cheating by examinees, who have learned the features maximally rated by the computer program. (For additional examples and discussion regarding automated scoring, see Bennett, 1998; Braun, Bennett, Frye, & Soloway, 1990; Dodd & Fitzpatrick, 1998; and Martinez & Bennett, 1992).

3. Conclusion

While today's automobiles are a vast improvement over the original horseless carriages of yesteryear, they share some basic elements. These include wheels, an engine, control devices operated by the driver, and seating for passengers. They also all still perform the same basic function-transporting people from point A to point B-albeit with more comfort, reliability, and speed. Just as we and others contemplate how innovative items might change computerized testing, so must early automotive pioneers have wondered what future cars might be like. Hindsight allows us to draw some conclusions in their case as to what innovations succeeded and what failed. By and large, the buying public has embraced innovations that have made cars truly better in the sense of being easier and more comfortable to drive. Automatic transmissions, power steering and air conditioning are examples of things that were once exotic and are now all but standard. A second class of innovations may not have appealed directly to consumers, but have been mandated by government regulation. Seat belts, air bags and emissions control devices are examples. What has not enjoyed certain or permanent success is innovation that did not further the car's basic purpose. Styling provides many examples. The huge tail fins popular in the 50s have disappeared, perhaps never to return. Vinyl roofs, chrome bumpers and elaborate grills are equally out of fashion.

There is probably a lesson here that we would do well to consider. Clearly, the main purpose of testing is to measure proficiencies (abilities, skills, or knowledge) in valid, reliable, and efficient ways. Innovative items that serve or further this purpose will surely succeed. In this chapter, we have delineated five dimensions along which innovative items might vary: 1) item format, 2) response action, 3) media inclusion, 4) level of interactivity, and 5) scoring algorithm. Many of the examples given here of items that take advantage of one or more of these aspects of innovation show promise for providing better measurement. However, innovation in and of itself does not ensure, or even promote, better assessment. It might even hinder the effort. Consider an item that requires extensive manipulations by the examinee through a complex computer interface. Examinees who should be thinking about what to respond might instead be focused on how to respond. The examinee's level of computer literacy may prove to be a confounding factor in this case. If it is, the validity and fairness of the item could be brought into question. Such items, like tail fins, may make a test look better to today's consumers, but unless they provide real and tangible measurement benefits, they are unlikely to succeed in the longer run. In a similar vein, consider innovative items that are more open-ended and accept various forms of the correct answer. Delineating all possible answers for these items can be a time-consuming task for the test developer, and the potential for item challenges can be increased substantially. Open-ended items can also be more time consuming for the examinee to answer, while delivering a similar amount of information about the examinee's proficiency as, for example, five-choice items. One category of innovations certain to gain acceptance are those things that can be done to accommodate disabled examinees. Like seat belts, these innovations are unlikely to attract consumer attention, but are important nonetheless. Computers offer considerable promise for interacting with nearly all examinees on a level footing. So, what other avenues look promising for future research on innovative items? A safe bet would be those innovative items that take advantage of the unique features of computers. Computerized assessments can provide highly interactive, media-rich, complex environments. They can also offer greater control over the testing process and provide increased measurement efficiency to boot. Innovative items that take advantage of these features to provide more valid measurements, while simultaneously being efficient for the test developers to produce and examinees to take, will have an excellent chance of succeeding.

References

ACT, Inc. (1995). *Work keys.* Iowa City, IA: Author.

ACT, Inc. (1998). *Assessing listening comprehension: A review of recent literature relevant to an LSAT Listening Component* [Unpublished manuscript]. Newton, PA: Law School Admission Council..

ACT, Inc. (1999). *Technical Manual for the ESL Exam.* Iowa City, IA: Author.

Baker, E.L., & O'Neil, H.F., Jr. (1995). Computer technology futures for the improvement of assessment. *Journal of Science Education and Technology, 4,* 37-45.

Bejar, I. I. (1991). A methodology for scoring open-ended architectural design problems. *Journal of Applied Psychology, 76,* 522-532.

Bennett, R.E. (1998). *Reinventing assessment.* Princeton, NJ: Educational Testing Service.

Bennett, R.E., & Bejar, I.I. (1999). Validity and automated scoring: It's not only the scoring. *Educational Measurement: Issues and Practices, 17,* 9-17.

Bennett, R. E., Goodman, M., Hessinger, J., Ligget, J., Marshall, G., Kahn, H., & Zack, J. (1997). *Using multimedia in large-scale computer-based testing programs.* (Research Rep. No. RR-97-3). Princeton, NJ: Educational Testing Service.

Bennett, R. E., & Sebrechts, M. M. (1997). A computer-based task for measuring the representational component of quantitative proficiency. *Journal of Educational Measurement, 34,* 64-77.

Bennett, R.E., Steffen, M., Singley, M.K., Morley, M., & Jacquemin, D. (1997). Evaluating an automatically scorable, open-ended response type for measuring mathematical reasoning in computer-adaptive tests. *Journal of Educational Measurement, 34,* 162-176.

Bennett, R. E., Morley, M., & Quardt, D. (1998, April). *Three response types for broadening the conception of mathematical problem solving in computerized-adaptive tests.* Paper presented at the annual meeting of the National Council of Measurement in Education, San Diego, CA.

Booth, J. (1991). The key to valid computer-based testing: The user interface. *European Review of Applied Psychology, 41,* 281-293.

Bosman, F., Hoogenboom, J., & Walpot, G. (1994). An interactive video test for pharmaceutical chemist's assistants. *Computers in Human Behavior, 10,* 51-62.

Braun, H.I. (1994). Assessing technology in assessment. In E. L. Baker & H. F. O'Neil (Eds.), *Technology assessment in education and training* (pp. 231-246). Hillsdale, NJ: Lawrence Erlbaum Associates

Braun, H.I., Bennett, R.E., Frye, D., & Soloway, E. (1990). Scoring constructed responses using expert systems. *Journal of Educational Measurement, 27,* 93-108.

Breland, H.M. (1998, April). *Writing assessment through automated editing.* Paper presented at the annual meeting of the National Council on Measurement in Education, San Diego.

Bugbee, A. C., Jr., & Bernt, F. M. (1990). Testing by computer: Findings in six years of use. *Journal of Research on Computing in Education, 23,* 87-100.

Burstein, J., Kukich, K., Wolff, S., Lu, C., & Chodorow, M. (1998, April). *Computer analysis of essays.* Paper presented at the annual meeting of the National Council on Measurement in Education, San Diego.

Buxton, W.A.S. (1987). There's more to interaction than meets the eye: Some issues in manual input. In R. M. Baecker & W. A. S. Buxton (Eds.) *Readings in human-computer interaction: A multidisciplinary approach* (pp.366-375). San Mateo, CA: Morgan Kaufmann.

Clauser, B.E., Margolis, M.J., Clyman, S.G., & Ross, L.P. (1997). Development of automated scoring algorithms for complex performance assessments: A comparison of two approaches. *Journal of Educational Measurement, 34,* 141-161.

Clauser, B.E., Ross, L.P., Clyman, S.G., Rose, K.M., Margolis, M.J., Nungester, R.J., Piemme, T.E., Chang, L., El-Bayoumi, G., Malakoff, G.L., & Pincetl, P.S. (1997). Development of a scoring algorithm to replace expert rating for scoring a complex performance-based assessment. *Applied Measurement in Education, 10,* 345-358.

Davey, T., Godwin, J., & Mittelholtz, D. (1997). Developing and scoring an innovative computerized writing assessment. *Journal of Educational Measurement, 34,* 21-41.

Dodd, B. G., & Fitzpatrick, S.J. (1998). *Alternatives for scoring computer-based tests.* Paper presented at the ETS Colloquium on Computer-based testing: Building the foundation for future assessments, Philadelphia, PA.

Drasgow, F., Olson-Buchanan, J. B., & Moberg, P. J. (1999). Development of an interactive video assessment: Trials and tribulations. In F. Drasgow & J. B. Olson-Buchanan, (Eds.), *Innovations in computerized assessment* (pp 177-196). Mahwah, NJ: Lawrence Erlbaum Associates.

ETS. (1998). Computer-Based TOEFL Score User Guide. Princeton, NJ: Author.

Fitch, W. T., & Kramer, G. (1994). Sonifying the body electric: Superiority of an auditory over a visual display in a complex, multivariate system. In G. Kramer (Ed.), *Auditory Display* (pp. 307-325). Reading, MA: Addison-Wesley.

French, A., & Godwin, J. (1996, April). *Using multimedia technology to create innovative items.* Paper presented at the annual meeting of the National Council on Measurement in Education, New York, NY.

Gaver, W. W. (1989). The SonicFinder: An interface that uses auditory icons. *Human-Computer Interaction, 4,* 67-94.

Gruber, J. S. (1998, October). Groupthink [Interview with James Kramer, Head of Virtual Technologies, Inc.]. *Wired,* 168-169.

Godwin, J. (1999, April). *Designing the ACT ESL Listening Test.* Paper presented at the annual meeting of the National Council on Measurement in Education, Montreal, Canada.

Koch, D. A. (1993). Testing goes graphical. *Journal of Interactive Instruction Development, 5,* 14-21.

Luecht, R. M., & Clauser, B. E. (1998, September). *Test models for complex computer-based testing.* Paper presented at the ETS Colloquium on Computer-based testing: Building the foundation for future assessments, Philadelphia, PA.

Martinez, M. E. (1991). A comparison of multiple choice and constructed figural response items. *Journal of Educational Measurement, 28,* 131-145.

Martinez, M. E. (1993). Item formats and mental abilities in biology assessment. *Journal of Computers in Mathematics and Science Teaching, 12,* 289-301.

Martinez, M.E., & Bennett, R.E. (1992). A review of automatically scorable constructed-response item types for large-scale assessment. *Applied Measurement in Education, 5,* 151-169.

Nissan, S. (1999, April). *Incorporating sound, visuals, and text for TOEFL on computer.* Paper presented at the annual meeting of the National Council on Measurement in Education, Montreal, Canada.

Olson-Buchanan, J. B., Drasgow, F., Moberg, P. J., Mead, A. D., Keenan, P.A., & Donovan, M.A. (1998). Interactive video assessment of conflict resolution skills. *Personnel Psychology, 51,* 1-24.

O'Neill, K., & Folk, V. (1996, April). *Innovative CBT item formats in a teacher licensing program.* Paper presented at the annual meeting of the National Council on Measurement in Education, New York, NY.

Page, E.B., & Petersen N.S. (1995). The computer moves into essay grading: Updating the ancient test. *Phi Delta Kappa, 76,* 561-565.

Parshall, C. G. (1999, February). *Audio CBTs: Measuring more through the use of speech and non-speech sound.* Paper presented at the annual meeting of the National Council on Measurement in Education, Montreal, Canada.

Parshall, C. G., Balizet, S., & Treder, D. W. (1998, July). *Using an audio computer-based progress exam to test the listening skills of non-native speakers.* Paper presented at the Simposio Sobre Tecnologias Educativas: Enseñanza para el Siglo XXI (Symposium on Educational Technology: Teaching for the 21st Century), Mèrida, Venezuela.

Parshall, C. G., Stewart, R, & Ritter, J. (1996, April). *Innovations: Sound, graphics, and alternative response modes.* Paper presented at the annual meeting of the National Council on Measurement in Education, New York.

Perlman, M., Berger, K., & Tyler, L. (1993). *An application of multimedia software to standardized testing in music.* (Research Rep. No. 93-36) Princeton, NJ: Educational Testing Service.

Shavelson, R. J., Baxter, G. P., & Pine, J. (1992). Performance assessments: Political rhetoric and measurement reality. *Educational Researcher, 21,* 22-27.

Shea, J. A., Norcini, J. J., Baranowski, R. A., Langdon, L. O., & Popp, R. L. (1992). A comparison of video and print formats in the assessment of skill in interpreting cardiovascular motion studies. *Evaluation and the Health Professions, 15,* 325-340.

Stone, B. (1998, March). Focus on technology: Are you talking to me? *Newsweek,* 85-86.

Taggart, W. R. (1995). Certifying pilots: Implications for medicine and for the future. In E. L. Mancall & P. G. Bashook (Eds.), *Assessing clinical reasoning: The oral examination and alternative methods* (pp. 175-182). Evanston, IL: American Board of Medical Specialties.

Vicino, F. L., & Moreno, K. E. (1997). Human factors in the CAT system: A pilot study. In W. A. Sands, B. K. Waters, & J. R. McBride (Eds.), *Computerized adaptive testing: From inquiry to operation* (pp. 157-160). Washington, DC: APA.

Vispoel, W. P., & Coffman, D. (1992). Computerized adaptive testing of music-related skills. *Bulletin of the Council for Research in Music Education, 112,* 29-49.

Vispoel, W. P., Wang, T., & Bleiler, T. (1997). Computerized adaptive and fixed-item testing of music listening skill: A comparison of efficiency, precision, and concurrent validity. *Journal of Educational Measurement, 34,* 43-63.

Williams, V. S. L., Sweeny, S. F., & Bethke, A. D. (1999, April). *The development and cognitive laboratory evaluation of an audio-assisted computer-adaptive test for eight-grade mathematics.* Paper presented at the annual meeting of the National Council on Measurement in Education, Montreal, Canada.

Chapter 8
Designing Item Pools for Computerized Adaptive Testing

Bernard P. Veldkamp & Wim J. van der Linden
University of Twente, The Netherlands

1. Introduction

In existing computerized adaptive testing (CAT) programs, each successive item in the test is chosen to optimize an objective function. Examples of well-known objectives in CAT are maximizing the information in the test at the ability estimate for the examinee and minimizing the deviation of the information in the test from a target value at this estimate. In addition, item selection is required to realize a set of content specifications for the test. For example, item content may be required to follow a certain taxonomy, or, if the items have a multiple-choice format, their answer key distribution should deviate not too much from uniformity. Content specifications are generally defined in terms of combinations of attributes the items in the test should have. They are typically realized by imposing a set of constraints on the item-selection process. The presence of an objective function and constraints in CAT leads to the notion of CAT as constrained (sequential) optimization. For a more formal introduction to this notion, see van der Linden (this volume).

In addition to content constraints, item selection in CAT is often also constrained with respect to the exposure rates of the items in the pool. These constraints are necessary to maintain item pool security. Sympson and Hetter (1985) developed a probabilistic method for item-exposure control. In their method, after an item is selected, a probability experiment is run to determine whether the item is or is not administered. By manipulating the (conditional) probabilities in this experiment, the exposure rates of the items are kept below their bounds. Several modifications of this method have been developed (Davey & Nering, 1998; Stocking & Lewis, 1998). For a review of these methods, see Stocking and Lewis (this volume).

Though current methods of item-exposure control guarantee upper bounds on the exposure rates of the items, they do not imply any lower bounds on these rates. In fact, practical experience with CAT shows that item pools often have surprisingly large subsets of items that are seldom administered. The reason for this phenomenon is that

W.J. van der Linden and C.A.W. Glas (eds.),
Computerized Adaptive Testing: Theory and Practice, 149–162.
© 2000 *Kluwer Academic Publishers. Printed in the Netherlands.*

such items contribute poorly to the objective function optimized in the CAT algorithm or have attribute values that are overrepresented in the pool relative to the requirements in the constraints on the test. Since item production usually involves a long and costly process of writing, reviewing, and pretesting the items, the presence of unused items in the pool forms an undesired waste of resources.

Though CAT algorithms could be developed to guarantee a lower bound on the exposure rates for the items in the pool as well, a more efficient approach to over- or underexposure of items is trying to prevent the problem at all and *design* the item pool to produce a more uniform item usage for the population of examinees. It is the purpose of this chapter to propose a method of item pool design that addresses this target. The main product of the method is an optimal blueprint for the item pool, that is, a document specifying what attributes the items in the CAT pool should have and how many items of each type are needed.

The blueprint should be used as a starting point for the item writing process. As will be shown below, if "the identity of the item writer" is used as a categorical item attribute in the design process, the blueprint can also lead to an optimal division of labor among the item writers. However, since some quantitative item attributes, in particular those that depend on statistical parameters estimated from empirical data, are difficult to realize exactly, a realistic approach is to use the method proposed in this chapter as a tool for continuous management of the item writing process. Repeated applications of it can then help to adapt the next stage in the item writing process to the part of the pool that has already been written.

2. Review of Item Pool Design Literature

The subject of item pool design has been addressed earlier in the literature, both for pools for use with CAT and the assembly of linear test forms. A general description of the process of developing item pools for CAT is presented in Flaugher (1990). This author outlines several steps in the development of an item pool and discusses current practices at these steps. A common feature of the process described in Flaugher and the method in the present paper is the use of computer simulation. However, in Flaugher's outline, computer simulation is used to evaluate the performance of an item pool once the items have been written and field tested whereas in the current chapter computer simulation is used to design an optimal blueprint for the item pool.

Methods of item pool design for the assembly of linear test forms are presented in Boekkooi-Timminga (1991) and van der Linden, Veldkamp and Reese (2000). These methods, which are based on the technique of integer programming, can be used to optimize the design of item pools that have to support the assembly of a future series of test forms. The method in Boekkooi-Timminga follows a sequential approach calculating the numbers of items needed for these test forms maximizing their information functions. The method assumes an item pool calibrated under the one-parameter logistic (1PL) or Rasch model. The method in van der Linden, Veldkamp and Reese directly calculates a blueprint for the entire pool minimizing an estimate of the costs involved in actually writing the items. All other test specifications, including those related to the information functions of the test forms, are represented by constraints in the integer programming model that produces the blueprint. This method can be used for item pools calibrated under any current IRT model. As will become clear below, the current proposal shares some of its logic with the latter method. However, integer programming is not used for direct calculation of the numbers of items needed in the pool—only to simulate constrained CAT.

Both Swanson and Stocking (1998) and Way, Steffen and Anderson (1998; see also Way, 1998) address the problem of designing a system of rotating item pools for CAT. This system assumes the presence of a master pool from which operational item pools are generated. A basic quantity is the number of operational pools each item should be included in (degree of item-pool overlap). By manipulating the number of pools items are included in, their exposure rates can be controlled. A heuristic based on Swanson and Stocking's (1993) weighted deviation model (WDM) is used to assemble the operational pools from the master pool such that, simultaneously, the desired degree of overlap between the operational pools is realized and they are as similar as possible. The method proposed in this paper does not assume a system of rotating item pools. However, as will be shown later, it can easily be adapted to calculate a blueprint for such a system.

3. Designing a Blueprint for CAT Item Pools

The process of designing an optimal blueprint for a CAT item pool involves the following stages: First, the set of specifications for the CAT is analyzed and all item attributes figuring in the specifications are identified. The result of this stage is a (multivariate) classification table defined as the product of all categorical and quantitative item attributes. Second, using this table, an integer programming model for the assembly of the shadow tests in a series of CAT simulations is formulated. (The notion of CAT with shadow tests will be explained below.) Third, the population of examinees is identified and an estimate of its ability distribution is obtained. In principle, the distribution is unknown but an accurate estimate may be obtained, for example, from historical data. Fourth, the CAT simulations are carried out by sampling examinees randomly from the ability distribution. Counts of the number of times items from the cells in the classification table are administered in the simulations are cumulated. Fifth, the blueprint is calculated from these counts adjusting them to obtain optimal projections of the item exposure rates. Some of these stages are now explained in more detail.

3.1. SETTING UP THE CLASSIFICATION TABLE

The classification table for the item pool is set up distinguishing the following three kinds of constraints that can be imposed on the item selection by the CAT algorithm: (1) constraints on categorical item attributes, (2) constraints on quantitative attributes, and (3) constraints needed to deal with inter-item dependencies (van der Linden, 1998).

Categorical item attributes, such as content, format, or item author, partition an item pool into a collection of subsets. If the items are coded by multiple categorical attributes, their Cartesian product induces a partitioning of the pool. A natural way to represent a partitioning based on categorical attributes is as a classification table. For example, let C1, C2, and C3 represent three levels of an item content attribute and let F1 and F2 represent two levels of a item format attribute. Table 1 shows the classification table for a partition that has six different cells, where n_{ij} represents the number of items in the pool that belong to cell (i,j).

Examples of possible quantitative item attributes in CAT are: word counts, values for the item difficulty parameters, and item response times. Classifications based on quantitative attributes are less straightforward to deal with. Some of them may have a continuous range of possible values. A possible way to overcome this obstacle is to pool adjacent values. For example, the difficulty parameter in the three parameter logistic IRT model takes real values in the interval $(-\infty, \infty)$. This interval could be partitioned into the collection of the following subintervals: $((-\infty, -2.5), (-2.5, -2), \ldots, (2, 2.5), (2.5, \infty))$. The larger the number of intervals, the more precise the approximation to the true item parameter values. After such partitioning, quantitative attributes can be used in setting up classification tables as if they were categorical. If single numbers are needed to represent intervals of attribute values, their midpoints are an obvious choice.

Table 1. Classification table

	F1	F2
C1	n_{11}	n_{21}
C2	n_{12}	n_{22}
C3	n_{13}	n_{23}

Inter-item dependencies deal with relations of exclusion and inclusion between the items in the pool. An example of an exclusion relation is the one between items in "enemy sets". Such items can not be included in the same test because they have clues to each other's solution. However, if previous experience has shown that enemies tend to be items with certain common combinations of attributes, constraints can be included in the CAT algorithm to prevent such combinations from happening. The problem of CAT from item pools with exclusion relations between the items will be addressed later in this chapter. An example of an inclusion relation is the one between set-based items in a test, that is, sets of items organized around common stimuli. When designing pools for the assembly of linear test forms, relations between set-based items can be dealt with by setting up a separate classification table based on the stimulus attributes and then assigning stimuli to item sets. An example of this approach is given in van der Linden, Veldkamp and Reese (2000). In this chapter, the problem of CAT from pools with item sets is not addressed.

The result of this stage is thus a classification table, $C \times Q$, that is the Cartesian product of a table C based upon the categorical attributes and a table Q based upon the quantitative attributes. Each cell of the table represents a possible subset of items in the pool that have the same values for their categorical attributes and belong to the same (small) interval of values for their quantitative attributes.

3.2. Constrained CAT Simulations

To find out how many items an optimal pool should contain in each cell in table $C \times Q$, simulations of the CAT procedure are carried out. Each cell in the $C \times Q$ table is represented by a decision variable in the integer programming model for the shadow test. The variables represents the number of times an item from a cell is selected for the shadow test. The method of constrained CAT with shadow tests (van der Linden, this volume; van der Linden & Reese, 1998) is briefly explained and a general formulation of an integer programming model for selecting a shadow test is given.

3.2.1. *Constrained Adaptive Testing with Shadow Tests*

In constrained adaptive testing with shadow tests, at each step a full test ("shadow test") is assembled. The shadow test is assembled to have an optimal value for the objective function and is required to meet a set of constraints that represents all test specifications. The item actually to be administered is selected from this shadow test; it is the item with the optimal contribution to the value of the objective function for the test. As a result of this procedure, each actual adaptive test eventually meets all constraints and has items selected with optimal values for the objective function.

The algorithm for constrained CAT with shadow tests can be summarized as follows:

Step 1: Choose an initial value of the examinee's ability parameter θ.

Step 2: Assemble the first shadow test such that all constraints are met and the objective function is optimized.

Step 3: Administer an item from the shadow test with optimal properties at the current θ estimate.

Step 4: Update the estimate of θ as well as all other parameters in the test assembly model.

Step 5: Assemble a new shadow test fixing the items already administered.

Step 6: Repeat Steps 3-5 until all n items have been administered.

In an application of this procedure in a real-life CAT program, the shadow tests are calculated using a 0-1 linear programming (LP) model for test assembly. The variables in the model represent for each individual item in the pool the decision to select or not select the item in the shadow test. However, in the CAT simulations in the current application, the more general technique of integer programming is used. The integer variables represent the number of items needed from each cell in the $C \times Q$ table for each simulated examinee.

3.2.2. *Integer Programming Model*

Let x_{cq} be the integer variable for cell (c, q) in table $C \times Q$. This variable determines how many items are to be selected from cell (c, q) for each simulated examinee. Furthe let n be the length of the CAT, $\widehat{\theta}_{k-1}$ the estimate of θ_j after $k - 1$ items have been administered, and S_{k-1} the set of cells with nonzero decision variable after $k - 1$ items have been administered. Fisher's information in the response on item i for an examinee with ability θ_j is denoted as $I_i(\theta_j)$. Finally, V_g denotes the set of cells representing the combination of attributes in categorical constraint $g = 1, \ldots, G$, V_h the set of cells representing the combination of attributes levels in quantitative constraint $h = 1, \ldots, H$, and V_e the set of cells in enemy set $e = 1, \ldots, E$.

The model has an objective function for the shadow tests that minimizes an estimate of the costs involved in writing the items in the pool. The information on the ability parameter at the ability estimate in the CAT is bounded from below by a target value, T. Generally, item writing costs can be presented as quantities k_{cq}, $(c, q) \in C \times Q$. In the empirical example below, k_{cq} is chosen to be the inverse of the numbers of items in cell (c, q) in a representative previous item pool, the idea being that types of items that are written more frequently involve less efforts and, therefore, are likely to be less costly. Several suggestions for alternative estimates of item writing costs are given in van der Linden, Veldkamp and Reese (2000). Also, if these costs are dependent on the item writer and it is known which authors wrote which items in the previous item pool, a convenient option is to adopt the identity of the item writer as a categorical item attribute in the $C \times Q$ table. The blueprint then automatically assigns numbers of items to be written to individual writers.

The general model for the assembly of the shadow test for the selection of the kth item in the CAT can be presented as:

$$\min \sum_{cq \in C \times Q} k_{cq} x_{cq} \qquad \text{(objective function)} \qquad (1)$$

subject to

$$\sum_{cq \in C \times Q} I_{cq}\left(\widehat{\theta}_{k-1}\right) x_{cq} \geq T, \qquad \text{(information target)} \qquad (2)$$

$$\sum_{cq \in S_{k-1}} x_{cq} = k - 1, \qquad \text{(items already selected)} \quad (3)$$

$$\sum_{cq \in C \times Q} x_{cq} = n, \qquad \text{(test length)} \qquad (4)$$

$$\sum_{cq \in Vg} x_{cq} = n_g, \qquad g = 1, ..., G, \qquad \text{(categorical constraint)} \quad (5)$$

$$f_h\left(x_{cq}\right) = n_h, \qquad h = 1, ..., H, \qquad \text{(quantitative constraint)} (6)$$

$$\sum_{cq \in V_e} x_{cq} \leq 1, \qquad e = 1, ..., E, \qquad \text{(enemy sets)} \qquad (7)$$

$$x_{cq} \in \{0, 1, ..\}, \qquad (c, q) \in C \times Q. \qquad \text{(range of variables)} \qquad (8)$$

The objective function in (1) minimizes the estimated item-writing costs. The constraint in (2) requires the information in the shadow test at the examinees' current ability estimate to meet the prespecified target value T. The constraint in (3) requires the $k - 1$ previously administered items to be in the test. The attribute values of the previous items are automatically taken into account when selecting the kth item. In (4), the length of the CAT is fixed at n items. In (5) and (6), the categorical and quantitative constraints are imposed on the shadow test. The function f_h in (6) is assumed to be linear in the decision variables, for example, a (weighted) average or a sum. The constraints in (5) and (6) are formulated as equalities but can easily be turned into inequalities. The constraints in (7) allow the shadow test to have no more than one item from each enemy set.

Practical experience should guide the selection of the target value for the test information function. Generally, to emulate the process of a CAT that maximizes the information in the test, the target value should be chosen as high as possible without making the selection of the shadow tests infeasible. If the CAT program has been operational for some time, it should be known what targets are feasible. Alternatively, the operational CAT procedure can be based on the same target for the information function. The only thing needed to implement the latter option is to insert the constraint (2) into the model for the operational CAT. Other approaches are also possible. For instance, in the empirical example below, no target value for the information in the CAT was available, and the objective function in (1) and the information function in the constraint in (2) were combined into a linear combination optimized by the test assembly model. A review of options to deal with multi-objective decision problems in test assembly is given in Veldkamp (1999).

After the shadow test for the kth item is assembled, the item with maximum information at $\widehat{\theta}_{k-1}$ among the items not yet administered is selected for administration as the kth item in the CAT.

3.3. CALCULATING THE BLUEPRINT

The blueprint for the item pool is based on the counts of the number of times items from the cells in table $C \times Q$ are administered to the simulated examinees, N_{cq}. These numbers are adapted to guarantee that the target values for the item-exposures rates are realized for a prespecified number of examinees sampled from the ability distribution. It is assumed that the number of examinees sampled, S, is large enough to produce stability among the relative values of N_{cq}.

The blueprint is calculated from the values of N_{cq} according to following formula:

$$I_{cq} = \left\lceil \frac{N_{cq}}{M} * \frac{C}{S} \right\rceil , \qquad (9)$$

where I_{cq} is the number of items in cell (c, q) of the blueprint, M is the maximum number of times an item can be exposed before it is supposed to be known, S is the number of simulees in the CAT simulation, and C is the number of adaptive test administrations the item pool should support.

Application of this formula is justified by the following argument. If the ability distribution in the CAT simulations is a reasonable approximation to the true ability distribution in the population, N_{cq} predicts the number of items needed in cell (c, q). Because the numbers are calculated for S simulees and the item pool should support CAT for C examinees, N_{ij} has to be corrected by the factor $\frac{C}{S}$. This correction thus yields the numbers of items with attribute values corresponding to cell (c, q). However, to meet the required exposure rates, these numbers should be divided by M. The results from (9), rounded upwards to obtain integer values, define the optimal blueprint for the item pool.

3.3.1. Rotating Item Pools

The method can also be used to design a system of rotating item pools from a master pool. The general case is addressed in which overlap between rotating pools is allowed. Let G be the number of item pools the master pool has to support and n_{cq} the number of overlapping pools in which an item from cell (c, q) is allowed to occur. The number of items in a cell of the master pool is equal to:

$$\tilde{I}_{cq} = \left\lceil \frac{N_{cq}}{M} * \frac{C}{S} * \frac{G}{n_{cq}} \right\rceil . \qquad (10)$$

The number of items needed in every rotating item pool is I_{cq}. Because the master pool has to support G rotating item pools, I_{cq} has to be multiplied by G. Finally, since an item in cell (c, q) figures in n_{cq} pools, this number has to be divided by n_{cq}.

4. Empirical Example

As an empirical example, an item pool was designed for the CAT version of the Graduate Management Admission Test (GMAT). Five categorical item attributes were used which are labeled here as $C1, ..., C5$.

Each attribute had between two and four possible values. The product of these attributes resulted in a table C with 96 cells.

All items were supposed to be calibrated by the three-parameter logistic (3PL) model:

$$P_i(\theta_j) \equiv c_i + (1 - c_i)\frac{e^{a_i(\theta_j - b_i)}}{1 + e^{a_i(\theta_j - b_i)}}, \tag{11}$$

where $P_i(\theta_j)$ is the probability that a person $j = 1 \ldots J$ with an ability parameter θ_j gives a correct response to an item $i = 1 \ldots I$, a_i is the value for the discrimination parameter, b_i for the difficulty parameter, and c_i for the guessing parameter of item i. These item parameters were the quantitative attributes in the current example. The range of values for the discrimination parameter, a_i, is the interval $[0, \infty)$. This interval was split into nine subintervals, the ninth interval extending to infinity. The difficulty parameter, b_i, takes values in the interval $(-\infty, \infty)$. Likewise, this interval was divided into fourteen subintervals. A previous item pool of 397 items from the GMAT was available. In this pool, the value of the guessing parameter, c_i, was approximately the same for all items. Therefore, in the simulation, c_i was fixed at this common value. The product of the quantitative attributes resulted in a table, Q, with 124 cells. The Cartesian product of the tables C and Q was a table with $96 \times 124 = 12,096$ cells.

As estimates of the item-writing costs, reciprocals of the frequencies of the items in a previous item pool were used (for cells with zero frequencies an arbitrary large number was chosen).

The actual specifications for the GMAT were used to formulate the integer programming model for the shadow tests in the CAT simulation. The model had 30 constraints dealing with such attributes as item content and test length. Because no target for the test information function was available, the following linear combination of test information and item writing costs was optimized:

$$\max\{\lambda \sum_{cq \in C \times Q} I_{cq}\left(\widehat{\theta}_{k-1}\right) x_{cq} - (1 - \lambda) \sum_{cq \in C \times Q} k_{cq} x_{cq}\}.$$

$$\text{(objective function)} \tag{12}$$

The examinees were sampled from $N(1,1)$. The simulations were executed using software for constrained CAT with shadow tests developed at the University of Twente. The integer programming models for the shadow tests were calculated using calls to the linear-programming software package CPLEX 6.0 (ILOG, 1998). The initial estimate for each new simulee was set equal to $\widehat{\theta} = 1$. The estimate was updated using the method of EAP estimation with a uniform prior.

The blueprint was calculated using realistic estimates for C and M in (9). For security reasons the blueprint can not be revealed here.

The simulation study was time intensive. For each examinee the complete test of 28 items took 8-9 minutes. The main reason for this is the large number of decision variables in the model, one for each of the 12,096 cells in the $C \times Q$ table. Large numbers of variables is not typical of real CAT, though. A previous simulation of constrained CAT with shadow tests directly from the previous GMAT pool had only 397 variables and took 2-3 seconds per item. Both the current and previous study were carried out on a Pentium 133 MHz computer.

5. Use of Item Pool Blueprint

The method presented in this chapter produces a blueprint for an item pool that serves as the best goal available to guide the item writing process. Its primary goal is to prepare the instructions for the item writers. If the identity of the writers is used as an attribute in the $C \times Q$ table for which cost estimates were obtained, the blueprint automatically assigns these instructions to them. In the item writing process, both the categorical item attributes as well as some of the quantitative attributes (e.g., word counts) can easily be realized. However, as already discussed, other quantitative item attributes, in particular those of a statistical nature, are more difficult to realize. If an existing item pool is used to estimate item writing costs, the blueprint for the item pool is automatically based on the empirical correlations between the statistical attributes and the other attributes. For example, if the difficult items tended to have other values for their categorical attributes than the easy items, the blueprint takes this fact automatically into account. This feature may improve the item-writing results but exact realization of statistical item attributes remains an optimistic goal.

The best way to implement the blueprint is, therefore, not in a one-shot approach but in a sequential fashion, recalculating the blueprint after a certain portion of the items has been written and field tested so that their actual attribute values are known. Repeated applications of the method helps to adapt the item writing efforts to the actual numbers of items already present in the pool. The same practice has been proposed for item pool design for assembling multiple linear test forms (van der Linden, Veldkamp & Reese, 2000).

If the method is implemented sequentially, in each rerun of the CAT simulations the model for the shadow tests in (1)-(8) should be adapted to allow for the items already admitted to the item pool. The result is a mixed model, with 0-1 decision variables for the items already in the

pool model and full integer variables for the new items in the cells of
the $C \times Q$ table.

Let $i = 1, .., I$ be the index for the items already in the pool and
x_i the variable for the decision to include ($x_i = 1$) or not to include
item i in the shadow test ($x_i = 0$). For these items, the actual attribute
values should be included in the model. Also, the actual costs of the
writing of item i, k_i, should be specified on the same scale as k_{cq}. The
variables x_{cq} still represent the selection of items with attribute values
associated with cell (c, q) in the $C \times Q$ table needed.

The adapted model is as follows:

$$\min \sum_{cq \in C \times Q} k_{cq} x_{cq} + \sum_{i=1}^{I} k_i x_i \qquad \text{(objective function)} \quad (13)$$

subject to

$$\sum_{cq \in C \times Q} I_{cq} \left(\widehat{\theta}_{k-1} \right) x_{cq} +$$

$$\sum_{i=1}^{I} I_i \left(\widehat{\theta}_{k-1} \right) x_i \geq T, \qquad \text{(information target)} \quad (14)$$

$$\sum_{cq \in S_{k-1}} x_{cq} + \sum_{i=1}^{I} x_i = k - 1, \qquad \text{(items already selected)} \quad (15)$$

$$\sum_{cq \in C \times Q} x_{cq} + \sum_{i=1}^{I} x_i = n, \qquad \text{(test length)} \quad (16)$$

$$\sum_{cq \in V_g} x_{cq} + \sum_{i \in V_g} x_i = n_g, \qquad g = 1, ..., G,$$

$$\text{(categorical constraint)} \quad (17)$$

$$f_h(x_{cq}) + f_h(x_i) = n_h, \qquad h = 1, ..., H,$$

$$\text{(quantitative constraint)} \quad (18)$$

$$\sum_{cq \in V_e} x_{cq} + \sum_{i \in V_e} x_i \leq 1, \qquad e = 1, ..., E,$$

(enemy sets) (19)

$$x_{cq} \in \{0, 1, 2, ...\}, \qquad (c, q) \in C \times Q,$$

(range of variables) (20)

$$x_i \in \{0, 1\}, \qquad i \in I.$$

(range of variables) (21)

The proposed application is thus to run the model repeatedly during the item writing process. At each next application, the number of decision variables x_i grows whereas the the values of the variables x_{cq} decrease. If the complete item pool is realized and the stage of operational CAT is entered, the model for the shadow test contains only the variables x_i.

6. Concluding Remark

One of the reasons for proposing a optimal blueprint as a target for the item writing process is to create more even item exposure. However, the $C \times Q$ table in this chapter can also be used to realize this goal for a CAT pool that has not been developed using the proposed blue-print. Suppose the items in an existing pool are clustered in the cells (c, q) of table $C \times Q$. Then all items in the same cell have identical values for their categorical attributes and values for their quantitative attributes that differ only slightly. As a consequence, items in the same cell are approximately equally informative at the estimated θ values for the examinees that take the adaptive test. Nevertheless, the actual exposure rates of the items may vary considerably. Adaptive testing in-volves optimal item selection and therefore tends to capitalize on small differences between the items.

Why not overcome this capitalization by reformulating the CAT algorithm to select cells from the $C \times Q$ table instead of individual items and randomly select one of the actual items from the cells for administration? The result of this algorithm is even exposure of items in the same cell. A similar approach has been proposed by Holmes and Segall (1999). Differences between cells can further be leveled by applying a method for item-exposure control on the selection of the cells in the $C \times Q$ table rather than the individual items in the pool.

In addition to the use of more rational methods of item pool design, continuous attempts to fine tune item selection criteria may be needed to produce CATs that combine accurate measurement with a more uniform item exposure.

References

Boekkooi-Timminga, E. (1991, June). *A method for designing Rasch model-based item banks.* Paper presented at the annual meeting of the Psychometric Society, Princeton, NJ.

Davey, T., & Nering, M.L. (1998, September). *Controlling item exposure and maintaining item security.* Paper presented at the ETS Computer-Based Testing Colloquium, Philadelphia, PA.

Flaugher, R. (1990). Item Pools. In H. Wainer, *Computerized adaptive testing: A primer* (pp. 41-64). Hillsdale, NJ: Lawrence Erlbaum Associates.

Holmes, R. M. , & Segall, D. O. (1999, April). *Reducing item exposure without reducing precision (much) in computerized adaptive testing.* Paper presented at the annual meeting of the National Council on Measurement in Education, Montreal, Canada.

ILOG, Inc. (1998). *CPLEX 6.0* [Computer Program and Manual]. Incline Village, NV: ILOG.

Stocking, M. L., & Lewis, C. (1998). Controlling item exposure conditional on ability in computerized adaptive testing. *Journal of Educational and Behavioral Statistics, 23,* 57-75.

Stocking, M. L., & Swanson, L. (1998). Optimal design of item banks for computerized adaptive tests. *Applied Psychological Measurement, 22,* 271-280.

Swanson, L., & Stocking, M. L. (1993). A model and heuristic for solving very large item selection problems, *Applied Psychological Measurement, 17,* 151-166.

Sympson, J. B., & Hetter, R. D. (1985). Controlling item-exposure rates in computerized adaptive testing. *Proceedings of the 27th annual meeting of the Military Testing Association* (pp. 973-977). San Diego, CA: Navy Personnel Research and Development Center.

van der Linden, W. J. (1998). Optimal assembly of psychological and educational tests, *Applied Psychological Measurement, 22,* 195-211.

van der Linden, W. J., & Reese, L. M. (1998). A model for optimal constrained adaptive testing. *Applied Psychological Measurement, 22,* 259-270.

van der Linden, W. J., Veldkamp, B. P., and Reese, L. M. (2000). An integer programming approach to item pool design. *Applied Psychological Measurement, 24,* 139-150.

Veldkamp, B. P. (1999). Multiple objective test assembly problems. *Journal of Educational Measurement, 36,* 253-266.

Way, W. D. (1998). Protecting the integrity of computerized testing item pools. *Educational Measurement: Issues and Practice, 17,* 17-26.

Way, W. D., Steffen, M., & Anderson, G. S. (1998, September). *Developing, maintaining, and renewing the item inventory to support computer-based testing.* Paper presented at the ETS Computer-Based Testing Colloquium, Philadelphia, PA.

Chapter 9
Methods of Controlling the Exposure of Items in CAT

Martha L. Stocking and Charles Lewis
Educational Testing Service, USA

1. Introduction

An important security issue in any kind of testing can be framed quite simply: preknowledge of items used in testing should not have serious effects on test scores. Of course, having framed the issue in such a simple fashion, it must be expanded to incorporate the nuances that may be unstated in any practical situation. The importance of limiting preknowledge in any testing context is dependent upon the use to which test scores are put. If the test scores are to be used to make important consequential decisions about test takers, for example, admission to college, graduate school, or a particular job, then limiting preknowledge becomes an important concern. In this context, we might conclude that any effect of preknowledge on test scores is too great, and that the investment of substantial resources in limiting preknowledge is appropriate. In contrast, if test scores are used to make less consequential decisions or if the consequences of seriously incorrect test scores can be easily discovered and remedied, then restricting preknowledge may be less important. For example, if the test scores are to be used to place test takers into appropriate level courses, incorrect placements can be easily discovered and remedied. In this context we might conclude that the expenditure of substantial resources to restrict even moderate to large effects of preknowledge on test scores is inefficient. For this chapter, we will address security concerns in the former, "high stakes" context but it is useful to keep in mind that the methods described are not necessarily appropriate for all contexts.

1.1. SOURCES OF PREKNOWLEDGE

High stakes tests usually pretest items to determine quality from both a statistical and content perspective before tests are assembled and administered. This is itself a source of preknowledge about items and many strategies, such as the use in final tests of items not pretested together, have been developed historically to minimize this type of preknowledge risk. Such strategies work equally well for computerized tests and will not be addressed further in this chapter.

163

W.J. van der Linden and C.A.W. Glas (eds.),
Computerized Adaptive Testing: Theory and Practice, 163–182.
© 2000 Kluwer Academic Publishers. Printed in the Netherlands.

Computer administered tests are generally conceived as being available for administration on a continuous or nearly continuous basis. Continuous testing, whether paper and pencil or computer administered, presents substantial additional concerns about preknowledge since preknowledge can now conceivably be obtained from test takers who have taken the test very recently and therefore may remember a great deal about individual items. This increased risk can be characterized as perhaps comparable, in an environment of vigorous information sharing activities among test takers, to the risk of repeatedly administering the same test in the periodic paper and pencil environment. Various possible strategies to capitalize on the computer itself in order to mitigate the possible effects of preknowledge are the subject of this chapter.

1.2. HIGH STAKES COMPUTERIZED ADAPTIVE TESTING (CAT)

As pointed out by Davey and Parshall (1995) high stakes adaptive testing has at least three goals: (1) to maximize test efficiency by selecting the most appropriate items for a test taker, (2) to assure that the tests measure the same composite of multiple traits for each test taker by controlling the nonstatistical characteristics of items included in the test (such as content, format, and cognitive demands), and (3) to protect the security of the item pool by seeking to control the rates at which items can be administered. The achievement of each of these goals is compromised by the presence of the other two. For instance, the imposition of the second and third goals requires test lengths that are longer than if only the first goal had been operating in item selection. Attempts to achieve the third goal are typically called "exposure control" methodologies and must function in the context of the first two goals.

Different approaches to the three goals yield different algorithms for item selection in adaptive testing. In general, any CAT item selection algorithm can be used to create a list of desirable items, possibly ordered in terms of their attractiveness, as candidates for the next item to be administered. Differences in the list membership or ordering of items typically reflect different definitions of item optimality, methods of constructing the list, and methods of estimating test taker ability. Any attempt to satisfy the third goal of controlling the exposure of items can then be viewed as the imposition of conditions on the list of items.

At this most general level, a number of questions can be posed concerning any CAT item selection algorithm with exposure control. Consider a "phase of testing" to be the state in which the next item is chosen for administration. For a given phase of testing:

(1) Which items should be in the (possibly ordered) list of items? The answer to this question depends the particular decisions reflected in the details of the CAT item selection algorithm.

(2) How many items should be in the list? While it may be useful to think of the entire pool of items not yet administered as forming the list at each phase, this strategy may be inefficient or even damaging for some methods of controlling item exposure (see discussion below). It may be a wiser strategy to limit the size of the list in some fashion, perhaps even allowing the size of the list to vary depending upon the phase of testing.

(3) What is the probability of administering each item in the list? These probabilities must, of course, sum to one for the administration of an item. However, some exposure control strategies may allow or insure that these probabilities vary over lists constructed for successive phases of testing in order to enhance security.

(4) Can an item be in more than one list? If items appear in the list for a given phase but are not administered and are removed from any future consideration, these items have been drawn from the pool without replacement. If some items in the list are removed from further consideration but some are not, this may be described as "partial" replacement. If all items are allowed to appear in subsequent lists (except, of course, the administered item), the items have been drawn from the pool with replacement. Decisions such as these will affect final exposure rates of items.

2. Previous Methods of Controlling Item Exposure

2.1. Pre-specifying Probabilities of Administration

Early theoretical investigations of CAT ignored the problem of item exposure (see, for example, Lord, 1970). Procedures that seek to prevent the overexposure of initial items developed when the prospect of actual implementation became more certain. Lord (1977), McBride and Martin (1983), Stocking (1987) and Weiss (1978) implemented strategies typical of these first attempts. In these approaches, a group of items, that is, the list, is composed of a specified number of items that are roughly equal in desirability as defined by the particular CAT algorithm. The next item to be administered is chosen randomly from this list. In most implementations, initial list sizes are larger and are reduced as the test progresses. An important feature of these procedures is that the items for each list are chosen with replacement from the pool, i.e., an item in a list that is not administered may appear in the lists for subsequent phases of testing.

An assumption underlying this approach is that initially, when estimates of test taker ability contain the greatest amount of error, it is difficult to differentiate among test takers and uncontrolled item selection will administer the same items frequently. After some number of initial items, ability estimates will be more precise and test takers will be sufficiently differentiated so that subsequently administered items will vary a great deal. Many variations on this approach are possible, including the possibility of never choosing the next item optimally with certainty, that is, the minimum list size is always at least two. This latter approach recognizes that in spite of randomization on initial items, test takers with similar abilities may receive many of the same items subsequently unless attempts are made to control the exposure of items later in the test.

The advantage of these kinds of schemes is that they are simple to implement and easily understood. They can be elaborated to make the random selection process depend upon the estimated ability of the test taker by specifying different list sizes for different levels of estimated ability. However, the success of such schemes is difficult to predict with complex but realistic item pool structures and test specifications, and may not prevent overuse of some items (for an example of this, see Stocking (1993), Figure 1). This is because items can be repeatedly selected for inclusion in the list and therefore popular items will eventually be administered. Thus, while these procedures may shuffle the order in which items are presented, they do not actually control the rate of item exposure. Moreover, it may be difficult if not impossible to develop criteria against which the results of any particular sequence of list sizes could be evaluated that would simultaneously provide guidance about other sequences of list sizes that might be more optimal.

Further developments of such approaches to exposure control, which can be characterized as pre-specifying the probability that an item will actually be administered given that it is considered for administration, will not be pursued in this chapter. Interested readers are referred to Kingsbury and Zara (1989), Revuelta and Ponsoda (1998), Robin (1999), and Thomasson (1995) for a variety of such approaches.

2.2. USING EXPOSURE CONTROL PARAMETERS

Sympson and Hetter (1985) tackle the issue of controlling item exposure directly in a probabilistic fashion based on the behavior of items over repeated simulations of a test design with a sample drawn from a typical distribution of abilities. This procedure considers a test taker randomly sampled from a typical group of test takers and distinguishes between the probability $P(S)$ that an item is selected as the best next item to administer from an ordered list formed by a CAT item selection algorithm, and $P(A|S)$, the probability that an item is administered given that it has been selected. The procedure seeks to control $P(A)$, the overall probability that an item is administered, where $P(A) = P(A|S) * P(S)$, and to insure that the maximum value of $P(A)$ for all items in the pool is less than some value r. This r is the desired maximum rate of item usage.

The "exposure control parameters", $P(A|S)$, one for each item in the pool, are determined through a series of adjustment simulations using an already established adaptive test design and simulated examinees (simulees) drawn from a typical distribution of ability. Following each simulation, the proportion of times each item is selected as the best item, $\hat{P}(S)$, and the proportion of times each item is administered, $\hat{P}(A)$, are separately tallied. The $\hat{P}(A|S)$ for the next iteration are then computed as

$$\hat{P}(A|S) = \begin{cases} 1.0 & \text{if } \hat{P}(S) \leq r, \\ r/\hat{P}(S) & \text{otherwise.} \end{cases}$$

The simulations and adjustments continue until the procedure stabilizes and the maximum observed $\hat{P}(A)$ for all items is approximately equal to the desired value of r. Note that there is no guarantee that this procedure will eventually stabilize, and indeed, it may not.

Exposure control parameters are used in item selection, both during the adjustment simulations and after final values have been established, as follows:

(1) Form an ordered list of candidate items.

(2) Consider the most desirable item.

(3) Administer the most desirable item with probability $P(A|S)$.

(4) If the most desirable item is not administered, remove it from the pool of remaining items for this test taker.

(5) Repeat steps (3) and (4) for the next-most-desirable item. Continue until an item is administered.

Items appearing in the list before the administered item are considered "selected" and selected items are sampled without replacement. Items in the list appearing after the administered item are returned to the pool for further consideration. Thus the Sympson and Hetter procedure can be considered to be a partial replacement procedure. If the length of the ordered list is too long in step (1) and if the generation of random numbers is "unlucky" in step (3) a substantial number of items may be removed from the pool. However, if the adaptive test is of length n, there must be at least n items in the pool with exposure control parameters of one. If there were not, then for some test takers there would not be enough items in the pool to administer a complete adaptive test. In the case where there are not n such items at the end of an iteration, Sympson and Hetter suggest the reasonable procedure of sorting the values of the exposure control parameters and setting the n largest to one. This has the effect of increasing the exposure rate for the items that are least popular – a conservative approach.

The advantage of the Sympson and Hetter approach is that one obtains direct control of the probability that an item is administered, $P(A)$, in a typical population of test takers. However, the simulations required to obtain values of the $P(A|S)$ for each item are time-consuming for pools and test specifications with complex structures. If an item pool is changed, even by the addition or deletion of a single item, or if the distribution of abilities for the target population changes, the adjustment simulations should be repeated to guarantee continued exposure control for the new pool and/or target population.

Moreover, if the structure of the item pool is not a good match with the structure of the specifications, it is possible for the Sympson and Hetter procedure to diverge, that is, it may not be possible to obtain stable values of the $P(A|S)$ for each element in the pool (See Stocking, 1993, Figure 4, for example). This happens, in part, because of the "fixup" to insure complete adaptive tests – setting the n highest $P(A|S)$ to one. This fixup seems to work well if all n of the high $P(A|S)$s are not too different from one. However, if some are very different, this fixup can cause wild fluctuations by repeatedly setting low $P(A|S)$ back to one or alternating among several items with low $P(A|S)$. This prevents (smooth) convergence of the procedure. A solution to this problem is to construct a context in which there is a better match between pool structure and test specifications, either by enriching the pool or by simplifying test structure. Either of these may be difficult to accomplish.

Stocking and Lewis (1998) developed a multinomial model for the Sympson and Hetter approach which overcomes some of its practical disadvantages. They propose exploiting this more robust model to develop exposure control parameters conditional on ability level. The basic model considers, at each phase of testing, the list of items ordered from the most desirable to the least desirable and the associated

$P(A \mid S)$, one for each item in the list. As in the Sympson and Hetter procedure, the same adjustment simulations are required to obtain the exposure control parameters, the $P(A \mid S)$s.

In identifying the next item to be administered, the operant conditional probabilities of administration for each item, k_i, are not the simple $P_i(A \mid S)$, but rather as follows:

$k_1 = P_1(A \mid S) \equiv P_1$, the probability that the first item in the ordered list for this phase is administered given that it is selected,

$k_2 = (1 - P_1) * P_2$, the probability that the first item is rejected given that it is selected and the probability that the second item in the ordered list is administered given that it is selected,

$k_3 = (1 - P_1) * (1 - P_2) * P_3$, the probability that the first two items in the list are rejected given selection and that the third item is administered given that it is selected, and so forth.

The sum of these probabilities must equal one for some event to occur, that is, for some item to be administered. If the sum is not one, we can define adjusted probabilities whose sum is guaranteed to equal one by dividing each k_i by the sum. This adjustment of probabilities is the analog of the fixup recommended in the Sympson and Hetter procedure in that it guarantees that an item (and therefore a complete adaptive test) can always be found for administration. If the list of items is long enough the adjustment to individual operant probabilities may be quite small, thus increasing the chance for smooth convergence.

The operant probabilities thus obtained specify a multinomial distribution from which we wish to sample one event. To do this, we form the cumulative distribution, generate a random number between zero and one, and locate the item to be administered. All items appearing in the ordered list before the item to be administered are removed from further consideration for this adaptive test. This elimination accords with the definition of the operant probabilities given above in that the operant probability of selecting item i includes the probabilities of rejecting all items before item i in the ordered list. This partial sampling without replacement, as in the Sympson and Hetter procedure, requires an ordered list of appropriate size in order to avoid the depletion of the pool. An appropriate length for the list is approximated by the ratio of the number of items in the pool to the number of items in the CAT.

2.3. CONDITIONING ON ABILITY

The unconditional multinomial model described above results in an exposure control parameter for each item in the pool. The adjustment simulations develop these in reference to a particular distribution of ability for the relevant population of test takers. An approach to developing a version of this method conditional on ability θ is to consider the range of ability covered by the distribution of ability and divide this range into, say, M different discrete values of θ that cover the range of interest, similar to the approach used by Thomasson (1995). Consider one particular discrete value in this range, θ_m. We can perform the adjustment simulations to develop exposure control parameters for each item in the pool using the multinomial procedure and in reference to only those simulees drawn that have true ability equal to θ_m.

These adjustment simulations result in a vector of exposure control parameters appropriate for all items in the pool for individuals with ability θ_m. If the adjustment simulations are performed simultaneously for all values of θ_m, $m = 1, \ldots, M$, then we produce a matrix of conditional exposure control parameters with elements $P_i(A \mid S, \theta_m)$ for the ith row (one for each item in the pool) and the mth column (one for each of the θ_m that span the range of true abilities in the target population of interest). During the adjustment phase, the tallies of item selection and administration are kept separately by θ level. In identifying the next item to be administered, the simulee's estimated ability level, $\hat{\theta}$, is used to select the appropriate column in the matrix of conditional exposure control parameters.

The advantage of conditional multinomial exposure control is that it allows direct control of item exposure for different levels of ability. Moreover, it may be possible to choose different target maximum exposure rates for different ability levels, perhaps reflecting the availability of items in the pool. In addition, conditional parameters are now independent of any target population of test takers.

2.4. CONDITIONING ON ITEMS

Davey and Parshall (1995) developed a different conditional exposure control methodology that also has its origins in the Sympson and Hetter procedure. This approach seeks to control item exposure probabilities conditional on those items that have already appeared in the test. This can be viewed as a method of attempting to control test overlap in addition to limiting the frequency of item use.

This procedure is based on a $T \times T$ table of exposure parameters where T is the number of items in the pool. Diagonal elements of this table contain unconditional exposure control parameters similar to those used in the Sympson and Hetter procedure. The off-diagonal entries are the conditional parameters that control the frequency with which pairs or clusters of items occur. The Sympson and Hetter procedure can then be viewed as a special case of the Davey and Parshall procedure in which all of the off-diagonal values are set to one. This table is used as follows:

(1) Form an ordered list of candidate items and select the most desirable item.

(2) Enter the table and find the diagonal element for the selected item, denoted by e_{ii}. Also find all off-diagonal entries involving both the selected item and those items administered previously and denote these values by e_{ij}, where j ranges over the items administered so far.

(3) Compute a conditional probability of administering the selected item by taking the average of the e_{ij} values and multiplying this average by e_{ii}. Administer the selected item with this probability.

(4) If an item is not administered, remove it from the available pool of items. Removed items are not reconsidered for this adaptive test unless the pool contains no unconsidered items, a rare circumstance.

To estimate the entries in the exposure control table, a series of adjustment simulations is conducted with a large number of simulees from a typical distribution of abilities, as in the Sympson and Hetter procedure. However, the details of these adjustment simulations are slightly different. Upper limits are set on the frequency with which individual items and item pairs are allowed to appear, similar to the maximum target probability of administration in the Sympson and Hetter procedure. All entries are initialized to one and a large number of adaptive tests are simulated, keeping track of the number of times each item and pair of items appears. After this simulation, the diagonal entries in the table are decreased for any item that appears more frequently than the specified limit, by multiplying by .95. For any item that appears less frequently than the specified limit, the diagonal entry is increased by multiplying by 1.04 (to a maximum of 1.0).

The off-diagonal counts for pairs of items are used to construct a χ^2-like statistic. If this statistic indicates that the pair-wise counts exceed chance, the corresponding off-diagonal element in this table is adjusted downward by multiplying by .95; if the pair-wise counts are below chance, the corresponding off-diagonal element is adjusted upward by 1.04 (to a maximum of 1.0). This process of adjustment simulations is repeated until all entries in the table are stabilized.

Nering, Davey, and Thompson (1998) proposed yet another exposure control procedure that combines elements of the Stocking and Lewis conditional-on-ability approach and the Davey and Parshall conditional-on-items-already-administered approach. In this procedure, the diagonal elements of the Davey and Parshall table are replaced by separate vectors for each of the M ability levels in the stratification on θ in the Stocking and Lewis procedure. These values limit the frequency with which items can be administered to examinees at each ability level. The off-diagonal elements are identical to those in the Davey and Parshall method and control the frequency with which pairs of items are allowed to appear. The overall probability of administration given selection for an item is again the combination of the appropriate diagonal element, chosen on the basis of the current $\hat{\theta}$, and the mean of the off-diagonal elements for every item already appearing in the adaptive test. This expanded table is again developed through a series of adjustment simulations that continues until all exposure control parameters have stabilized.

2.5. EVALUATING METHODS

Nering, Davey, and Thompson also conducted an evaluation study comparing a number of different exposure control methods. Four observed properties of each procedure were of interest: (1) the extent to which procedures minimized test overlap, (2) the extent to which the overall or unconditional exposure rates were controlled, (3) the extent to which the control procedures forced a balanced use of the pool, and (4) the extent to which the procedure hindered the efficiency of the CAT by barring access to the most informative items in the pool. The results of their study suggest that, in general, the control of item security improves with increased conditionality. Conditioning on previous items and test taker abilities, as in Nering, Davey and Thompson, gave better results in terms of the first three criteria than conditioning on abilities, as in Stocking and Lewis, or conditioning on items, as in Davey and Parshall. However, differences were not great, and there was a diminishing return as conditioning became more complex.

Chang (1998) conducted an extensive study comparing the properties of five exposure control procedures: simple randomization, Sympson and Hetter, Davey and Parshall, Stocking and Lewis unconditional, and Stocking and Lewis conditional exposure control procedures. These procedures were evaluated using criteria similar to those used by Nering, Davey, and Thompson. Chang found that no exposure control procedure performed best on all criteria. The simple randomization procedure with replacement performed worst on criteria associated with item and test security, and best on criteria associated with the measurement efficiency of the resultant CATs. The Sympson and Hetter procedure and the Stocking and Lewis unconditional procedure performed similarly on all criteria. This is reassuring since the Stocking and Lewis procedure was intended to model the Sympson and Hetter procedure. The Davey and Parshall conditional procedure and the Stocking and Lewis conditional procedure also performed similarly, with the Stocking and Lewis conditional procedure performing slightly better when all evaluation factors were considered.

The Nering, Davey, and Thompson, and Chang evaluation studies highlight an important and vital aspect of the three goals of adaptive testing discussed earlier – measurement efficiency, measurement of the same composite of multiple traits, and enhancing security through the control of item exposure rates. These three goals compete directly with each other in a complex, and sometimes surprising, fashion. Whatever criteria are used to measure the satisfaction of the three goals form a system that can be likened, for a fixed item pool, to a three-sided balloon – a push in on one side will inevitably cause a bulge out on another side. Thus, it may be possible to satisfy security concerns, but at the expense of loss of measurement efficiency or consistency of the construct being measured or both. Much of the work involved in the establishment of a successful program of adaptive testing lies in achieving an appropriate balance among these three goals.

3. A New Approach

As Thomasson (1995) emphasized, it seems desirable that methods of controlling item exposure be independent of the distribution of ability in the population for whom the test is intended. In theory, both the Stocking and Lewis conditional multinomial approach and the Nering, Davey, and Thompson hybrid conditional approach have this property. These approaches can be characterized as follows:

(1) Perform adjustment simulations to develop exposure control parameters that are conditional on true ability levels θ_m, $m = 1, 2, \ldots M$.

(2) During actual CAT administrations (and also during the adjustment simulations) use a test taker's current estimated ability, $\hat{\theta}$, to locate the appropriate vector of conditional exposure control parameters in determining whether or not an item may be administered, given that it is selected.

This characterization highlights a potential problem – the development of the exposure control parameters conditional on true ability but their use based on estimated ability. In practice, this amounts to pretending that $\theta = \hat{\theta}$ at each item administration. It is required, of course, that $\hat{\theta}$ be a good approximation to θ by the end of an adaptive test. But it is less likely for this approximation to hold at the beginning of an adaptive tests when nothing may be known about test taker ability. The difference between θ and $\hat{\theta}$ can sometimes produce unexpected results in the context of the Stocking and Lewis conditional multinomial method. It is likely that similar unexpected results occur with the Nering, Davey, and Thompson hybrid approach. In this section we develop a new approach to exposure control in which the exposure control parameters are developed by conditioning on $\hat{\theta}$, rather than θ.

3.1. A NUMERICAL DEMONSTRATION

To demonstrate various aspects of both the original Stocking and Lewis conditional multinomial approach and the new approach, a pool of items designed to measure quantitative reasoning was obtained. This pool contained 348 "elements": 201 discrete items, 18 stimuli, and 129 items associated with the 18 stimuli. All items were calibrated using large (2000+) samples of test takers from a typical population and the computer program LOGIST (Wingersky, 1983). The estimated distribution of true ability in the population was computed using the approach developed by Mislevy (1984) and the responses from a large (6000+) sample of test takers on an exemplar form of a paper and pencil linear test which was demonstrated to measure the same construct measured by the adaptive test. For further details concerning this pool and adaptive test, see Eignor, Way, Stocking, and Steffen (1993).

The adaptive test used the Weighted Deviations (WD) algorithm of Swanson and Stocking (1993) and Stocking and Swanson (1993) to select items for administration subject to 28 nonstatistical constraints. The Stocking and Lewis multinomial method, both conditional and unconditional, was used to control the exposure of items.

An initial ability estimate for the start of an adaptive test may be obtained using a variety of methods. It is possible to employ Bayesian approaches, perhaps using collateral information about an individual test taker or information about the population of test takers of which the test taker is a member. In practice, these approaches may suffer from the criticism that the use of such information may give some test takers an advantage over other test takers. The perception of fairness to test takers may be enhanced if the same starting value is used for all test takers. In the particular adaptive test design studied here, the first item chosen was selected to be the most appropriate item for a test taker whose ability level was −1.0.

Adjustment simulations were performed to develop the Stocking and Lewis exposure control parameters. These simulations were performed using stratified sampling, with 1000 simulees at each of 11 discrete ability levels that spanned the range of interest. For evaluation with respect to the marginal distribution of ability in a typical population, the results of the simulations were weighted by the estimated distribution of true ability.

The adjustment simulations were evaluated with respect to the three goals of adaptive testing discussed earlier. A single number summary of measurement efficiency was computed using an approximation to reliability proposed by Green, Bock, Humphreys, Linn, and Reckase (1984), equation 6. The conformation to test specifications for a typical population was evaluated by the average total weighted deviations (TWD). This was computed by averaging the total weighted-by-importance deviations for adaptive tests administered to simulees with ability θ, and forming the weighted average of the results across all levels of θ, where these latter weights were given by the estimated distribution of true ability in the typical population of test takers. The third goal, item security, was evaluated by comparing the maximum observed $\hat{P}(A)$ over all elements in the pool with the target set for the adjustment simulations. $\hat{P}(A)$ was computed for each element by taking a weighted average of the results at each level of θ.

Table 1 displays the results of the adjustment simulations for five different experiments. The results for each experiment occupy two adjacent columns. The labels T and O stand for "Target" and "Observed", respectively. The first three rows report the values on the three unconditional criteria for each experiment. The remaining rows report the conditional maximum $\hat{P}(A \mid \theta)$ for each θ compared to the specified conditional targets. The two values of θ that bracket the initial starting point of $\hat{\theta} = -1.0$ for all test takers are highlighted with an asterisk. The adjustment simulations in the first set of columns used the Stocking and Lewis unconditional multinomial procedure to control the probability of administration with the unconditional target set to .20. Since the control was unconditional, no conditional targets are specified. The observed conditional maximum $\hat{P}(A \mid \theta)$ ranges from a low of .37 for $\theta = .05$ to a high of .98 for $\theta = 3.55$. These values are probably too large to be acceptable, although the unconditional observed maximum $P(A)$ is .21.

Note that the value of max $\hat{P}(A)$ is substantially less than any of the values of max $\hat{P}(A|\theta)$. To understand how this can occur, recall that, for any item, the value of $\hat{P}(A)$ is obtained by averaging the values of the $\hat{P}(A \mid \theta)$ for that item with respect to the estimated target distribution

Table 1. A comparison of experiments conditioning on θ.

	Uncon		Con Equal		Con High		Con Low		Con Real	
Reliability	.924		.912		.919		.933		.926	
Average TWD	.73		.86		.71		.71		.63	
max $\hat{P}(A)$	.21		.20		.89		.55		.48	
max $\hat{P}(A \mid \theta)$	T	O	T	O	T	O	T	O	T	O
$\theta = -3.84$	-	.54	.2	.26	.2	.89	.9	.94	.5	.61
$\theta = -2.18$	-	.56	.2	.25	.2	.89	.9	.77	.5	.55
$\theta = -1.38^*$	-	.66	.2	.24	.9	.90	.2	.61	.5	.58
$\theta = -.81^*$	-	.49	.2	.24	.9	.90	.2	.56	.5	.53
$\theta = -.35$	-	.39	.2	.23	.2	.89	.9	.79	.4	.48
$\theta = .05$	-	.37	.2	.24	.2	.88	.9	.92	.3	.49
$\theta = .43$	-	.38	.2	.23	.2	.89	.9	.93	.3	.48
$\theta = .81$	-	.49	.2	.25	.2	.89	.9	.92	.2	.48
$\theta = 1.24$	-	.88	.2	.24	.2	.89	.9	.93	.2	.48
$\theta = 1.88$	-	.96	.2	.24	.2	.89	.9	.93	.2	.48
$\theta = 3.55$	-	.98	.2	.24	.2	.89	.9	.92	.2	.47

of ability. In general the average of a set of conditional maxima will differ from the unconditional maximum of the averages, because the conditional maxima will be associated with different items.

The second set of columns contains the results of the adjustment simulations using the Stocking and Lewis conditional approach with equal targets set for all ability levels. Conditional control of item use is associated with a slight decrease in overall reliability (from .924 to .912) and a slight increase in average TWD (from .73 to .86). However, the conditional probabilities of administration are substantially lower, as expected. These results are typical for many such comparisons that have been made for different CATs.

The third set of columns, labeled Conditional High, present the results of the adjustment simulations when the conditional targets are unequal. The targets for all ability levels are .2, except for those ability levels that bracket the initial value of $\hat{\theta}$. This produces unconditional and conditional observed maximum exposure rates near .9, even though most of the conditional targets are .2.

This phenomenon occurs because of the discrepancy between true theta and estimated theta, particularly at the beginning of the adaptive test.

Every test taker, regardless of their true ability, starts the adaptive test with an estimated ability of -1.0. While $\hat{\theta}$ controls the choice of which vector of exposure control parameters is used in selecting the first item, the results of the selection are recorded using a simulee's true ability. Thus all conditional observed maximum exposure rates are about the same as the one closest to the initial $\hat{\theta}$ of -1.0.

The next set of columns, Conditional Low, has all the targets equal to .9 except those that bracket -1.0, which are set to .2. The conditional observed maximum $\hat{P}(A \mid \theta)$ are lowest for these two values, which represent common values of $\hat{\theta}$ early in the test, but do not approximate the target very well. Conditional observed maximum $\hat{P}(A \mid \theta)$ are larger for those ability levels farther away from the initial starting value.

The final set of columns summarizes the results of using a more realistic set of conditional target values. These values reflect the belief that test takers with low true ability are less able to remember and convey accurately information about items and therefore exposure can be greater at these abilities than at higher levels of ability. Again, most observed values are close to the targets set for θ's close to initial values of $\hat{\theta}$ and do not match the conditional targets very well.

These three experiments setting different targets for different ability levels illustrate that it is not possible to achieve these targets using the conditional multinomial approach to exposure control. They also suggest that the problem arises because of the discrepancy between θ and $\hat{\theta}$ early in the adaptive tests. This is true regardless of how the first item is chosen. It is possible that improvements would be gained by conditioning on $\hat{\theta}$ rather than θ.

3.2. CONDITIONING ON $\hat{\theta}$

This approach begins by constructing an artificial sampling space, based on our simulation of adaptive test administrations. The random variables in this space are the true ability θ of a simulee, the phase of testing p for the simulee, the current estimate $\hat{\theta}$ of the true ability of the simulee at the given phase of testing, a variable S indicating whether or not a particular item is selected for the simulee, and a variable A indicating whether or not this item is administered to the simulee.

The probabilities in the artificial sampling space are defined in a stepwise manner. First a uniform distribution for θ is defined over the discrete values θ_m introduced earlier: $P(\theta_m) = 1/M$, for $m = 1, \dots, M$. Note that no claim is made that an actual population of test takers has a discrete uniform distribution of abilities. Rather this definition is introduced to provide a framework within which more flexible control of item exposure can be achieved.

The phase of testing in the current framework simply refers to the number of items that have been administered to the simulee, plus the next one to be administered. Thus, before testing begins the simulee is in the first phase of testing. Given that the focus is on control of item exposure, there is no interest in the situation when testing is complete. Thus the last phase of interest occurs when the last item in the test must still be administered, and the number of phases of interest equals the number of items in the test.

In this development, only fixed-length tests will be considered. If the number of items in a test is denoted by n, then the probability of being in phase p is defined as $P(p \mid \theta) = P(p) = 1/n$, for $p = 1, \dots, n$. Note that p and θ are taken to be independent.

The reason for introducing p into the sampling space is that $\hat{\theta}$ will, in general, vary with p for a given simulee. For purposes of constructing the artificial sampling space, each obtained value of $\hat{\theta}$ will be rounded to the nearest of the discrete values of θ_m introduced earlier. The result will be denoted by $\hat{\theta}_m$. The joint distribution for $\hat{\theta}$, S, and A, given θ and p, is defined by the simulation process (including the rules governing the selection of an item and its administration given that it has been selected).

Suppose the simulation is carried out in the stratified manner described earlier, with N simulees at each value of θ for a total of $M * N$ simulated adaptive tests and $M * N * n$ equally likely points from the sample space just constructed. If Freq(.) is used to denote the observed frequency in the simulation of the event described within the parentheses, then expressions for estimating the corresponding probability may be obtained in a straight-forward manner. Thus $P(\hat{\theta})$ may be estimated by

$$\hat{P}(\hat{\theta}) = \text{Freq}(\hat{\theta})/(M * N * n), \tag{1}$$

and $P(S, \hat{\theta})$ by

$$\hat{P}(S, \hat{\theta}) = \text{Freq}(S, \hat{\theta})/(M * N * n). \tag{2}$$

Combining (1) and (2) using the definition of conditional probabilities gives

$$\hat{P}(S \mid \hat{\theta}) = \text{Freq}(S, \hat{\theta})/\text{Freq}(\hat{\theta}). \tag{3}$$

The adjustment simulations used to develop exposure control parameters conditional on $\hat{\theta}$ are identical to those described earlier with

Table 2. A comparison of experiments conditioning on $\hat{\theta}$.

	Con Equal		Con High		Con Low		Con Real	
Reliability	.903		.909		.927		.915	
Average TWD	1.04		.87		.76		.72	
max $\hat{P}(A)$	.24		.28		.42		.34	
max $\hat{P}(A \mid \theta)$	T	O	T	O	T	O	T	O
$\theta, \hat{\theta}$	$\hat{\theta}$	$\hat{P}$	$\hat{\theta}$	$\hat{P}$	$\hat{\theta}$	$\hat{P}$	$\hat{\theta}$	$\hat{P}$
-3.84	.2	.25	.2	.31	.9	.92	.5	.53
-2.18	.2	.26	.2	.38	.9	.67	.5	.64
-1.38*	.2	.25	.9	.56	.2	.37	.5	.54
-.81*	.2	.27	.9	.55	.2	.41	.5	.57
-.35	.2	.25	.2	.40	.9	.65	.4	.46
.05	.2	.27	.2	.30	.9	.72	.3	.33
.43	.2	.25	.2	.27	.9	.78	.3	.30
.81	.2	.25	.2	.29	.9	.75	.2	.27
1.24	.2	.26	.2	.28	.9	.82	.2	.24
1.88	.2	.26	.2	.26	.9	.88	.2	.24
3.55	.2	.30	.2	.31	.9	1.00	.2	.31

two exceptions. The first is that all conditioning occurs with respect to $\hat{\theta}$, rather than θ. Thus the estimate for $P(S \mid \hat{\theta})$ given in (3) is obtained for each item in the pool. During the adjustment simulations this quantity is compared with the specified target and the exposure control parameter for the item at the value of $\hat{\theta}$ is adjusted accordingly, as described earlier.

The second difference with the procedure described earlier is related to the setting of an appropriate target. Here it must be remembered that all probabilities refer to what happens at a randomly sampled phase of testing. This was a necessary complication because $\hat{\theta}$ can be expected to change as testing proceeds through successive phases, so conditioning on $\hat{\theta}$ rather than θ implies taking the phase of testing into account. The goal of the procedure, however, is to control exposure for the entire test, not just for a single phase. Since there are n phases in the test, controlling exposure for a randomly selected phase at a target of r/n should produce control at the target level of r for the test as a whole.

3.3. Results

The four experiments described earlier in which exposure control was conditioned on θ were repeated with conditioning on $\hat{\theta}$. The results are displayed in Table 2 and may be compared with those in Table 1. Conditioning on $\hat{\theta}$ produces adaptive tests with slightly lower estimated reliability and slightly higher estimated average TWD than when the conditioning is on θ. The unconditional evaluation of the maximum observed $\hat{P}(A)$ ranges from slightly higher (.24 vs .20) to substantially improved when the conditioning is on $\hat{\theta}$.

The conditional evaluation shows that conditioning on $\hat{\theta}$ substantially improves the match between the observed maximum $\hat{P}(A \mid \theta)$ and the target at each ability level. (Note that the targets are set for $P(A \mid \hat{\theta})$ but evaluated with $\hat{P}(A \mid \theta)$ and that the actual targets used were those in the table divided by the number of items in the test. In Table 2, the column labels $\hat{\theta}$ and $\hat{P}$ stand for $\max P(A|\hat{\theta})$ and $\max \hat{P}(A|\theta)$, respectively.) The improvement in exposure properties comes at the expense of lowered reliabilities and increased average TWDs, but since the differences are small, may well be worth this cost.

4. Discussion

The role that item security concerns play in adaptive test design depends upon the purpose of the test. In this chapter we have focussed on the environment of continuous high stakes adaptive testing in which item security concerns play an important role. An important goal of such concerns is to prevent test taker preknowledge of items from having too large an effect on test scores. In the future it may be possible to capitalize on the computer to generate items with known properties, including statistical characteristics, on-the- fly in this environment, thus mitigating the issues dealt with here. Until such generation is possible, it is likely that capitalizing on the computer to control the rate of item usage will continue to be a fruitful avenue of further research.

Sophisticated methods of statistical control of item exposure rates began with the efforts of Sympson and Hetter. These methods were elaborated upon by other researchers in the latter half of the 1990's to include various kinds of conditioning in this statistical control. Conditional methods such as those of Stocking and Lewis or of Nering, Davey, and Thompson use θ as the conditioning variable. Such models can sometimes produce disappointing results, as shown in Table 1.

In this chapter, we have focussed on developing a method of statistically controlling the rate of item exposure conditional on estimated ability, rather than true ability. Such methods are difficult not only because of the difference between $\hat{\theta}$ and θ, but also because the best estimate of $\hat{\theta}$ changes as the CAT adapts to the test taker's previous responses. The method developed here appears to at least partially solve the problems of conditioning on θ shown in Table 1, although the results reported in Table 2 could possibly be improved upon by further refinement of the approach. Future research should focus upon the application of conditioning on $\hat{\theta}$ in controlling exposure in a wide variety of contexts with many different item pools before this approach could be recommended for operational use. In particular, it seems important to compare the results predicted from simulation experiments with those obtained from real test takers.

In addition, it should be noted that although exposure rate is controlled statistically by the methods discussed in this chapter, it is not controlled with respect to candidate volume. Any exposure control methodology that seeks to control exposure rates as opposed to absolute exposure suffers from this criticism. This suggests that future research might profitably begin to focus on the effects of absolute exposure for expected candidate volumes. The development of pool rotation schedules and partial or complete pool replacement methods are crucial to this effort. Finally, it seems important that the consequences for test scores of administering items about which test takers may have some preknowledge must be thoroughly understood.

References

Chang, S.W. (1998). *A comparative study of item exposure control methods in computerized adaptive testing.* Unpublished doctoral dissertation, Iowa City, IA: University of Iowa.

Davey, T., & Parshall, C. G. (1995). *New algorithms for item selection and exposure control with computerized adaptive testing.* Paper presented at the annual meeting of the American Educational Research Association, San Francisco, CA.

Eignor, D. R., Way, W. D., Stocking, M. L., & Steffen, M. (1993). *Case studies in computer adaptive test design through simulations.* Research Report 93-56, Princeton, NJ: Educational Testing Service.

Green, B. F., Bock, R. D., Humphreys, L. G., Linn, R. L., & Reckase, M. D. (1984). Technical guidelines for assessing computerized adaptive tests. *Journal of Educational Measurement,21,* 347-360.

Kingsbury, G. G., & Zara, A. R. (1989). Procedures for selecting items for computerized adaptive tests. *Applied Measurement in Education,2,* 359-375.

Lord, F. M. (1970). Some test theory for tailored testing. In W. H. Holtzman,(Ed.), *Computer assisted instruction, testing, and guidance,* (pp.139-183), New York, NY: Harper and Row.

Lord, F. M. (1977). A broad-range tailored test of verbal ability. *Applied Psychological Measurement, 1,* 95-100.

McBride, J. R., & Martin, J. T. (1983). Reliability and validity of adaptive ability tests in a military setting. In D. J. Weiss, (Ed.), *New Horizons in Testing*, (pp. 223-236), New York, NY: Academic Press.

Mislevy, R. J. (1984). Estimating latent distributions. *Psychometrika, 49*, 359-381.

Nering, M. L., Davey, T., & Thompson, T. (1998). *A hybrid method for controlling item exposure in computerized adaptive testing*. Paper presented at the annual meeting of the Psychometric Society, Urbana, IL.

Revuelta, J., & Ponsoda, V. (1998). A comparison of item exposure control methods in computerized adaptive testing. *Journal of Educational Measurement, 35*, 311-327.

Robin, F. (1999). *Alternative item selection strategies for improving test security and pool usage in computerized adaptive testing*. Paper presented at the annual meeting of the National Council on Measurement in Education, Montreal, Quebec.

Stocking, M. L. (1993). *Controlling item exposure rates in a realistic adaptive testing program*. Research Report 93-2, Princeton, NJ: Educational Testing Service.

Stocking, M. L. (1987). Two simulated feasibility studies in computerized adaptive testing, *Applied Psychology: An International Review, 36*, 263-277.

Stocking, M. L., & Lewis, C. (1998). Controlling item exposure conditional on ability in computerized adaptive testing. *Journal of Educational and Behavioral Statistics, 23*, 57-75.

Stocking, M. L., & Swanson, L. (1993). A method for severely constrained item selection in adaptive testing. *Applied Psychological Measurement, 17*, 277-292.

Swanson, L., & Stocking, M. L. (1993). A model and heuristic for solving very large item selection problems. *Applied Psychological Measurement, 17*, 151-166.

Sympson, J. B., & Hetter, R. D. (1985). Controlling item-exposure rates in computerized adaptive testing. *Proceedings of the 27th annual meeting of the Military Testing Association*, (pp. 973-977). San Diego, CA: Navy Personnel Research and Development Center.

Thomasson, G. L. (1995). *New item exposure control algorithms for computerized adaptive testing*. Paper presented at the annual meeting of the Psychometric Society, Minneapolis, MN.

Weiss, D. J. (1978). *Proceedings of the 1977 Computerized Adaptive Testing Conference*. Minneapolis, MN: University of Minnesota.

Wingersky, M. S. (1983). LOGIST: A program for computing maximum likelihood procedures for logistic test models. In R. K. Hambleton, (Ed.), *Applications of item response theory*, (pp.44-45). Vancouver, BC: Educational Research Institute of British Columbia.

Chapter 10
Item Calibration and Parameter Drift

Cees A.W. Glas*
University of Twente, The Netherlands

1. Introduction

Computer-based testing (CBT), as computerized adaptive testing (CAT), is based on the availability of a large pool of calibrated test items. Usually, the calibration process consists of two stages.

(1) A pre-testing stage: In this stage, subsets of items are administered to subsets of respondents in a series of pre-test sessions, and an item response theory (IRT) model is fit to the data to obtain item parameter estimates to support computerized test administration.

(2) An on-line stage: In this stage, data are gathered in a computerized assessment environment, proficiency parameters for examinees are estimated, and the incoming data may also be used for further item parameter estimation.

The topic of this chapter is the evaluation of differences in item parameter values in the pre-test and on-line stage, that is, the evaluation of parameter drift. Essentially, evaluation of parameter drift boils down to checking whether the pre-test and on-line data comply with the same IRT model. Parameter drift may have different sources. Security is one major problem in adaptive testing. If adaptive testing items are administered to examinees on an almost daily basis, after a while some items may become known to new examinees. In an attempt to reduce the risk of overexposure, several exposure control methods have been developed. All these procedures prevent items from being administered more often than desired. Typically, this goal is reached by modifying the item selection criterion so that "psychometrically optimal" items are not always selected. Examples of methods of exposure control are the random-from-best-n method (see e.g. Kingsbury & Zara, 1989, p. 369-370), the count-down random method (see e.g. Stocking & Swanson,

* This study received funding from the Law School Admission Council (LSAC). The opinions and conclusions contained in this paper are those of the author and do not necessarily reflect the position or policy of LSAC.

W.J. van der Linden and C.A.W. Glas (eds.),
Computerized Adaptive Testing: Theory and Practice, 183–199.
© 2000 *Kluwer Academic Publishers. Printed in the Netherlands.*

1993, p. 285-286), and the method of Sympson and Hetter (1985; see also Stocking, 1993). With relatively low exposure rates items will probably become known later than with high exposure rates. Still, sooner or later some items may become known to some future examinees.

Differences between the pretest and the on-line stage may also result in other forms of parameter drift. One might, for instance, think of differences in item difficulty resulting from the different modes of presentation (computerized or paper-and-pencil administration) or differences in item difficulty resulting from a changing curriculum. Also, differences in motivation of the examinees between the pre-test and on-line stage might result in subtle shifts of the proficiency that is measured by the test. Response behavior in these stages might not be properly modeled by the same set of IRT parameters when examinees in the pre-test stage are significantly less motivated than those in the high-stakes on-line stage.

In this chapter, two methods for the evaluation of parameter drift are proposed. The first method is based on a global item-oriented test for parameter drift using a Lagrange multiplier statistic. The method can be viewed as a generalization to adaptive testing of the modification indices for the 2PL model and the nominal response model introduced by Glas (1998, 1999). The second method is targeted at parameter drift due to item disclosure. It addresses the one-sided hypothesis that the item is becoming easier and is loosing its discriminative power. The test for this hypothesis is based on a so-called cumulative sum (CUSUM) statistic. Adoption of this approach in the framework of IRT-based adaptive testing was first suggested by Veerkamp (1996) for use with the Rasch model. The present method is a straightforward generalization of this work.

This chapter is organized as follows. First the most common method of item calibration, marginal maximum likelihood, will be explained. Then the Lagrange multiplier test and the CUSUM test for parameter drift will be explained. Finally, the power of the two classes of tests will be examined in a number of simulation studies.

2. Calibration

2.1. MML ESTIMATION

Marginal maximum likelihood (MML) estimation is probably the most used technique for item calibration. For the 1PL, 2PL and 3PL models, the theory was developed by such authors as Bock and Aitkin (1981), Thissen (1982), Rigdon and Tsutakawa (1983), and Mislevy

(1984,1986), and computations can be made using the software package Bilog-MG (Zimowski, Muraki, Mislevy, & Bock, 1996). MML estimation procedures are also available for IRT models with a multidimensional ability structure (see, for instance, Segall, this volume). Under the label "Full Information Factor Analysis", a multidimensional version of the 2PL and 3PL normal ogive model was developed by Bock, Gibbons, and Muraki (1988) and implemented in TESTFACT (Wilson, Wood, and Gibbons, 1991). A comparable model using a logistic rather than a normal-ogive representation was studied by Reckase (1985, 1997) and Ackerman (1996a and 1996b). In this section, a general MML framework will be sketched, and then illustrated by its application to the 3PL model.

Let $\mathbf{u}_j$ be the response pattern of respondent j, $j = 1, ..., m$, and let $\mathbf{U}$ be the data matrix. In the MML approach, it is assumed that the possibly multidimensional ability parameters $\boldsymbol{\theta}_j$ are independent and identically distributed with density $g(\boldsymbol{\theta}; \boldsymbol{\lambda})$. Usually, it is assumed that ability is normally distributed with population parameters $\boldsymbol{\lambda}$ (which are the mean μ and the variance σ^2 for the unidimensional case, or the mean vector $\boldsymbol{\mu}$ and the covariance matrix Φ for the multidimensional case). Item parameters $\boldsymbol{\beta}$ consist of discrimination parameters (a_i, or $\mathbf{a}_i$ for the unidimensional and the multidimensional cases, respectively), item difficulties b_i, and guessing parameters c_i.

In applications of IRT to CAT, testees seldom respond to all available items. In the calibration stage, a calibration design is used where samples of testees respond to subsets of items which are often called booklets. In the on-line stage, every testee is administered a virtually unique test by the very nature of the item selection mechanism of CAT. Both these test administration designs are captured by introducing a test administration vector $\mathbf{d}_j$, which has elements d_{ij}, $i = 1, ..., I$, where I is the number of items in the item pool. The item administration variable d_{ij} is equal to one if testee j responded to item i, and zero otherwise. The design for all testees is represented by an $m \times I$ design matrix $\mathbf{D}$. The definition of the response variable is extended: the vector $\mathbf{u}_j$ has I elements, which are equal to one if a correct response is observed, equal to zero if an incorrect response is observed and equal to an arbitrary constant if no response is observed. In this context, it is an interesting question whether estimates can be calculated treating the design as fixed and maximizing the likelihood of the parameters conditional on $\mathbf{D}$. If so, the design is called ignorable (Rubin, 1976). Using Rubin's theory on ignorability of designs, this question is extensively studied by Mislevy and Wu (1996). They conclude that the administration design is ignorable in adaptive testing. Further, their conclusions also have consequences for the calibration stage, because the design is usually also ignorable in estimation using data from tests with time limits and tests targeted to the ability level of the testees. MML estimation derives its name from maximizing the log-likelihood that is marginalized with respect to θ, rather than maximizing the joint log-likelihood of all person parameters θ and item parameters β. Let η be a vector of all item and population parameters. Then the marginal likelihood of η is given by

$$\log L(\,\eta; \mathbf{U}, \mathbf{D}\,) = \sum_j \log \int \cdots \int p(\mathbf{u}_j \mid \mathbf{d}_j, \theta_j, \beta_i) g(\theta_j; \lambda) d\theta_j \ . \quad (1)$$

The reason for maximizing the marginal rather than the joint likelihood is that maximizing the latter does not lead to consistent estimates. This is related to the fact that the number of person parameters grows proportional with the number of observations, and, in general, this leads to inconsistency (Neyman & Scott, 1948). Simulation studies by Wright and Panchapakesan (1969) and Fischer and Scheiblechner (1970) show that these inconsistencies can indeed occur in IRT models. Kiefer and Wolfowitz (1956) have shown that marginal maximum likelihood estimates of structural parameters, say the item and population parameters of an IRT model, are consistent under fairly reasonable regularity conditions, which motivates the general use of MML in IRT models.

The marginal likelihood equations for η can be easily derived using Fisher's identity (Efron, 1977; Louis 1982; also see, Glas, 1992, 1998). The first order derivatives with respect to η can be written as

$$\mathbf{h}(\eta) = \frac{\partial}{\partial \eta} \log L(\eta; \mathbf{U}, \mathbf{D}) = \sum_j E(\omega_j(\eta) \mid \mathbf{u}_j, \mathbf{d}_j, \eta) \ , \quad (2)$$

with

$$\omega_j(\eta) = \frac{\partial}{\partial \eta} \log p(\mathbf{u}_j, \theta_j \mid \mathbf{d}_j, \eta), \quad (3)$$

where the expectation is with respect to the posterior distribution $p(\theta_j \mid \mathbf{u}_j, \mathbf{d}_j; \eta)$. The identity in (2) is closely related to the EM-algorithm (Dempster, Laird and Rubin, 1977), which is an algorithm for finding the maximum of a likelihood marginalized over unobserved data. The present application fits this framework when the response patterns are viewed as observed data and the ability parameters as unobserved data. Together they are referred to as the complete data. The EM algorithm is applicable in situations where direct inference based on the marginal likelihood is complicated, and the complete data likelihood equations, i.e., equations based on $\omega_j(\eta)$, are easily solved. Given some estimate of η, say η^*, the estimate can be improved by solving $\sum_j E(\omega_j(\eta) \mid \mathbf{u}_j, \mathbf{d}_j, \eta^*) = 0$ with respect to η. Then this new estimate becomes η^* and the process is iterated until convergence.

Application of this framework to deriving the likelihood equations of the structural parameters of the 3PL model proceeds as follows. The likelihood equations are obtained upon equating (2) to zero, so explicit expressions are needed for (3). Given the design vector $\mathbf{d}_j$, the ability parameter θ_j and the item parameters of the 3PL model, the probability of response pattern $\mathbf{u}_j$ is given by

$$p(\mathbf{u}_j \mid \mathbf{d}_j, \theta_j, a_i, b_i, c_i) = \prod_i P_i(\theta_j)^{d_{ij}u_{ij}}(1 - P_i(\theta_j))^{d_{ij}(1-u_{ij})} ,$$

where P_{ij} is the probability of a correct response to item i, as defined in van der Linden and Pashley (this volume), formula (1). Let S_{ij} be defined by $P_{ij} = c_i + (1 - c_i)S_{ij}$, so S_{ij} is the logistic part of the probability P_{ij}. By taking first order derivatives of the logarithm of this expression, the expressions for (3) are found as

$$\omega_j(a_i) = \frac{(u_{ij} - P_{ij})(1 - c_i)S_{ij}(1 - S_{ij})(\theta_j - b_i)}{P_{ij}(1 - P_{ij})}, \tag{4}$$

$$\omega_j(b_i) = \frac{(P_{ij} - u_{ij})(1 - c_i)S_{ij}(1 - S_{ij})a_i}{P_{ij}(1 - P_{ij})}, \tag{5}$$

and

$$\omega_j(c_i) = \frac{(u_{ij} - P_{ij})(1 - S_{ij})}{P_{ij}(1 - P_{ij})}. \tag{6}$$

The likelihood equations for the item parameters are found upon inserting these expressions into (2) and equating the resulting expressions to zero. To derive the likelihood equations for the population parameters, the first order derivatives of the log of the density of the ability parameters $g(\theta; \mu, \sigma)$ are needed. In the present case, $g(\theta; \mu, \sigma)$ is the well-known expression for the normal distribution with mean μ and standard deviation σ, so it is easily verified that these derivatives are given by

$$\omega_j(\mu) = \frac{(\theta_j - \mu)}{\sigma^2}$$

and

$$\omega_j(\sigma) = \frac{(\theta_j - \mu)^2 - \sigma^2}{\sigma^3}.$$

The likelihood equations are again found upon inserting these expressions in (2) and equating the resulting expressions to zero.

Also the standard errors are easily derived in this framework: Mislevy (1986) points out that the information matrix can be approximated as

$$\mathbf{H}(\,\boldsymbol{\eta},\,\boldsymbol{\eta}) \approx \sum_j E(\,\omega_j(\boldsymbol{\eta}) \mid \mathbf{u}_j, \mathbf{d}_j, \boldsymbol{\eta})E(\,\omega_j(\boldsymbol{\eta}) \mid \mathbf{u}_j, \mathbf{d}_j, \boldsymbol{\eta})', \quad (7)$$

and the standard errors are the diagonal elements of the inverse of this matrix.

The basic approach presented so far can be generalized in two ways. First, the assumption that all respondents are drawn from one population can be replaced by the assumption that there are multiple populations of respondents. Usually, it is assumed that each population has a normal ability distribution indexed by a unique mean and variance parameter. Bock and Zimowski (1997) point out that this generalization together with the possibility of analyzing incomplete item-administration designs provides a unified approach to such problems as differential item functioning, item parameter drift, non-equivalent groups equating, vertical equating and matrix-sampled educational assessment. Item calibration for CAT also fits within this framework.

A second extension of this basic approach is Bayes modal estimation. This approach is motivated by the fact that item parameter estimates in the 3PL model are sometimes hard to obtain because the parameters are poorly determined by the available data. In these instances, item characteristic curves can be appropriately described by a large number of different item parameter values over the ability scale region where the respondents are located. As a result, the estimates of the three item parameters in the 3PL model are often highly correlated. To obtain "reasonable" and finite estimates, Mislevy (1986) considers a number of Bayesian approaches. Each of them entails the introduction of prior distributions on item parameters. Parameter estimates are then computed by maximizing the log-posterior density of $\boldsymbol{\eta}$, which is proportional to $\log L(\,\boldsymbol{\eta};\, \mathbf{U}) + \log p(\,\boldsymbol{\eta} \mid \boldsymbol{\zeta}) + \log p(\boldsymbol{\zeta})$, where $p(\,\boldsymbol{\eta} \mid \boldsymbol{\zeta})$ is the prior density of the $\boldsymbol{\eta}$, characterized by parameters $\boldsymbol{\zeta}$, which in turn follow a density $p(\,\boldsymbol{\zeta})$. In one approach, the prior distribution is fixed; in another approach, often labeled empirical Bayes, the parameters of the prior distribution are estimated along with the other parameters. In the first case, the likelihood equations in (1) change to $\partial \log L(\eta; \mathbf{U})/\partial \eta + \partial \log p(\eta \mid \zeta)/\partial \eta = 0$. In the second case, in addition to these modified likelihood equations, the additional equations $\partial \log p(\zeta)/\partial \zeta = 0$ must also be solved. For details refer to Mislevy (1986). In the following sections, two methods for parameter drift in the framework of the 3PL model and MML estimation will be presented.

3. Evaluation of Parameter Drift

3.1. A Lagrange Multiplier Test for Parameter Drift

The idea behind the Lagrange Multiplier (LM) test (Aitchison and Silvey, 1958), and the equivalent efficient score test (Rao, 1947), can be summarized as follows. Consider some general parameterized model and a special case of the general model, the so-called restricted model. The restricted model is derived form the general model by imposing constraints on the parameter space. In many instances, this is accomplished by setting one or more parameters of the general model to constants. The LM test is based on evaluating a quadratic function of the partial derivatives of the log-likelihood function of the general model evaluated at the ML estimates of the restricted model. The LM test is evaluated using the ML estimates of the parameters of the restricted model. The unrestricted elements of the vector of the first-order derivatives are equal to zero because their values originate from solving the likelihood equations. The magnitude of the elements of the vector of first-order derivatives corresponding with restricted parameters determine the value of the statistic: the closer they are to zero, the better the model fits.

More formally, the principle can be described as follows. Consider a null hypothesis about a model with parameters ϕ_0. This model is a special case of a general model with parameters ϕ. In the case discussed here, the special model is derived from the general model by setting one or more parameters to zero. So if the parameter vector ϕ_0 is partitioned as $\phi_0 = (\phi_{01}, \phi_{02})$, the null hypothesis entails $\phi_{02} = 0$. Let $\mathbf{h}(\phi)$ be the partial derivatives of the log-likelihood of the general model, so $\mathbf{h}(\phi) = \partial \log L(\phi)/\partial \phi$. This vector of partial derivatives gauges the change of the log-likelihood as a function of local changes in ϕ. The test will be based on the statistic

$$LM = \mathbf{h}(\phi_{02})^t \Sigma^{-1} \mathbf{h}(\phi_{02}) , \qquad (8)$$

where

$$\Sigma = \Sigma_{11} - \Sigma_{10}\Sigma_{00}^{-1}\Sigma_{01}$$

and

$$\Sigma_{pq} = \sum_j \mathbf{h}_j(\phi_{0p})\mathbf{h}_j(\phi_{0q})^t.$$

The statistic has an asymptotic χ^2-distribution with degrees of freedom equal to the number of parameters in ϕ_{02} (Aitchison & Silvey, 1958, Rao, 1948).

Recently, the LM principle has been applied in the framework of IRT for evaluating differential item functioning (Glas, 1998) and the axioms of unidimensionality and local stochastic independence (Glas, 1999). Though originally presented in the framework of a fixed item administration design, these tests can also be applied in the framework of the stochastic design characteristics for CAT. However, the result with respect to the asymptotic distribution of the statistics does not automatically apply the to case of a stochastic design. The ignorability principle ensures consistency of estimators in a CAT design, but it does not apply to sample inferences, such as confidence intervals and the distributions of statistics for evaluation of model fit (Mislevy & Chang, 1998). Chang and Ying (1999) have addressed these problems for the ability estimates, but for the item parameter estimates the impact of the problems is, as yet, unclear. Therefore, for the applications presented here, a power study under the null model will be part of the example to be presented. The results will show that the asymptotic distribution of the LM statistics is hardly affected by CAT.

Above, it was already noted that parameter drift can be evaluated by checking whether pre-test and on-line data can be properly described by the same IRT model. Consider G groups labeled $g = 1, ..., G$. It is assumed that the first group partakes in the pre-testing stage, and the following groups partake in the on-line stage. The application of the LM tests to monitoring parameter drift is derived from the LM test for differential item functioning proposed by Glas (1998) for the 2PL model. This is a test of the hypothesis that the item parameters are constant over groups, that is, the hypothesis $a_{ig} = a_i$ and $b_{ig} = b_i$, for all g. To see the relation with the LM framework, consider two groups, and define a variable y_j, that is equal to one if j belongs to the first group and zero if j belongs to the second group. Defining $a_{iy} = a_i + y_j\delta_1$ and $b_{iy} = b_i + y_j\delta_2$, the hypothesis given by $\delta_1 = 0$ and $\delta_2 = 0$ can be evaluated using the LM test. For more than two groups more dummy variables y_j are needed to code group membership. This approach can of course also be used to monitor parameter drift in CAT. Further, generalization to the 3PL model entails adding $\delta_3 = 0$, with $c_{iy} = c_i + y_j\delta_3$ to the null hypothesis.

For actual implementation of this approach using adaptive testing data, the high correlation of estimates of the three item parameters discussed in the previous section must be taken into account. Another parameter estimation problem arises specifically in the context of adaptive testing. Guessing (which may be prominent in the calibration stage) may hardly occur in the on-line stage because items are tailored to the ability level of the respondents. Therefore, a test focussed on all three parameters simultaneously often proves computationally unstable. In the present paper, three approaches will be studied. In the first, the LM test will be focussed on simultaneous parameter drift in a_i and b_i, in the second approach, the LM test will be focussed on parameter drift in c_i. These two tests will be labeled $LM(a_i, b_i)$ and $LM(c_i)$, respectively. In the third approach, the guessing parameter will be fixed to some plausible constant, say, to the reciprocal of the number of response alternatives of the items, and the LM statistic will be used to test whether this fixed guessing parameter is appropriate in the initial stage and remains so when the adaptive testing data are introduced. So the hypothesis considered is that $c_{ig} = c_i$ for all g. Using simulation studies, it will be shown that the outcomes of these three approaches are quite comparable.

3.2. A CUSUM TEST FOR PARAMETER DRIFT

The CUSUM chart is an instrument of statistical quality control used for detecting small changes in product features during the production process (see, for instance, Wetherill, 1977). The CUSUM chart is used in a sequential statistical test, where the null hypothesis of no change is never accepted. In the present application, loss of production quality means that the item is becoming easier and less discriminating.

Contrary to the case of the LM test, the CUSUM test needs estimation of the item parameters for every group of testees $g = 1, ..., G$. As above, the first group partakes in the pre-testing stage, and the following groups take an adaptive test. However, estimation of the guessing parameter is problematic in a CAT situation because, as already mentioned above, guessing may be prominent in the calibration stage, while it may hardly occur in the on-line stage, where the items are tailored to the ability level of the respondents. Two possible solutions include: fixing the guessing parameter to some plausible constant such as the reciprocal of the number of response options, or concurrent estimation of the item guessing parameter using all available data. In either approach, the null hypothesis is $a_{ig} - a_{i1} \geq 0$ and $b_{ig} - b_{i1} \geq 0$, for the respondent groups $g = 1, ..., G$. Therefore, a one-sided CUSUM chart will be based on the quantity

$$S_i(g) = \max\left\{ S_i(g-1) + \frac{a_{i1} - a_{ig}}{Se(a_{ig} - a_{i1})} + \right.$$

$$\left. \frac{b_{i1} - b_{ig}}{Se(b_{i1} - b_{ig} \mid a_{i1} - a_{ig})} - k, \ 0 \right\} \qquad (9)$$

where $Se(a_{ig} - a_{i1}) = \sigma_a$ and $Se(b_{i1} - b_{ig} \mid a_{i1} - a_{ig}) = \sqrt{\sigma_b^2 - \sigma_{ab}^2/\sigma_a^2}$, with σ_a^2, σ_b^2 and σ_{ab} the appropriate elements of the covariance matrix of the parameter estimates given by (7). Further, k is a reference value determining the size of the effects one aims to detect. The CUSUM chart starts with $S_i(1) = 0$ and the null hypothesis is rejected as soon as $S_i(g) > h$, where h is some constant threshold value. The choice of the constants k and h determines the power of the procedure. In the case of the Rasch model, where the null hypothesis is $b_{ig} - b_{i1} \geq 0$, and the term involving the discrimination indices is lacking from (9), Veerkamp (1996) successfully uses $k = 1/2$ and $h = 5$. This choice was motivated by the consideration that the resulting test has good power against the alternative hypothesis of a normalized shift in item difficulty of approximately half a standard deviation. In the present case, one extra normalized decision variable is employed, namely, the variable involving the discrimination indices. So, for instance, a value $k = 1$ can be used to have power against a shift of one standard deviation of both normalized decision variables in the direction of the alternative hypothesis. However, there are no compelling reasons for this choice; the attractive feature of the CUSUM procedure is that the practitioner can chose the effect size k to meet the specific characteristics of the problem. Also, the choice of a value for h is determined by the targeted detection rate, especially by the trade off between Type I and II errors. In practice, the values of h and k can be set using simulation studies. Examples will be given below.

4. Examples

In this section, the power of the procedures suggested above will be investigated using a number of simulation studies. Since all statistics involve an approximation of the standard error of the parameter estimates using (7), first the precision of the approximation will be studied by assessing the power of the statistics under the null model, that is, by studying the Type I error rate. Then the power of the tests will be studied under various model violations. These two topics will first be studied for the LM tests, then for the CUSUM test.

Table 1. Type I error rate of LM test

K	L	N_g	$LM(c_i)$	$LM(a_i, b_i)$
			Percentage Significant at 10 %	
50	20	500	8	9
		1000	10	10
	40	500	9	10
		1000	11	8
100	20	500	12	10
		1000	8	9
	40	500	10	12
		1000	10	10

In all simulations, the ability parameters θ were drawn from a standard normal distribution. The item difficulties b_i were uniformly distributed on $[-1.5, 1.5]$, the discrimination indices a_i were drawn from a log-normal distribution with a zero mean and a standard deviation equal to 0.25, and the guessing parameters were fixed at 0.20, unless indicated otherwise. In the on-line stage, item selection was done using the maximum information principle. The ability parameter was estimated by its expected a-posteriori value (EAP); the initial prior was standard normal.

The results of eight simulation studies with respect to the Type I error rate of the LM test are shown in Table 1. The design of the study can be inferred from the first three columns of the table. It can be seen that the number of items K in the item bank was fixed at 50 for the first four studies and at 100 for the next four studies. Both in the pre-test stage and the on-line stage, test lengths L of 20 and 40 were chosen. Finally, as can be seen in the third column, the number of respondents per stage, N_g, was fixed at 500 and 1000 respondents. So summed over the pre-test and on-line stage, the sample sizes were 1000 and 2000 respondents, respectively. For the pre-test stage, a spiraled test administration design was used. For instance, for the $K = 50$ studies, for the pre-test stage, five subgroups were used; the first subgroup was administered the items 1 to 20, the second the items 11 to 30, the third the items 21 to 40 the fifth the items 31 to 50, and the last group received the items 1 to 10 and 41 to 50. In this manner, all items drew the same number of responses in the pre-test stage. For the $K = 100$ studies, for the pre-test stage four subgroups administered

Table 2. Power of LM test

Model Violation		Percentage Significant at 10 %	
		$LM(a_i, b_i)$	$LM(c_i)$
$c_i = 0.25$	Hits	25	15
	False Alarm	08	10
$c_i = 0.30$	Hits	45	35
	False Alarm	13	11
$c_i = 0.40$	Hits	95	85
	False Alarm	17	20
$b_i = -0.20$	Hits	25	30
	False Alarm	13	12
$b_i = -0.40$	Hits	55	70
	False Alarm	15	20
$b_i = -0.60$	Hits	80	95
	False Alarm	13	27

50 items were formed, so here the design was 1 - 50, 26 - 75, 51 - 100 and 1 - 25 and 76 - 100. For each study, 100 replications were run. The results of the study are shown in the last two columns of Table 1. These columns contain the percentages of $LM(c_i)$ and $LM(a_i, b_i)$ tests that were significant at the 10% level. It can be seen that the Type I error rates of the tests conform to the nominal value of 10%. These results support the adequacy of the standard error approximations for providing accurate Type I error rates.

The second series of simulations pertained to the power of the LM statistics under various model violations. The setup was the same as in the above study with $K = 100$ items in the item bank, a test length $L = 50$, $N_1 = 1000$ simulees in the pre-test stage and $N_2 = 1000$ simulees in the on-line stage. Two model violations were simulated. In the first, the guessing parameter c_i went up in the on-line stage, in the second, the item difficulty b_i went down in the on-line stage. Six conditions were investigated: c_i rose from 0.20 to 0.25, 0.30 and 0.40, respectively, and b_i changed from the initial value by -0.20, -0.40 and -0.60, respectively. These model violations were imposed on the items 5, 10, 15, etc. So 20 out of the 100 items were affected by this form of parameter drift. 100 replications were made for each condition. Both the $LM(c_i)$ and $LM(a_i, b_i)$ tests were used. The results are shown

Table 3. Type I error rate of CUSUM test

Effect Size	$h = 2.5$	$h = 5.0$	$h = 7.5$	$h = 10.0$
$k = 0.50$	17	04	01	00
$k = 1.00$	09	06	01	00
$k = 2.00$	01	00	00	00

in Table 2. This table displays both the percentage of "hits" (correctly identified items with parameter drift) and "false alarms" (items without parameter drift erroneously identified as drifting). Three conclusions can be drawn. Firstly, it can be seen that the power of the tests increases as the magnitude of the model violation grows. Secondly, the power of the test specifically aimed at a model violation is always a little larger than the power of the other test, but the differences are quite small. For instance, in the case $b_i = -0.60$, the power of $LM(a_i, b_i)$ is 0.95, while the power of $LM(c_i)$ is 0.85. The third conclusion that can be drawn from the table is that the percentage of "false alarms" is clearly higher than the nominal 10% error rate. A plausible explanation might be that the improper parameter estimates of the 20% items with parameter drift influence the estimates of the 80% non-affected items. Finally, it can be noted that the agreement between the two tests with respect to the flagged items was high; agreement between the two tests was always higher than 0.84.

As mentioned above, the power of the CUSUM procedure is governed by choosing an effect size k and a critical value h. A good way to proceed in a practical situation is to calibrate the procedure when the pre-test data have become available. First the practitioner must set an effect size k of interest. Then, assuming no parameter drift, on-line data can be simulated using the parameter estimates of the pre-test stage. Finally, CUSUM statistics can be computed to find a value for h such that an acceptable Type I error rate is obtained. An example will be given using the same set-up as above: there were $K = 100$ items in the item bank, test length was $L = 50$, and the pre-test data consisted of the responses of $N_1 = 1000$ simulees. Then, four batches responses of $N_g = 1000$, $(g = 2, ..., 5)$ simulees were generated as on-line data, and CUSUM statistics $S_i(g)$ were computed for the iterations $g = 2, ..., 5$. This procedure was carried out for three effect sizes k and four thresholds h; the values are shown in Table 3.

In the table, the percentages items flagged in the fifth iteration ($g = 5$) of the procedure are shown for the various combinations of k and

h. Since no parameter drift was induced, the percentages shown can be interpreted as Type I error rates. For an effect size $k = 0.50$, it can be seen that a value $h = 2.5$ results in 17% flagged items, which is too high. A value $h = 5.0$ results in 4% flagged items, which might be considered an acceptable Type I error rate. Also for an effect size $k = 1.00$ a critical value $h = 5.0$ seems a good candidate. Finally, for $k = 2.00$ all four values of h produce low Type I error rates. So it must be concluded that, given the design and the sample size, detection of parameter drift with an effect size of two standard deviations may be quite difficult.

This result was further studied in a set of simulations where model violations were introduced. These studies used the setup $K = 100$, $L = 50$, and $N_g = 1000$, for $g = 1, ..., 5$. The model violations were similar to the ones imposed above above. So in six conditions, the guessing parameter c_i rose from 0.20 to 0.25, 0.30 and 0.40, respectively, and b_i changed from the initial value by -0.20, -0.40 and -0.60, respectively. Again, for each condition, 20 of the 100 items were affected by the model violation. The results are shown in Table 4. For the simulation studies with effect sizes $k = 0.50$ and $k = 1.00$, a critical value $h = 5.0$ was chosen, for the studies with effect size $k = 2.00$, the critical value was $h = 2.5$. For every combination of effect size and model violation 20 replications were made. The last four columns of Table 4 give the percentages of "hits" (flagged items with parameter drift) and "false alarms" (erroneously flagged items per condition) for the iterations $g = 2, ..., 5$. The percentages are aggregated over the 20 replications per condition. As expected, the highest percentages of "hits" were obtained for the smaller effect sizes $k = 0.50$ and $k = 1.00$, and the larger model violations. The top is the combination $k = 1.00$ and $b_i = -0.60$, which, for $g = 5$, has an almost perfect record of 99% "hits". In this condition, the percentage of "false alarms" remained at a 10% level. The worst performances were obtained for combinations of $k = 0.50$ and $k = 2.00$ with small violations as $c_i = 0.25$, $c_i = 0.30$ and $b_i = -0.20$.

These conditions both show a low "hit"-rate and a "false alarm"-rate of approximately the same magnitude, which is relatively high for a "false alarm"-rate.

5. Discussion

In this chapter, it was shown how to evaluate whether the IRT model of the pre-test stage also fits the on-line stage. Two approaches were presented. The first was based on LM statistics. It was shown that the approach supports the detection of specific model violations and has

Table 4. Power of CUSUM test

Effect Size	Model Violation		Iteration			
			$g = 2$	$g = 3$	$g = 4$	$g = 5$
$k = 0.50$	$c_i = 0.25$	Hits	00	00	05	15
		False Alarm	00	04	05	13
	$c_i = 0.30$	Hits	00	05	10	20
		False Alarm	00	03	05	06
	$c_i = 0.40$	Hits	00	30	75	85
		False Alarm	00	00	01	03
$k = 1.00$	$c_i = 0.25$	Hits	15	25	30	45
		False Alarm	05	13	17	21
	$c_i = 0.30$	Hits	15	35	55	50
		False Alarm	03	03	03	06
	$c_i = 0.40$	Hits	30	75	90	85
		False Alarm	03	04	06	09
$k = 2.00$	$c_i = 0.25$	Hits	00	05	15	15
		False Alarm	00	00	03	00
	$c_i = 0.30$	Hits	05	15	15	20
		False Alarm	03	01	04	04
	$c_i = 0.40$	Hits	15	30	55	60
		False Alarm	00	01	01	01
$k = 0.50$	$b_i = -0.20$	Hits	00	00	10	15
		False Alarm	00	00	06	05
	$b_i = -0.40$	Hits	00	15	45	60
		False Alarm	01	06	09	15
	$b_i = -0.60$	Hits	05	35	65	80
		False Alarm	00	00	04	04
$k = 1.00$	$b_i = -0.20$	Hits	00	20	40	35
		False Alarm	00	01	03	05
	$b_i = -0.40$	Hits	25	50	55	65
		False Alarm	01	04	06	09
	$b_i = -0.60$	Hits	20	75	95	99
		False Alarm	03	06	10	10
$k = 2.00$	$b_i = -0.20$	Hits	00	00	05	05
		False Alarm	00	01	03	03
	$b_i = -0.40$	Hits	05	10	30	35
		False Alarm	00	00	03	01
	$b_i = -0.60$	Hits	00	25	75	75
		False Alarm	01	03	04	03

the advantage of known asymptotic distributions of the statistics on which it is based. In the present paper, two specific model violations were considered, but the approach also applies to other model violations, such as violation of local independence and multidimensionality (see Glas, 1999). The second approach is based on CUSUM statistics. The distribution of these statistics is not known, but an appropriate critical value h can be found via simulation. An advantage, however, is that the practitioner can tune the procedure to the needs of the specific situation. When choosing h, the subjective importance of making "hits" and of avoiding "false alarms" can be taken into account, and the effect size k can be chosen to reflect the magnitude of parameter drift judged relevant in a particular situation. Summing up, both approaches provide practical tools for monitoring parameter drift.

References

Ackerman, T.A. (1996a). Developments in multidimensional item response theory. *Applied Psychological Measurement 20*, 309-310.

Ackerman, T.A. (1996b). Graphical representation of multidimensional item response theory analyses. *Applied Psychological Measurement 20*, 311-329.

Aitchison, J., & Silvey, S.D. (1958). Maximum likelihood estimation of parameters subject to restraints. *Annals of Mathematical Statistics 29*, 813-828.

Bock, R.D., & Aitkin, M. (1981). Marginal maximum likelihood estimation of item parameters: an application of an EM-algorithm. *Psychometrika, 46*, 443-459.

Bock, R.D., Gibbons, R.D. & Muraki, E. (1988). Full-information factor analysis. *Applied Psychological Measurement 12*, 261-280.

Bock, R.D. & Zimowski, M.F. (1997). Multiple group IRT. In W.J.van der Linden and R.K.Hambleton (Eds.). *Handbook of modern item response theory.* (pp. 433-448). New York: Springer.

Chang, H.-H., & Ying, Z. (1999). Nonlinear sequential designs for logistic item response theory models, with applications to computerized adaptive tests. *Annals of Statistics.* (In press).

Dempster, A. P., Laird, N. M., & Rubin, D. B. (1977). Maximum likelihood from incomplete data via the EM algorithm (with discussion). *J. R. Statist. Soc. B. 39*, 1-38.

Efron, B. (1977). Discussion on maximum likelihood from incomplete data via the EM algorithm (by A. Dempster, N. Laird, and D. Rubin). *J. R. Statist. Soc. B. 39*, 1-38.

Fischer, G.H. & Scheiblechner, H.H. (1970). Algorithmen und Programme für das probabilistische Testmodell von Rasch. *Psychologische Beiträge, 12*, 23-51.

Glas, C.A.W. (1992). A Rasch model with a multivariate distribution of ability. In M. Wilson, (Ed.), *Objective measurement: Theory into practice* (Volume 1) (pp. 236-258). Norwood, NJ: Ablex Publishing Corporation.

Glas, C.A.W. (1998) Detection of differential item functioning using Lagrange multiplier tests. *Statistica Sinica, 8*, 647-667.

Glas, C. A. W. (1999). Modification indices for the 2-PL and the nominal response model. *Psychometrika, 64*, 273-294.

Kiefer, J., & Wolfowitz, J. (1956). Consistency of the maximum likelihood estimator in the presence of infinitely many incidental parameters. *Annals of Mathematical Statistics, 27*, 887-903.

Kingsbury, G.G., & Zara, A.R. (1989). Procedures for selecting items for computerized adaptive tests. *Applied Measurement in Education, 2*, 359-375.

Louis, T.A. (1982). Finding the observed information matrix when using the EM algorithm. *Journal of the Royal Statistical Society, Series B, 44*, 226-233.

Mislevy, R.J. (1984). Estimating latent distributions. *Psychometrika, 49*, 359-381.

Mislevy, R.J. (1986). Bayes modal estimation in item response models. *Psychometrika, 51*, 177-195.

Mislevy, R.J., & Chang, H.-H. (1998). *Does adaptive testing violate local independence?* (Research Report RR-98-33). Princeton, NJ: Educational Testing Service.

Mislevy, R.J. & Wu, P.K. (1996). *Missing responses and IRT ability estimation: omits, choice, time limits, and adaptive testing* (Research Report RR-96-30-ONR). Princeton, NJ: Educational Testing Service.

Neyman, J., and Scott, E.L. (1948). Consistent estimates, based on partially consistent observations. *Econometrica, 16*, 1-32.

Rao, C.R. (1947). Large sample tests of statistical hypothesis concerning several parameters with applications to problems of estimation. *Proceedings of the Cambridge Philosophical Society, 44*, 50-57.

Reckase, M.D. (1985). The difficulty of test items that measure more than one ability. *Applied Psychological Measurement, 9*, 401-412.

Reckase, M.D. (1997). A linear logistic multidimensional model for dichotomous item response data. In W.J.van der Linden and R.K.Hambleton (Eds.). *Handbook of modern item response theory* (pp.271-286). New York: Springer.

Rigdon S.E., & Tsutakawa, R.K. (1983). Parameter estimation in latent trait models. *Psychometrika, 48*, 567-574.

Rubin, D.B. (1976). Inference and missing data. *Biometrika, 63*, 581-592.

Stocking, M.L. (1993). *Controlling exposure rates in a realistic adaptive testing paradigm* (Research Report 93-2). Princeton, NJ: Educational Testing Service.

Stocking, M.L., & Swanson, L. (1993). A method for severely constrained item selection in adaptive testing. *Applied Psychological Measure-ment, 17*, 277-292.

Sympson, J.B., & Hetter, R.D. (1985). Controlling item-exposure rates in computerized adaptive testing. *Proceedings of the 27th annual meeting of the Military Testing Association* (pp. 973-977). San Diego, CA: Navy personnel Research and Devel-op-ment Center.

Thissen D. (1982). Marginal maximum likelihood estimation for the one-parameter logistic model. *Psychometrika, 47*, 175-186.

Veerkamp, W.J.J. (1996). *Statistical methods for computerized adaptive testing.* Unpublished doctoral thesis, Twente University, the Netherlands.

Wilson, D.T., Wood, R., & Gibbons, R.D. (1991) *TESTFACT: Test scoring, item statistics, and item factor analysis* (computer software). Chicago: Scientific Software International, Inc.

Wetherill, G.B. (1977). *Sampling inspection and statistical quality control* (2nd edition). London: Chapman and Hall.

Wright, B.D., & Panchapakesan, N. (1969). A procedure for sample- free item analysis. *Educational and Psychological Measurement, 29*, 23-48.

Zimowski, M.F., Muraki, E., Mislevy, R.J., & Bock, R.D. (1996). *Bilog MG: Multiple-group IRT analysis and test maintenance for binary items.* Chicago: Scientific Software International, Inc.

Chapter 11
Detecting Person Misfit in Adaptive Testing Using Statistical Process Control Techniques

Edith M. L. A. van Krimpen-Stoop & Rob R. Meijer*
University of Twente, The Netherlands

1. Introduction

The aim of a computerized adaptive test (CAT) is to construct an optimal test for an individual examinee. To achieve this, the ability of the examinee is estimated during test administration and items are selected that match the current ability estimate. This is done using an item response theory (IRT) model that is assumed to describe an examinee's response behavior. It is questionable, however, whether the assumed IRT model gives a good description of each examinee's test behavior. For those examinees for whom this is not the case, the current ability estimate may be inadequate as a measure of the ability level and as a result the construction of an optimal test may be flawed.

There are all sorts of factors that may cause an ability estimate to be invalidated. For example, examinees may take a CAT to familiarize themselves with the questions and randomly guess the correct answers on almost all items in the test. Or examinees may have pre-knowledge of some of the items in the item pool and correctly answer these items independent of their trait level and the item characteristics. These types of aberrant behavior invalidate the ability estimate and it therefore seems useful to investigate the fit of an item score pattern to the test model. Research with respect to methods that provide information about the fit of an individual item score pattern to a test model is usually referred to as appropriateness measurement or person-fit measurement. Most studies in this area are, however, in the context of paper-and-pencil (P&P) tests. As will be argued below, the application of person-fit theory presented in the context of P&P tests cannot simply be generalized to CAT.

The aim of this article is first to give an introduction to existing person-fit research in the context of P&P tests, then to discuss some

* This study received funding from the Law School Admission Council (LSAC). The opinions and conclusions contained in this report are those of the authors and not necessarily reflect the position or policy of LSAC.

W.J. van der Linden and C.A.W. Glas (eds.),
Computerized Adaptive Testing: Theory and Practice, 201–219.

specific problems regarding person fit in the context of CAT, and finally to explore some new methods to investigate person fit in CAT.

2. Review of Existing Literature

2.1. PERSON FIT IN PAPER-AND-PENCIL TESTING

Several statistics have been proposed to investigate the fit of an item score pattern to an IRT model.

In IRT, the probability of obtaining a correct answer on item i ($i = 1, ..., n$) is explained by an examinee's latent trait value (θ) and the characteristics of the item (Hambleton & Swaminathan, 1985). Let U_i denote the binary $(0, 1)$ response to item i, a_i the item discrimination parameter, b_i the item difficulty parameter, and c_i the item guessing parameter. The probability of correctly answering an item according to the three-parameter logistic IRT model (3PLM) is defined by

$$P\left(U_i = 1 \,|\theta\right) \equiv P_i\left(\theta\right) = c_i + \left(1 - c_i\right) \frac{\exp\left[a_i\left(\theta - b_i\right)\right]}{1 + \exp\left[a_i\left(\theta - b_i\right)\right]}.$$

When $c_i = 0$, the 3PLM becomes the two-parameter logistic IRT model (2PL model):

$$P_i\left(\theta\right) = \frac{\exp\left[a_i\left(\theta - b_i\right)\right]}{1 + \exp\left[a_i\left(\theta - b_i\right)\right]}. \tag{1}$$

Most person-fit research has been conducted using fit statistics that are designed to investigate the probability of an item score pattern under the null hypothesis of fitting response behavior. A general form in which most person-fit statistics can be expressed is

$$\sum_{i=1}^{n} U_i w_i\left(\theta\right) - w_0\left(\theta\right), \tag{2}$$

where $w_i\left(\theta\right)$ and $w_0\left(\theta\right)$ are suitable weights (see Snijders, 2000). Or person-fit statistics can be expressed in a centered form, that is, with expectation 0, as

$$W = \sum_{i=1}^{n} \left[U_i - P_i\left(\theta\right)\right] w_i\left(\theta\right). \tag{3}$$

Note that, as a result of binary scoring $U_i^2 = U_i$. Thus, statistics of the form

$$W^* = \sum_{i=1}^{n} \left[U_i - P_i\left(\theta\right)\right]^2 v_i\left(\theta\right),$$

can be re-expressed as statistics of the form in Equation 2. The expected value of the statistic equals 0 and often the variance is taken into account to obtain a standardized version of the statistic. For example, Wright and Stone (1979) proposed a person-fit statistic based on standardized residuals, with the weight

$$w_i(\theta) = \frac{1}{nP_i(\theta)[1 - P_i(\theta)]},$$

resulting in

$$V = \sum_{i=1}^{n} \frac{[U_i - P_i(\theta)]^2}{nP_i(\theta)[1 - P_i(\theta)]}. \tag{4}$$

V can be interpreted as the corrected mean of the squared standardized residuals based on n items; relatively large values of V indicate deviant item score patterns.

Most studies in the literature have been conducted using some suitable form of the log-likelihood function

$$l = \sum_{i=1}^{n} \{U_i \ln P_i(\theta) + [1 - U_i] \ln[1 - P_i(\theta)]\}.$$

This statistic, first proposed by Levine and Rubin (1979), was further developed by Drasgow, Levine, and Williams (1985). Two problems exist when using l as a fit statistic. The first problem is that the numerical values of l depend on the trait level. As a result, item score patterns may or may not be classified as aberrant depending on the trait level. The second problem is that in order to classify an item score pattern as aberrant, a distribution of the statistic under the null hypothesis of fitting response behavior, the null distribution, is needed. Using the null distribution, the probability of exceedance or significance probability can be determined. Because large negative values of l indicate aberrance, the significance probabilities in the left tail of the distribution are of interest.

As a solution for these two problems Drasgow, Levine, and Williams (1985) proposed a standardized version of l, l_z, that was less confounded with the trait level,

$$l_z = \frac{l - E(l)}{[Var(l)]^{\frac{1}{2}}},$$

where $E(l)$ and $Var(l)$ denote the expectation and variance of l, respectively. These quantities are given by

$$E(l) = \sum_{i=1}^{n} \{P_i(\theta) \ln P_i(\theta) + [1 - P_i(\theta)] \ln[1 - P_i(\theta)]\},$$

and

$$Var\left(l\right) = \sum_{i=1}^{n} P_i\left(\theta\right)\left[1 - P_i\left(\theta\right)\right]\left[\ln \frac{P_i(\theta)}{1-P_i(\theta)}\right]^2.$$

For long tests, the null distribution was assumed to be the standard normal distribution, due to the central limit theorem. Molenaar and Hoijtink (1990) argued that l_z is only standard normally distributed for long tests when true θ is used, but in practice θ is replaced by its estimate $\hat{\theta}$, which affects the distributional characteristics of l_z. This was shown in Molenaar and Hoijtink (1990) and also in van Krimpen-Stoop and Meijer (1999). In both studies it was found that when maximum likelihood estimation was used to estimate θ, the variance of l_z was smaller than expected under the standard normal distribution. In particular for short tests and tests of moderate length (tests of 50 items or less), the variance was found to be seriously reduced. As a result, the statistic was very conservative in classifying score patterns as aberrant. In the context of the Rasch (1960) model, Molenaar and Hoijtink (1990) proposed three approximations to the distribution of l conditional on the total scores: using (1) complete enumeration, (2) Monte Carlo simulation, or (3) a chi-square distribution, where the mean, standard deviation, and skewness of l were taken into account. These approximations were all conditional on the total score which in the Rasch model is a sufficient statistic for θ. Recently, Snijders (2000) derived the asymptotic null distribution for a family of person-fit statistics that are linear in the item responses and in which θ is replaced by $\hat{\theta}$.

Trabin and Weiss (1983) proposed to use the person response function (PRF) as a measure of person-fit. At a fixed θ value a PRF gives the probability of a correct response as a function of b. In IRT, the item response function is assumed to be a nondecreasing function of θ, whereas the PRF is assumed to be nonincreasing in b. The fit of an examinee's item score pattern can be investigated by comparing the observed and expected PRF. Large differences between observed and expected PRFs may indicate nonfitting response behavior. To construct an observed PRF, Trabin and Weiss (1983) ordered items by increasing estimated item difficulties $\hat{b}$ and then formed subtests of items by grouping items according to $\hat{b}$ values. For fixed $\hat{\theta}$, the observed PRF was constructed by determining, in each subtest, the mean probability of a correct response. The expected PRF was constructed by estimating according to the 3PL model, in each subtest, the mean probability of a correct response. A large difference between the expected and observed PRF was interpreted as an indication of nonfitting item scores for that examinee.

Let n items be ordered by their b values and let item rank numbers be assigned accordingly, such that

$$b_1 < b_2 < ... < b_k$$

Furthermore, let $\hat{\theta}$ be the maximum likelihood estimate of θ under the 3PL model. Assume that A_q ($q = 1, ..., Q$) ordered classes can be formed, each containing m items; thus, $A_1 = \{1, ..., m\}, A_2 = \{m+1, ..., 2m\}, ..., A_Q = \{k-m+1, ..., k\}$. To construct the expected PRF, an estimate of the expected proportion of correct responses on the basis of the 3PL model in each subset

$$m^{-1} \sum_{i \in A_q} P_i(\hat{\theta}), \text{ for } q = 1, 2, ..., Q.$$

is taken. This expected proportion is compared with the observed proportion of correct responses, given by

$$m^{-1} \sum_{i \in A_q} U_i, \text{ for } q = 1, 2, ..., Q.$$

Thus, within each subtest, for a particular $\hat{\theta}$ the difference of observed and expected correct scores is taken and this difference is divided by the number of items in the subtest. This yields

$$D_q(\hat{\theta}) = m^{-1} \sum_{i \in A_q} \left[U_i - P_i(\hat{\theta}) \right], \text{ for all } q = 1, ..., Q.$$

Next, the Ds are summed across subtests, which yields

$$D(\hat{\theta}) = \sum_{q=1}^{Q} D_q(\hat{\theta}).$$

$D(\hat{\theta})$ was taken as a measure of the fit of an examinee's score pattern. For example, if an examinee copied the answers on the most difficult items, for that examinee the scores on the most difficult subtests are likely to be substantially higher than predicted by the expected PRF. Plotting the expected and the observed PRF may reveal such differences. Nering and Meijer (1998) compared the PRF approach with the l_z statistic using simulated data and found that the detection rate of l_z in most cases was higher than using the PRF. They suggested that the PRF approach and l_z can be used in a complementary way: aberrant item score patterns can be detected using l_z, and differences between expected and observed PRFs can be used to retrieve more information at the subtest level. Klauer and Rettig (1990) statistically refined the methodology of Trabin and Weiss (1983).

Statistics like V or l_z can only be used to investigate the probability of an item score pattern under the null hypothesis of normal response behavior. An alternative is to test this null hypothesis against an a priori specified alternative model of aberrant response behavior. Levine and Drasgow (1988) proposed a method for the identification of aberrant item score patterns which is statistical optimal; that is, no other method can achieve a higher rate of detection at the same Type I error rate. They calculated a likelihood-ratio statistic that provides the most powerful test for the null hypothesis that an item score pattern is normal versus the alternative hypothesis that it is aberrant. In this test, the researcher has to specify a model for normal behavior and for a particular type of aberrant behavior in advance. Klauer (1991, 1995) followed the same strategy and used uniformly most powerful tests in the context of the Rasch model to test against person-specific item discrimination as well as violations of local independence and unidimensionality.

For a review of person-fit statistics, see Meijer and Sijtsma (1995).

2.2. PERSON FIT IN COMPUTERIZED ADAPTIVE TESTING

To investigate the fit of an item score pattern in a CAT one obvious option is to use one of the statistics proposed for P&P tests. However, person-fit research conducted in this context has shown that this is not straightforward. Nering (1997) evaluated the first four moments of the distribution of l_z for a CAT. His results were in concordance with the results using P&P tests: the variance and the mean were smaller than expected and the null distributions were negatively skewed. As a result the normal approximation in the tails of the null distribution was inaccurate. Van Krimpen-Stoop and Meijer (1999) simulated the distributions of l_z and l_z^*, an adapted version of l_z in which the variance was corrected for $\hat{\theta}$ according to the theory presented in Snijders (2000). They simulated item scores with a fixed set of administered items and item scores generated according to a stochastic design, where the choice of the administered items depended on the responses to the previous items administered. Results indicated that the distribution of l_z and l_z^* differed substantially from the standard normal distribution although the item characteristics and the test length determined the magnitude of the difference. Glas, Meijer, and van Krimpen-Stoop (1998) adapted the person-fit statistics discussed by Klauer (1995) to the 2PL model and investigated the detection rate of these statistics in a CAT. They found small detection rates for most simulated types of aberrant item score patterns: the rates varied between 0.01 and 0.24 at significance level $\alpha = 0.10$ (one-sided).

A possible explanation for these results is that the characteristics of CATs are generally unfavorable for the assessment of person fit using existing person-fit statistics. The first problem is that CATs contain relatively few items compared to P&P tests. Because the detection rate of a person-fit statistic is sensitive to test length, and longer tests result in higher detection rates (e.g., Meijer, Molenaar, & Sijtsma, 1994), the detection rate in a CAT will, in general, be lower than that of a P&P test. A second problem is that almost all person-fit statistics assume a spread in the item difficulties: generally speaking, aberrant response behavior consists of many 0 scores for easy items and many 1 scores for difficult items. In CAT, the spread in the item difficulties is relatively modest: in particular, towards the end of the test when $\hat{\theta}$ is close to θ, items with similar item difficulties will be selected and as a result it is difficult to distinguish normal from aberrant item scores.

An alternative to the use of P&P person-fit statistics is to use statistics that are especially designed for CATs. Few examples of research using such statistics exist. McLeod and Lewis (1998) discussed a Bayesian approach for the detection of a specific kind of aberrant response behavior, namely for examinees with preknowledge of the items. They proposed to use a posterior log-odds ratio as a method for detecting item preknowledge in CAT. However, the effectiveness of this ratio to detect nonfitting item score patterns was unclear because of the absence of the null distribution and thus the absence of critical values for the log-odds ratio.

Below we will explore the usefulness of new methods to investigate person-fit in CATs. We will propose new statistics that can be used in CATs and that are based on theory from Statistical Process Control (see Veerkamp, 1997, chap. 5, for an application of Statistical Process Control in the context of item exposure). Also, the results of simulation studies investigating the critical values and detection rates of the statistics will be presented.

3. New Approaches

3.1. STATISTICAL PROCESS CONTROL

In this section, theory from Statistical Process Control (SPC) is introduced. SPC is often used to control production processes. Consider, for example, the process of producing tea bags, where each tea bag should have a certain weight. Too much tea in each bag is undesirable because of the increasing costs of material (tea) for the company. On the other hand, the customers will complain when there is too little tea in the bags. Therefore, the weight of the tea bags needs to be controlled during the production process. This can be done using techniques from SPC.

A (production) process is in a state of statistical control if the variable being measured has a stable distribution. A technique from SPC that is effective in detecting small shifts in the mean of the variable being measured, is the cumulative sum (CUSUM) procedure, originally proposed by Page (1954). In a CUSUM procedure, sums of statistics are accumulated, but only if they exceed 'the goal value' by more than d units. Let Z_t be the value of a standard normally distributed statistic Z (e.g., the standardized mean weight of tea bags) obtained from a sample of size N at time point t. Furthermore, let d be the reference value. Then, a two-sided CUSUM procedure can be written in terms of C_t^+ and C_t^-, where

$$
\begin{aligned}
C_1^+ &= \max\left[0, Z_1 - d\right] \\
C_2^+ &= \max\left[0, (Z_1 - d) + (Z_2 - d)\right] \\
&= \max\left[0, (Z_2 - d) + C_1^+\right] \\
C_3^+ &= \max\left[0, (Z_3 - d) + C_2^+\right] \\
&\cdots \\
C_t^+ &= \max\left[0, (Z_t - d) + C_{t-1}^+\right],
\end{aligned}
$$

and, analogously,

$$
C_t^- = \min\left[0, (Z_t + d) + C_{t-1}^-\right],
$$

with starting values $C_0^+ = C_0^- = 0$. Note that the sums are accumulating on both sides concurrently. The sum of consecutive positive values of $Z_t - d$ is reflected by C_t^+ and the sum of consecutive negative values of $Z_t + d$ by C_t^-. Thus, as soon as $|Z_t| > d$, Z_t values are taken into account. Let h be some threshold. The process is 'out-of-control' when $C^+ > h$ or $C^- < -h$ and 'in-control' otherwise.

One assumption underlying the CUSUM procedure is that the Z_t-values are asymptotically standard normally distributed; the values of d and h are based on this assumption. The value of d is usually selected as one-half of the mean shift (in Z_t-units) one wishes to detect; for example, $d = 0.5$ is the appropriate choice for detecting a shift of one times the standard deviation of Z_t. In practice, CUSUM-charts with $d = 0.5$ and $h = 4$ or $h = 5$ are often used (for a reference of the underlying rationale of this choice, see Montgomery, 1997, p.322). Setting these values for d and h results in a significance level of approximately

$\alpha = 0.0027$ (two-sided). Note that in person-fit research α is fixed and critical values are derived from the null distribution of the statistic. In this study, we will also use a fixed α and will derive critical values from simulations.

3.2. CAT AND STATISTICAL PROCESS CONTROL

CUSUM procedures are sensitive to strings of positive and negative values of a statistic. Person-fit statistics are often defined in terms of the difference between observed and expected scores (Equation 3). A commonly used statistic is V, the mean of the squared standardized residuals based on n items (Equation 4). One of the drawbacks of V is that negative and positive residuals can not be distinguished. The distinction is of interest in a CAT because a string of negative or positive residuals may indicate aberrant behavior. For example, suppose an examinee with an average θ-value responds to a test and the examinee has preknowledge of the items in the last part of the test. As a result, in the first part of the test the responses will alternate between zero and one, whereas in the second part of the test more and more items will be correctly answered due to item preknowledge; thus, in the second part of the test, consecutive positive differences will tend to occur.

Sums of consecutive negative or positive residuals can be investigated using a CUSUM procedure. This can be explained as follows. A CAT can be viewed as a multistage test, where each item is a stage and each stage can be seen as a point in time; at each stage a response to one item is given. Let i_k denote the kth item in the CAT; that is, k is the stage of the CAT. Further, let the statistic T_k be a function of the residuals at stage k, n the final test length, and let, without loss of generality, the reference value d be equal to 0. Below, some examples of the statistic T are proposed. For each examinee, at each stage k of a CAT, the CUSUM procedure can be determined as

$$C_k^+ = \max\left[0, T_k + C_{k-1}^+\right], \tag{5}$$
$$C_k^- = \min\left[0, T_k + C_{k-1}^-\right], \text{ and} \tag{6}$$
$$C_0^+ = C_0^- = 0, \tag{7}$$

where C^+ and C^- reflect the sum of consecutive positive and negative residuals, respectively. Let UB and LB be some appropriate upper and lower bound, respectively. Then, when $C^+ > UB$ or $C^- < LB$ the item score pattern can be classified as not fitting the model, otherwise, the item score pattern can be classified as fitting the model.

3.2.1. *Some Person-Fit Statistics*

Let S_k denote the set of items administered as the first k items in the CAT and $R_k = \{1, ..., I\} \setminus S_{k-1}$ the set of remaining items in the pool; from R_k the kth item in the CAT is administered. A principle of CAT is that θ is estimated at each stage k based on the responses to the previous administered items; that is, the items in set S_{k-1}. Let $\hat{\theta}_{k-1}$ denote the estimated θ at stage $k - 1$. Thus, based on $\hat{\theta}_{k-1}$, the item for the next stage, k, is selected from R_k. The probability of correctly answering item i_k, according to the 2PL model, evaluated at $\hat{\theta}_{k-1}$ can be written as

$$P_{i_k}\left(\hat{\theta}_{k-1}\right) = \frac{\exp\left[a_{i_k}\left(\hat{\theta}_{k-1} - b_{i_k}\right)\right]}{1 + \exp\left[a_{i_k}\left(\hat{\theta}_{k-1} - b_{i_k}\right)\right]}. \tag{8}$$

Two sets of four statistics, all corrected for test length and based on the difference between observed and expected item scores, are proposed. The first four statistics, T^1 through T^4, are proposed to investigate the sum of consecutive positive or negative residuals in an on-line situation, when the test length of the CAT is fixed. These four statistics use as the expected score the probability of correctly answering the item, evaluated at the updated ability estimate, $\hat{\theta}_{k-1}$, defined in Equation 8. The other four statistics, T^5 through T^8, use as the expected score the probability of correctly answering the item, evaluated at the final ability estimate $\hat{\theta}_n$. As a result of using $\hat{\theta}_n$ instead of $\hat{\theta}_k$, the development of the accumulated residuals can not be investigated during the test.

All of these statistics are based on the general form defined in Equation 3: a particular statistic is defined by choosing a particular weight.

Statistics T_k^1 and T_k^5 are defined by choosing $w_i(\theta) = 1$. Furthermore, in line with the literature on person fit in P&P tests, statistics T_k^2 and T_k^6 are defined by weighting the residual $U - P(\cdot)$ by the estimated standard deviation, $[P(\cdot)(1 - P(\cdot))]^{-\frac{1}{2}}$. In order to construct person-fit statistics that are sensitive to nonfitting item scores in the first part of a CAT, in T_k^3 and T_k^7 the residual is weighted by the reciprocal of the square root of the test information function containing the items administered up to and including stage k. This weight is a monotonically decreasing function of the stage of the CAT and, as a result, the residuals in the beginning of the test receive a larger weight than the residuals in the last part of the CAT. Finally, T_k^4 and T_k^8 are proposed to detect nonfitting item scores at the end of a CAT. Here, the residuals are multiplied by the square root of the stage of the CAT, $\sqrt{k}$. Because $\sqrt{k}$ is a monotonically increasing function of the test length, the residuals at the beginning of the CAT are also weighted less than the residuals at the end of the CAT.

Define

$$T_k^1 = \frac{1}{n} \left\{ U_{i_k} - P_{i_k} \left(\hat{\theta}_{k-1} \right) \right\},$$

$$T_k^2 = T_k^1 \times \left\{ P_{i_k} \left(\hat{\theta}_{k-1} \right) \left[1 - P_{i_k} \left(\hat{\theta}_{k-1} \right) \right] \right\}^{-\frac{1}{2}},$$

$$T_k^3 = T_k^1 \times \left\{ I \left(\hat{\theta}_{k-1} \right) \right\}^{-\frac{1}{2}}, \text{ and}$$

$$T_k^4 = \sqrt{k} \times T_k^1,$$

where $I \left(\hat{\theta}_{k-1} \right)$ is the test information function according to the 2PLM, of a test containing the items administered up to and including stage $k - 1$, evaluated at $\hat{\theta}_{k-1}$. That is,

$$I \left(\hat{\theta}_k \right) = \sum_{g=1}^{k-1} I_{i_g} \left(\hat{\theta}_{k-1}, a_{i_g}, b_{i_g} \right) = \sum_{g=1}^{k-1} a_{i_g}^2 P_{i_g} \left(\hat{\theta}_{k-1} \right) \left[1 - P_{i_g} \left(\hat{\theta}_{k-1} \right) \right].$$

Thus, T_k^1 is the residual between the observed response and the probability of a correct response to item i_k, where the probability is evaluated at the estimated ability at the previous stage. T_k^2, T_k^3, and T_k^4 are functions of these residuals.

Define

$$T_k^5 = \frac{1}{n} \left\{ U_{i_k} - P_{i_k} \left(\hat{\theta}_n \right) \right\},$$

$$T_k^6 = T_k^5 \times \left\{ P_{i_k} \left(\hat{\theta}_n \right) \left[1 - P_{i_k} \left(\hat{\theta}_n \right) \right] \right\}^{-\frac{1}{2}},$$

$$T_k^7 = T_k^5 \times \left\{ I \left(\hat{\theta}_n \right) \right\}^{-\frac{1}{2}}, \text{ and}$$

$$T_k^8 = \sqrt{k} \times T_k^5,$$

where $I \left(\hat{\theta}_n \right)$ is the test information function of a test containing the items administered up to and including stage i, evaluated at the final estimated ability, $\hat{\theta}_n$. Thus,

$$I \left(\hat{\theta}_n \right) = \sum_{g=1}^{k} I_{i_g} \left(\hat{\theta}_n, a_{i_g}, b_{i_g} \right) = \sum_{g=1}^{k} a_{i_g}^2 P_{i_g} \left(\hat{\theta}_n \right) \left[1 - P_{i_g} \left(\hat{\theta}_n \right) \right].$$

The statistics T^5 through T^8 are proposed to investigate the sum of consecutive negative or positive residuals, evaluated at the final estimate $\hat{\theta}_n$. Again, due to the use of $\hat{\theta}_n$ instead of $\hat{\theta}_{k-1}$, the accumulated residuals can no longer be investigated during the test.

These eight statistics can be used in the CUSUM procedure described in Equations 5 through 7. To determine upper and lower bounds in a CUSUM procedure it is assumed that the statistic computed at each stage is asymptotically standard normally distributed. However, the null distributions of T^1 through T^8 are far from standard normal. T^1 and T^5 follow a binomial distribution with only one observation, while the other statistics are weighted versions of T^1 and T^5, also based only on one observation. As a result, setting $d = 0.5$ and the upper and lower bound to 5 and -5, respectively, might not be appropriate in this context. Therefore, in this study, the numerical values of the upper and lower bound are investigated through simulation, with $\alpha = 0.05$ and $d = 0$.

4. A Simulation Study

As already noted, a drawback of the CUSUM procedure is the absence of guidelines for determining the upper and lower bounds for statistics that are not normally distributed. Therefore, in Study 1, a simulation study was conducted to investigate the numerical values of the upper and lower thresholds of the CUSUM procedures using statistics T^1 through T^8 across θ-levels. When these bounds are independent of θ, a fixed upper and lower bound for each statistic can be used. In Study 2, the detection rate of the CUSUM procedures with the statistics T^1 through T^8 for several types of nonfitting response behavior were investigated.

In these two simulation studies we used the 2PL model because it is less restrictive with respect to empirical data than the one-parameter logistic model and does not have the estimation problems involved in the guessing parameter in the 3PL model (e.g., Baker, 1992, pp.109-112). Furthermore, true item parameters were used. This is realistic when item parameters are estimated using large samples: Molenaar and Hoijtink (1990) found no serious differences between true and estimated item parameters for samples consisting of $1,000$ examinees or more.

4.1. Study 1

4.1.1. *Method*
Five data sets consisting of 10,000 fitting adaptive item score vectors each were constructed at five different θ-levels; $\theta = -2, -1, 0, 1,$ and 2. An item pool of 400 items fitting the 2PLM with $a_i \sim N(1; 0.2)$ and $b_i \sim U(-3; 3)$ was used to generate the adaptive item score patterns.

Fitting item score vectors were simulated as follows. First, the true θ of a simulee was set to a fixed θ-level. Then, the first CAT item selected was the item with maximum information given $\theta = 0$. For this item, $P_i(\theta)$ was determined according to Equation 1. To simulate the answer (1 or 0), a random number y from the uniform distribution on the interval $[0, 1]$ was drawn; when $y < P_i(\theta)$ the response to item i was set to 1 (correct response), 0 otherwise. The first four items of the CAT were selected with maximum information for $\theta = 0$, and based on the responses to these four items, $\widehat{\theta}$ was obtained using weighted maximum likelihood estimation (Warm, 1989). The next item selected was the item with maximum information given $\widehat{\theta}$ at that stage. For this item, $P_i(\theta)$ was computed, a response was simulated, θ was estimated and another item was selected based on maximum information given $\widehat{\theta}$ at that stage. This procedure was repeated until the test attained the length of 30 items.

For each simulee, eight different statistics, T^1 through T^8, were used in the CUSUM procedure described in Equations 5 through 7. For each simulee and for each statistic,

$$\max C^+ = \max_k \left(C_k^+ \right) \text{ and}$$
$$\min C^- = \min_k \left(C_k^- \right)$$

were determined, resulting in 10,000 values of $\max C^+$ and $\min C^-$ for each data set and for each statistic. Then, for each data set and each statistic, the upper bound, UB, was determined as the value of $\max C^+$ for which 2.5% of the simulees had higher $\max C^+$-values and the lower bound, LB, was determined as the value of $\min C^-$ for which 2.5% of the simulees had lower $\min C^-$-values. That is, a two-sided test at $\alpha \leq 0.05$ was conducted, where $P(\max C^+ \geq UB) = P(\min C^- \leq LB) = 0.025$. So, for each data set and for each statistic two bounds (the upper and lower bounds) were determined.

When the bounds are stable across θ, it is possible to use one fixed upper and one fixed lower bound for each statistic. Therefore, to construct fixed bounds, the weighted averages of the upper and lower bound were calculated across θ for each statistic, with different weights for different θ-values; weights 0.05, 0.2, and 0.5, for $\theta = \pm 2, \pm 1$, and 0, respectively were used. These weights represent a realistic population distribution of abilities.

4.1.2. *Results*

In Table 1, the upper and lower bounds for $\alpha \leq 0.05$ (two-sided) for the statistics T^1 through T^8 are tabulated at five different θ-levels. Table 1 shows that, for most statistics the upper and lower bounds were stable across θ-levels. For statistic T^7, the bounds were less stable across θ-values. Table 1 also shows that, for most statistics except T^4 and T^8,

Table 1. Bounds of CUSUM using statistics T^1 through T^8

θ	Weights	T^1 LB	T^1 UB	T^2 LB	T^2 UB	T^3 LB	T^3 UB	T^4 LB	T^4 UB
-2	.05	$-.23$	.19	$-.47$	.40	$-.12$	.09	$-.13$	1.81
-1	.20	$-.20$	.19	$-.42$	.40	$-.08$	.07	$-.13$	1.83
0	.50	$-.20$	.20	$-.41$	.42	$-.07$	.07	$-.13$	1.86
1	.20	$-.20$	.20	$-.41$	.43	$-.07$	.09	$-.13$	1.86
2	.05	$-.18$	.23	$-.41$	.47	$-.09$	.11	$-.13$	2.02
	Average	$-.20$	.20	$-.41$	.42	$-.07$	.07	$-.13$	1.86

θ	Weights	T^5 LB	T^5 UB	T^6 LB	T^6 UB	T^7 LB	T^7 UB	T^8 LB	T^8 UB
-2	.05	$-.13$	.13	$-.27$	.29	$-.11$	.30	$-.10$	1.72
-1	.20	$-.13$	.13	$-.28$	.28	$-.07$	.13	$-.10$	1.73
0	.50	$-.13$	.14	$-.29$	.29	$-.07$	.06	$-.11$	1.76
1	.20	$-.13$	.13	$-.28$	.28	$-.12$	.07	$-.10$	1.73
2	.05	$-.13$	.13	$-.29$	.27	$-.28$	.11	$-.10$	1.85
	Average	$-.13$	.13	$-.28$	.28	$-.09$	.09	$-.10$	1.75

the weighted average bounds were approximately symmetrical around 0.

As a result of the stable bounds for almost all statistics across θ, one fixed upper and one fixed lower bound can be used as the bounds for the CUSUM procedures.

4.2. STUDY 2

4.2.1. *Method*
To examine the detection rates for the eight proposed statistics, different types of nonfitting item score patterns were simulated. Nonfitting item scores were simulated for all items, or for the first or second part of the item score pattern. Nine data sets containing 1,000 nonfitting adaptive item score patterns were constructed; an item pool of 400 items with $a_i \sim N(1.0; 0.2)$ and $b_i \sim U(-3; 3)$ was used to construct a test of 30 items. The detection rate was defined as the proportion of a priori defined nonfitting item score patterns that were classified as nonfitting. For each item score vector, CUSUM procedures using T^1 through T^8 were used. An item score vector was classified as nonfitting when

$$\max C^+ (T^q) > UB(T^q) \text{ or}$$

$$\min C^- (T^q) \quad < \quad LB (T^q)$$

for $q = 1, ...8$. The upper and lower bounds for each statistic were set equal to the weighted average of the values presented in Table 1. To facilitate comparisons, a data set containing 1,000 adaptive item score patterns fitting the 2PLM was simulated and the percentage of fitting item score vectors classified as nonfitting was determined.

4.2.2. Types of Aberrant Response Behavior

4.2.2.1. Random Response Behavior

To investigate the detection rate of nonfitting item scores to all items of the test, random response behavior to all items was simulated. This type of response behavior may be the result of guessing the answers to the items of a test and was empirically studied by van den Brink (1977). He described examinees who took a multiple-choice test only to familiarize themselves with the questions that would be asked. Because returning an almost completely blank answering sheet may focus a teacher's attention on the ignorance of the examinee, each examinee was found to randomly guess the answers on almost all items of the test. In the present study, "guessing" simulees were assumed to answer the items by randomly guessing the correct answers on each of the 30 items in the test with a probability of 0.2. This probability corresponds to the probability of obtaining the correct answer by guessing in a multiple-choice test with five alternatives per item.

4.2.2.2. Non-invariant Ability

To investigate nonfitting response behavior in the first or in the second half of the CAT, item score vectors were simulated with a two-dimensional θ (Klauer, 1991). It was assumed that during the first half of the test an examinee had another θ-value than during the second half. Carelessness, fumbling, or preknowledge of some items can be the cause of non-invariant abilities. Two data sets containing item score vectors with a two-dimensional θ were simulated by drawing the first θ_1 from the standard normal distribution. The second ability, θ_2, was set to $\theta_2 = \theta_1 + r$ with $r = -2, -1, 1$, and 2. Thus, during the first half of the test $P(\theta_1)$ was used and during the second half $P(\theta_2)$ was used to simulate the responses to the items.

Examples of higher ability ($r > 0$) in the second half of the CAT are item preknowledge of items in the second half of the CAT. Examples of a lower ability ($r < 0$) in the second half of the CAT are preknowledge of the items in the first half of the test, or carelessness in the second part of the CAT.

4.2.2.3. *Violations of Local Independence.* To examine aberrance in the second part of the CAT, item score vectors with violations of local independence between all items in the test were simulated. When previous items provide new insights useful for answering the next item or when the process of answering the items is exhausting, the assumption of local independence between the items may be violated. A generalization of a model proposed by Jannarone (1986) (see Glas, Meijer, and van Krimpen, 1998) was used to simulate response vectors with local independence between all subsequent items

$$P\left(U_i = u_i, U_{i+1} = u_{i+1} \mid \theta\right) \propto \exp\left[\sum_{j=i}^{i+1} u_j a_j \left(\theta - b_j\right) + u_i u_{i+1} \delta_{i,i+1}\right],$$

where $\delta_{i,i+1}$ was a parameter modeling association between items (see Glas et al., 1998, for more details). Using this model, the probability of correctly answering an item was determined by the item parameters a and b, the person parameter θ and the association parameter δ. If $\delta = 0$ the model equals the 2PL model. Compared to the 2PL model, positive values of δ result in a higher probability of a correct response (e.g., learning-effect), and negative values of δ result in a lower probability of correctly answering an item (e.g., carelessness). The values $\delta = -2$, -1, 1, and 2 were used to simulate nonfitting item score patterns.

4.2.3. *Results*

In Table 2, the detection rates for the eight different CUSUM procedures are given for several types of nonfitting item score patterns and for model-fitting item score patterns. For most statistics the percentage of fitting item score patterns classified as nonfitting was around the expected 0.05. For T^3, however, this percentage was 0.13. The detection rates for guessing simulees were high for most statistics, except for T^5 and T^8. For example, the detection rate was 0.19 for T^5, whereas for T^7 it was 0.97.

For non-invariant abilities and all values of r, and for most statistics (except T^3), the detection rates were higher when $|r|$ was larger. However, the detection rates were different across statistics: statistic T^3 had the lowest detection rates for all values of r, whereas the detection rates for statistics T^4, T^7, and T^8 were all higher than statistic T^3 for each r. The highest (and similar) detection rates were observed for statistics T^1, T^2, T^5, and T^6. For example, for $r = 2$ the detection rates for T^1, T^2, T^5, and T^6 were 0.76, 0.79, 0.78, and 0.78, respectively.

Table 2 also shows that, for violations of local independence, the highest detection rates were found for $\delta = 2$. For $\delta = 1$ the detection rates were approximately 0.10 for all statistics, whereas for $\delta = 2$ the

Table 2. Detection rates of CUSUM procedures

	T^1	T^2	T^3	T^4	T^5	T^6	T^7	T^8
Normal	.05	.06	.13	.04	.04	.04	.07	.04
Guessing	.66	.72	.89	.59	.19	.59	.97	.21
$r =$ -2	.71	.76	.12	.53	.72	.74	.37	.49
-1	.23	.28	.14	.14	.21	.23	.18	.15
1	.25	.28	.09	.29	.22	.24	.17	.25
2	.76	.79	.12	.80	.75	.78	.44	.77
$\delta =$ -2	.03	.05	.11	.01	.00	.01	.12	.00
-1	.03	.04	.11	.01	.01	.01	.10	.01
1	.10	.13	.19	.11	.11	.12	.11	.12
2	.22	.27	.33	.28	.29	.32	.18	.34

detection rates were approximately 0.30 for most statistics, except for T^1 (0.22) and T^7 (0.18).

5. Discussion

The results of Study 1 showed that the bounds of the CUSUM procedures were stable across θ-values for all statistics except T^7. As a result, for most statistics, one fixed UB and one fixed LB can be used as thresholds for the CUSUM procedures. In Study 2 these fixed bounds were used to determine the detection rates for the eight CUSUM procedures. The results showed that for statistic T^3 the percentage of fitting item score patterns classified as nonfitting was larger than the expected 0.05 (Type I error); that is, the CUSUM procedure using statistic T^3 may result in a liberal classification of nonfitting item score patterns.

In this paper, several statistics for detecting person misfit were proposed. It was expected that statistics T^3 and T^7 were more sensitive to nonfitting item scores in the beginning of the test and that statistics T^4 and T^8 were more sensitive to nonfitting item scores at the end of the test. The results showed that when $r > 0$ (indicating aberrant response behavior at the end of the test) the detection rates of T^4 and T^8 were larger than T^3 and T^7; however when $r < 0$ (aberrant response behavior in the beginning of the test) detection rates of T^4 and T^8 were equal or larger than T^3 and T^7. Moreover, the detection rates of T^4 and T^8 were comparable and sometimes smaller than the unweighted statistic T^1. Thus weighing the residual with $[I(.)]^{-\frac{1}{2}}$ does not seem to have the desired effect. For guessing, however, the detection rates of T^3 and T^7 were highest. On the basis of our results it may tentatively be concluded that some statistics are more sensitive to particular types of aberrant response behavior than others. Future research could build on these findings and investigate person fit with different types of statistics.

Because few other studies on person fit in CAT have been conducted, it is difficult to compare the detection rates found in this study with the detection rates from other studies. An alternative is to compare the results with those from studies using P&P data. For example, in the Zickar and Drasgow (1996) study, real data were used in which some examinees had been instructed to distort their own responses. In their study, detection rates between 0.01 and 0.32 were found for P&P data. In the present study it was shown that such detection rates can also be found in CAT. It was shown that most statistics were sensitive to random response behavior to all items in the test and that most statistics were sensitive to response behavior simulated using a two-dimensional θ. None of the proposed statistics was shown to be sensitive to local dependence between all items in the test when the probability of a correct response decreased during the test (negative values of δ). However, most statistics were shown to be sensitive to strong violations of local independence when the probability of a correct response increased during the test ($\delta = 2$).

References

Baker, F. B. (1992). *Item response theory: Parameter estimation techniques*. New York: Dekker.

Drasgow, F., Levine, M. V., & Williams, E. A. (1985). Appropriateness measurement with polychotomous item response models and standardized indices. *British Journal of Mathematical and Statistical Psychology 38*, 67-86.

Glas, C. A. W., Meijer, R. R., & van Krimpen-Stoop, E. M. L. A. (1998). *Statistical tests for person misfit in computerized adaptive testing* (Reseach Report RR 98-01). Enschede, The Netherlands: University of Twente.

Hambleton, R. K., & Swaminathan, H. (1985). *Item response theory: Principles and applications*. Boston: Kluwer-Nijhoff.

Jannarone, R. J. (1986). Conjunctive item response theory kernels. *Psychometrika 51*, 357-373.

Klauer, K. C. (1991). An exact and optimal standardized person test for assessing consistency with the Rasch model. *Psychometrika 56*, 213-228.

Klauer, K. C.(1995). The assessment of person fit. In G. F. Fischer & I. W. Molenaar (Eds.): *Rasch models: Foundations, recent developments, and applications* (pp. 97-110). New York: Springer-Verlag.

Klauer, K. C. & Rettig, K. (1990). An approximately standardized person test for assessing consistency with a latent trait model. *British Journal of Mathematical and Statistical Psychology, 43*, 193-206

Levine, M. V., & Drasgow, F. (1988). Optimal appropriateness measurement. *Psychometrika 53*, 161-176.

Levine, M. V., & Rubin, D. B. (1979). Measuring the appropriateness of multiple-choice test scores. *Journal of Educational Statistics 4*, 269-290.

McLeod, L. D., & Lewis, C. (1998, April). *A Bayesian approach to detection of item preknowledge in a CAT*. San Diego, CA: Paper presented at the Annual Meeting of the National Council of Measurement in Education.

Meijer, R. R., Molenaar, I. W., & Sijtsma, K. (1994). Item, test, person and group characteristics and their influence on nonparametric appropriateness measurement. *Applied Psychological Measurement 18*, 111-120.

Meijer, R. R., & Sijtsma, K. (1995). Detection of aberrant item score patterns: a review of recent developments. *Applied Measurement in Education 8*, 261-272.

Molenaar, I. W., & Hoijtink, H. (1990). The many null distributions of person fit indices. *Psychometrika 55*, 75-106.

Montgomery, D. C.(1997). *Introduction to statistical quality control (3rd ed.)*. New York: John Wiley and Sons.

Nering, M. L. (1997). The distribution of indexes of person fit within the computerized adaptive testing environment. *Applied Psychological Measurement 21*, 115-127.

Nering, M. L., & Meijer, R. R. (1998). A comparison of the person response function and the lz statistic to person-fit measurement. *Applied Psychological Measurement 22*, 53-69.

Page, E. S. (1954). Continuous inspection schemes. *Biometrika 41*, 100-115.

Rasch, G. (1960). *Probabilistic models for some intelligent and attainment tests*. Copenhagen: Nielsen and Lydiche.

Snijders, T. A. B. (2000). Asymptotic null distribution of person fit statistics with estimated person parameter. *Psychometrika,* in press.

Trabin, T. E., & Weiss, D. J. (1983). The person response curve: Fit of individuals to item response theory models. In D. J. Weiss (Ed.): *New horizons in testing: Latent trait test theory and computerized adaptive testing*. New York: Academic Press.

van den Brink, W. P. (1977). Het verken-effect [The scouting effect]. *Tijdschrift voor Onderwijsresearch 2*, 253-261.

van Krimpen-Stoop, E. M. L. A., & Meijer, R. R. (1999). Simulating the null distribution of person-fit statistics for conventional and adaptive tests. *Applied Pychological Measurement, 23*, 327-345.

Veerkamp, W. J. J.(1997). Statistical methods for computerized adaptive testing. Ph.D. thesis, University of Twente, Enschede, The Netherlands.

Warm, T. A. (1989). Weighted likelihood estimation of ability in item response theory. *Psychometrika 54*.

Wright, B. D., & Stone, M. H. (1979). *Best test design: Rasch measurement*. Chicago: Mesa Press.

Zickar, M. J., & Drasgow, F. (1996). Detecting faking on a personality instrument using appropriateness measurement. *Applied Psychological Measurement 20*, 71-87.

Chapter 12

The Assessment of Differential Item Functioning in Comput Adaptive Tests

Rebecca Zwick*

University of California at Santa Barbara, USA

1. Introduction

Differential item functioning (DIF) refers to a difference in item performance between equally proficient members of two demographic groups. From an item response theory (IRT) perspective, DIF can be defined as a difference between groups in item response functions. The classic example of a DIF item is a mathematics question that contains sports jargon that is more likely to be understood by men than by women. An item of this kind would be expected to manifest DIF against women: They are less likely to give a correct response than men with equivalent math ability. In reality, the causes of DIF are often far more obscure. The recent book by Camilli and Shepard (1994) and the volume edited by Holland and Wainer (1993) provide an excellent background in the history, theory, and practice of DIF analysis.

There are several reasons that DIF detection may be more important for computerized adaptive tests (CATs) than it is for nonadaptive tests. Because fewer items are administered in a CAT, each item response plays a more important role in the examinees' test scores than it would in a nonadaptive testing format. Any flaw in an item, therefore, may be more consequential. Also, an item flaw can have major repercussions in a CAT because the sequence of items administered to the examinees depends in part on their responses to the flawed item. Finally, administration of a test by computer creates several potential sources of DIF which are not present in conventional tests, such as differential computer familiarity, facility, and anxiety, and differential preferences for computerized administration. Legg and Buhr (1992) and Schaeffer, Reese, Steffen, McKinley, and Mills (1993) both reported ethnic and gender group differences in some of these attributes. Powers and O'Neill (1993) reviewed the literature on this topic.

* Acknowledgments: the author is grateful to Daniel Eignor, Kathy O'Neill, Donald E. Powers, Louis Roussos, Dorothy T. Thayer, and Denny Way, who provided thoughtful comments on an earlier draft, and to Joyce Wang, who offered the conjecture about a possible source of DIF involving non-native speakers of English.

221

W.J. van der Linden and C.A.W. Glas (eds.),
Computerized Adaptive Testing: Theory and Practice, 221–244.
© 2000 Kluwer Academic Publishers. Printed in the Netherlands.

The investigation of DIF in CATs can be conducted using several different administration schemes: First, the items to be assessed for DIF can be administered adaptively. Second, the items to be assessed can be "seeded" throughout the exam and administered nonadaptively. Finally, the potential DIF items can be administered in an intact non-adaptive section. This chapter focuses on the first situation. DIF analysis for adaptively administered items involves two major technical challenges: First, an appropriate matching variable for DIF analysis must be determined. Clearly, number-right score, typically used in large-scale applications of DIF analyses by major testing companies, is not appropriate; on the other hand, matching on a scale score based on an IRT model is not entirely straightforward. Second, a method is needed which can provide stable results in small samples: Even if the total number of examinees for a given CAT is large, the number of responses for some items may be very small. This chapter presents the methods that have been developed to analyze DIF in CATs, along with results of applications to simulated data. In the final section, fruitful directions for future research are outlined.

2. Methods for Assessing DIF in CATS

Formal discussion of DIF procedures for CATs appears to have begun in the late eighties. Steinberg, Thissen, & Wainer (1990) recommended the application of a likelihood-ratio test approach that involves determining whether the fit of an IRT model to the data is impaired by constraining item parameters to be the same for two groups of examinees. While this approach has become a well-established DIF analysis method for non-adaptive tests (e.g., see Thissen, Steinberg, & Wainer, 1993), it does not appear to have been applied to CATs, possibly because of the complexities introduced by the incomplete data that result from CAT administration.

In another early proposal for assessing DIF in CATs, Holland suggested comparing examinee groups in terms of item percents correct, basing the analysis on only those test-takers who received the item late in their CATs (Holland & Zwick, 1991). However, analyses of simulated data (Zwick, Thayer, & Wingersky, 1993; 1994a) did not support the assumption underlying this procedure–that "examinees" who receive a particular item late in the CAT will be well-matched in ability. This key assumption would likely be violated even more severely in actual CATs, which involve many non-psychometric constraints on item selection.

In addition to recommending the IRT likelihood-ratio approach, Steinberg et al. (1990) suggested that DIF assessment procedures for

CATs might be developed by matching examinees on expected true score and then applying existing DIF methods. The CAT DIF methods of Zwick, Thayer, and Wingersky (ZTW; 1994a; 1994b; 1995) and the CAT version of the empirical Bayes DIF method of Zwick, Thayer, and Lewis (ZTL; 1997; 1999; 2000; Zwick & Thayer, in press) are consistent with this recommendation; these methods are discussed in the subsequent sections. The only other CAT DIF method that appears in the literature–the CATSIB procedure of Nandakumar and Roussos (in press; see also Roussos, 1996; Roussos & Nandakumar, 1998)–is also discussed below. In addition to the publications that propose specific methods, Miller (1992) and Way (1994) have addressed the general data analysis issues involved in conducting DIF analyses in CATs.

The ZTW methods are modifications of the Mantel-Haenszel (MH; 1959) DIF procedure of Holland and Thayer (1988) and of the standardization method of Dorans and Kulick (1986); the ZTL approach, originally developed for nonadaptive tests and later modified for CATs, is an enhancement of the MH method; and CATSIB is a modification of the SIBTEST procedure of Shealy and Stout (1993a; 1993b). The original "nonadaptive" versions of the Mantel-Haenszel, standardization, and SIBTEST methods are reviewed in the next section; the CAT analogues are then described.

2.1. A REVIEW OF THE MANTEL-HAENSZEL, STANDARDIZATION, AND SIBTEST PROCEDURES

In the MH procedure of Holland and Thayer (1988), which is widely used by testing companies for DIF screening, a $2 \times 2 \times K$ table of examinee data is constructed based on item performance (right or wrong), group membership (the *focal group*, which is of primary interest, or the *reference group*), and score on an overall proficiency measure (with K levels), used to match examinees. The two examinee groups are then compared in terms of their odds of answering the item correctly, conditional on the proficiency measure. The odds ratio is assumed to be constant over all levels of the proficiency measure.

Assume that there are T_k examinees at the kth level of the matching variable. Of these, n_{Rk} are in the reference group and n_{Fk} are in the focal group. Of the n_{Rk} reference group members, A_k answered the studied item correctly while B_k did not. Similarly C_k of the n_{Fk} matched focal group members answered the studied item correctly, whereas D_k did not. The MH measure of DIF can then be defined as

$$MH\ D\text{-}DIF = -2.35 \ln(\hat{\alpha}_{MH}) \tag{1}$$

where $\hat{\alpha}_{MH}$ is the Mantel-Haenszel (1959) conditional odds-ratio estimator given by

$$\hat{\alpha}_{MH} = \frac{\sum_k A_k D_k / T_k}{\sum_k B_k C_k / T_k}. \tag{2}$$

In (1), the transformation of $\hat{\alpha}_{MH}$ places $MH\ D\text{-}DIF$ (which stands for 'Mantel-Haenszel delta difference') on the ETS delta scale of item difficulty (Holland & Thayer, 1985). The effect of the minus sign is to make $MH\ D\text{-}DIF$ negative when the item is more difficult for members of the focal group than it is for comparable members of the reference group. Phillips and Holland (1987) derived an estimated standard error for $\ln(\hat{\alpha}_{MH})$; their result proved to be identical to that of Robins, Breslow and Greenland (1986).

The Mantel-Haenszel chi-square test provides an approximation to the uniformly most powerful unbiased test of the null hypothesis of no DIF (common odds ratio equal to one) versus the hypothesis of constant DIF (common odds ratio not equal to one). Rejection of the null hypothesis suggests that item performance and group membership are associated, conditional on the matching variable.

The results of a MH DIF analysis typically include $MH\ D\text{-}DIF$ (or some equivalent index based on the estimated odds ratio), along with its estimated standard error. In making decisions about whether to discard items or flag them for review, however, testing companies may rely instead on categorical ratings of the severity of DIF. Several testing companies have adopted a system developed by ETS for categorizing the severity of DIF based on both the magnitude of the DIF index and the statistical significance of the results (see Zieky, 1993). According to this classification scheme, a "C" categorization, which represents moderate to large DIF, requires that the absolute value of $MH\ D\text{-}DIF$ be at least 1.5 and be significantly greater than 1 (at $\alpha = .05$). A "B" categorization, which indicates slight to moderate DIF, requires that $MH\ D\text{-}DIF$ be significantly different from zero (at $\alpha = .05$) and that the absolute value of $MH\ D\text{-}DIF$ be at least 1, but not large enough to satisfy the requirements for a C item. Items that do not meet the requirements for either the B or the C categories are labeled "A" items, which are considered to have negligible DIF. Items that fall in the C category are subjected to further scrutiny and may be eliminated from tests. For most purposes, it is useful to distinguish between negative DIF (DIF against the focal group, by convention) and positive DIF (DIF against the reference group). This distinction yields a total of five DIF classifications: C-, B-, A, B+, and C+.

In the standardization DIF procedure (Dorans & Kulick, 1986), data are organized the same way as in MH DIF analysis. The standardization index, often called *STD P-DIF* (which stands for 'standardized proportion difference'), compares the item proportions correct for the reference and focal groups, after adjusting for differences in the distribution of members of the two groups across the levels of the matching variable.

More specifically,

$$STD \; P\text{-}DIF = \sum w_k \hat{p}_{Fk} - \sum w_k \hat{p}_{Rk} \qquad (3)$$

where w_k is a weight associated with the kth level of the matching variable. In typical applications of *STD P-DIF*, including those described below,

$$w_k = \frac{n_{Fk}}{n_F} \qquad (4)$$

where $n_F = \sum_k n_{Fk}$ is the total number of examinees in the focal group. Under this weighting scheme the term before the minus sign in (3) is simply the proportion of the focal group that answers the studied item correctly, and the term following the minus sign is an adjusted proportion correct for the reference group. Although a standard error formula for *STD P-DIF* was developed by Holland (see Dorans & Holland, 1993) and two alternative formulations were derived by Zwick (1992; see Zwick & Thayer, 1996), *STD P-DIF* is usually used as a descriptive measure and not as the basis for a formal hypothesis test.

The original versions of the MH and standardization DIF procedures involve matching examinees from two groups on the basis of observed test score–typically, the number correct. Under the classical test theory model, it can be shown that reference and focal group members who are matched in terms of observed scores will not, in general, be matched in terms of true score (see also Shealy & Stout, 1993a; 1993b; Zwick, 1990; Zwick, Thayer & Mazzeo, 1997). The measurement error problem vanishes under certain Rasch model conditions because of the sufficiency of number-correct score for θ, but, except in that special case, the severity of the problem increases as the difference between the reference and focal group ability distributions increases and as the test reliability decreases. To address this problem, the SIBTEST procedure developed by Shealy and Stout (1993a; 1993b) matches examinees on an estimated true score obtained by applying a "regression correction" to the observed score. The SIBTEST measure of DIF, $\hat{\beta}$, can be defined as follows:

$$\hat{\beta} = \sum_k w_k (\overline{Y}^*_{Fk} - \overline{Y}^*_{Rk}). \qquad (5)$$

where $\overline{Y}^*_{Fk}$ and $\overline{Y}^*_{Rk}$ are adjusted mean scores (described below) on the studied item for the focal and reference groups, respectively, and w_k is a weight. Although the weight can in principle be defined as in (4), Shealy and Stout recommend defining it as

$$w_k = (n_{Rk} + n_{Fk})/N, \tag{6}$$

where N is the total sample size. (Note that SIBTEST can be applied to a set of studied items simultaneously, in which case the $\overline{Y}^*_k$ values in (5) are the adjusted means for the set of items.)

The steps involved in obtaining the adjusted means, which are described in detail by Shealy and Stout (1993a), are as follows: (1) Assuming a classical test theory model for the regression of true score on observed score (in this case, the number-right score on the matching items, excluding the studied item), obtain the expected true score for each group at each of the K levels of the matching variable. For each group, the slope of this regression is the reliability of the set of matching items in that group (Shealy & Stout, 1993a, pp. 190-193). This adjustment is equivalent to the correction proposed by T. L. Kelley (1923) as a means of adjusting an observed test score for measurement error. (Newer versions of SIBTEST formulate the correction somewhat differently.) (2) For each of the K levels of the matching variable, average the expected true score for the reference and focal groups and regard that average as the true score corresponding to the kth level. (3) For each level of the matching variable, estimate the expected item score, given true score, for each group, assuming that the regression of item score on true score is locally linear. This expected item score is the adjusted item mean $\overline{Y}^*_k$ for that group.

If the weighting function in (4) rather than (6) is chosen, and if the test is either very reliable or yields similar score distributions and reliabilities for the two groups, then the value of $\hat{\beta}$ will be close to that of STD P-DIF (3). The SIBTEST test statistic, which is obtained by dividing $\hat{\beta}$ by its standard error, is approximately standard normal under the null hypothesis of no DIF. Under some conditions, SIBTEST has been shown to provide better Type I error control than the MH (Roussos & Stout, 1996).

2.2. A MODIFICATION OF THE MH AND STANDARDIZATION APPROACHES FOR CATs (ZTW)

The ZTW CAT DIF approach requires that IRT item parameter estimates be available for all items. After responding to the CAT, examinees are matched on the expected true score for the entire CAT pool, and the MH or standardization procedures applied. Specifically, the matching variable is

$$Expected\ true\ score\ on\ CAT = \sum_{j=1}^{J} \hat{p}_j(\hat{\theta}_{CAT}), \qquad (7)$$

where $\hat{p}_j(\hat{\theta}_{CAT})$ is the estimated item response function for item j, evaluated at $\hat{\theta}_{CAT}$, the maximum likelihood estimate (MLE) of ability based on responses to the set of items received by the examinee, and J is the number of items in the pool. In the original ZTW studies, one-unit intervals of expected true score were used for matching; in our more recent application to very sparse data (Zwick & Thayer, in press), two-unit intervals were found to work better.

In the initial ZTW simulation study (1993; 1994a) that evaluated the performance of these methods, the pool consisted of 75 items, 25 of which were administered to each examinee using an information-based CAT algorithm. Item responses were generated using the three-parameter logistic (3PL) model, in which the probability of a correct response on item i in group $G(G = R$ or F, denoting the reference or focal group) can be represented as

$$p_{iG}(\theta) = c_i + (1 - c_i)\{1 + \exp[-(1.7a_i(\theta - b_{iG})]\}^{-1}, \qquad (8)$$

where θ is the examinee ability parameter, a_i is the discrimination parameter for item i, c_i is the probability of correct response for a very low-ability examinee (which was constant across items in our simulation), and b_{iG} is the item difficulty in group G. The focal group difficulty, b_{iF}, is equal to $b_{iR} - d_i$. Hence, d_i is the difference between reference and focal group difficulties.

A simple relation between item parameters and MH DIF exists only in the Rasch model (Fischer, 1995; Holland & Thayer, 1988; Zwick, 1990), in which the $MH\ D\text{-}DIF$ statistic provides an estimate of $4\ a_i d_i$ under certain assumptions (see Donoghue, Holland, & Thayer, 1993). Even when the Rasch model does not hold, however, $MH\ D\text{-}DIF$ tends to be roughly proportional to $a_i d_i$ (ZTW, 1993; 1994a). Therefore, in this study, we used $a_i d_i$ as an index of the magnitude of DIF present in item i.

In practice, the true item parameters, are, of course, unavailable for estimating abilities and calculating item information within the CAT algorithm. To produce more realistic predictions about the functioning of the DIF methods in applications to actual examinee data, item parameter estimates, rather than the generating parameters, were used for these purposes. (This simulation design issue is discussed further in a later section.) A calibration sample was generated which consisted of 2,000 simulated examinees who responded to all 75 items in the pool under non-DIF conditions. Item calibration, based on the 3PL model, was conducted using LOGIST (Wingersky, Patrick, & Lord, 1988).

The main simulation included 18 conditions. In half the conditions, the number of examinees per group was 500, while in the other half, the reference group had 900 members and the focal group had 100. The simulation conditions also varied in terms of focal group ability distribution (same as or different from reference group) and pattern of DIF. A detailed analysis of the accuracy of the CAT DIF estimates and of the classification of items into the A, B, and C categories (ZTW, 1993; 1994a) showed that the methods performed well. A small portion of the validity evidence is included here.

2.3. CORRELATIONS BETWEEN CAT DIF STATISTICS, NONADAPTIVE DIF STATISTICS AND GENERATING DIF

For six of the 18 simulation conditions (all of which included reference and focal sample sizes of 500), two nonadaptive versions of the *MH D-DIF* and *STD P-DIF* statistics were computed for comparison to the CAT results. For both nonadaptive approaches, all 75 pool items were "administered" to all examinees. In the first procedure (referred to as "$\hat{\theta}$-75"), examinees were matched on an expected true score calculated using the MLE of ability based on all 75 responses. That is, the matching variable in equation 7 was replaced by

$$Expected\ true\ score\ based\ on\ all\ pool\ items = \sum_{j=1}^{J} \hat{p}_j(\hat{\theta}_J), \qquad (9)$$

where $\hat{\theta}_J$ is the MLE of ability based on all $J = 75$ items. The second nonadaptive approach ("Number Right") was a conventional DIF analysis in which examinees were matched on number-right score. Correlations (across the 75 items in the pool) between the CAT-based DIF statistics, the DIF statistics based on nonadaptive administration, and the DIF magnitude index, $a_i d_i$, are presented in Table 1. (A more sophisticated measure of an item's true DIF, given in equation 14, was used in the ZTL studies.)

Because of a complex estimation procedure used only for the CAT-based DIF analyses, the CAT DIF statistics were much more precisely determined than were the DIF statistics for the other two matching variables. (This estimation procedure was used only within the context of the simulation and is not involved in ordinary applications;

Table 1. Correlations between DIF estimates and DIF magnitude index values (from Zwick, Thayer, & Wingersky, 1993; 1994a)

Type of DIF Measure		Type of Correlation	Median Correlation MH D	STD P
$\hat{\theta}-CAT$	$\hat{\theta}-75$	Uncorrected	.89	.86
		Corrected	.99	.95
$\hat{\theta}-CAT$	Number Right	Uncorrected	.88	.87
		Corrected	.99	.95
$\hat{\theta}-CAT$	Magnitude Index	Uncorrected	.96	.96
		Corrected	.97	.97
$\hat{\theta}-75$	Number Right	Uncorrected	.99	.98
		Corrected	>1.00	>1.00
$\hat{\theta}-75$	Magnitude Index	Uncorrected	.87	.87
		Corrected	.97	.95
Number Right	Magnitude Index	Uncorrected	.88	.87
		Corrected	.98	.95

see ZTW, 1994.) To avoid giving a spuriously inflated impression of the performance of the CAT analyses, correlations were corrected for unreliability (see ZTW, 1993; 1994a for details). These corrected correlations (which occasionally exceed one) provide a more equitable way of comparing the three sets of DIF statistics than do the uncorrected correlations (also shown).

Table 1 [1] shows that the CAT, $\hat{\theta}$-75, and Number Right analyses produced results that were highly correlated with each other and with the true DIF indexes. In particular, the two analyses based on all 75 item responses produced virtually identical results. (The similarity between these approaches may be substantially less for shorter tests.) The median (over conditions) of the corrected correlations with the true DIF index were very similar for the CAT, $\hat{\theta}$-75, and Number Right analyses, which is somewhat surprising since the CAT DIF approach uses ability estimates based on only 25 item responses. Correlations with the true DIF index tended to be slightly higher for MH D-DIF than for STD

[1] The two left-most columns refer to the variables used to compute the correlations, i.e., $\hat{\theta}$-CAT refers to the DIF statistics (MH D-DIF or STD P-DIF) obtained after matching examinees on the CAT-based ability estimate. The DIF magnitude index is defined as $a_i d_i$ (see text). "Corrected" correlations are Pearson correlations that have been corrected for attenuation due to unreliability. The two right-most columns give median correlations over six simulation conditions.

$P - DIF$, a finding that is probably an artifact of the metric of the true DIF index (i.e., the index is roughly proportional to the quantity estimated by $MH\ D\text{-}DIF$, whereas $STD\ P\text{-}DIF$ is in the proportion metric).

Several extensions of the initial ZTW research were conducted. In one study (ZTW, 1995), we examined the effect on ability and DIF estimation of applying the Rasch model to data that were generated using the 3PL model. Although the DIF statistics were highly correlated with the generating DIF, they tended to be slightly smaller in absolute value than in the 3PL analysis, resulting in a lower probability of detecting items with extreme DIF. This reduced sensitivity appeared to be related to a degradation in the accuracy of matching. In another study (ZTW, 1994b), we addressed the question of how to assess DIF in nonadaptively administered pretest items that have not yet been calibrated. A simple procedure that involved matching on the sum of the CAT-based expected true score (equation 7) and the score on the pretest item (0 or 1) was found to work as well as more sophisticated matching procedures that required calibration of the pretest items. In another spin-off of the ZTW research, we discovered that adaptive administration has a systematic effect on the standard errors of DIF statistics. For fixed group sample sizes, adaptive administration tends to lead to smaller standard errors for $MH\ D\text{-}DIF$ and larger standard errors for $STD\ P\text{-}DIF$ than does nonadaptive administration. Although this phenomenon seems counterintuitive at first, it appears to be related to the fact that item proportions correct are closer to .5 in adaptive than in nonadaptive tests; this has opposite effects on the standard error of $MH\ D\text{-}DIF$, which is in the logit metric, and the standard error of $STD\ P\text{-}DIF$, which is in the proportion metric (see Zwick, 1997; ZTW, 1994b).

2.4. An Empirical Bayes (EB) Enhancement of the MH Approach (ZTL)

Zwick, Thayer and Lewis (1997; 1999; 2000) developed an empirical cal Bayes (EB) approach to Mantel-Haenszel DIF analysis which yields more stable results in small samples than does the ordinary MH approach and is therefore well suited to adaptive testing conditions. The computations, which involve only the $MH\ D\text{-}DIF$ indexes and their standard errors, are detailed in the original references.

The model can be expressed as follows. Because $ln(\hat{\alpha}_{MH})$ has an asymptotic normal distribution (Agresti, 1990), it is reasonable to assume that

$$MH_i \,|\, \omega_i \sim N\left(\omega_i, \sigma_i^2\right), \tag{10}$$

where MH_i denotes the MH $D\text{-}DIF$ statistic for item i, $E\left(MH_i\right) = \omega_i$ represents the unknown parameter value corresponding to MH_i, and σ_i^2 is the sampling variance of MH_i.

The following prior distribution is assumed for ω_i :

$$\omega_i \sim N\left(\mu, \tau^2\right), \tag{11}$$

where μ is the across-item mean of ω_i and τ^2 is the across-item variance. The parameters of the prior are estimated from the data. The posterior distribution of ω_i , given the observed MH $D - DIF$ statistic, can be expressed as

$$f\left(\omega_i \,|\, MH_i\right) \propto f\left(MH_i \,|\, \omega_i\right) f\left(\omega_i\right). \tag{12}$$

Standard Bayesian calculations (see, e.g., Gelman, Carlin, Stern & Rubin, 1995) show that this distribution is normal with mean $W_i MH_i + (1 - W_i)\mu$ and variance $W_i \sigma_i^2$, where

$$W_i = \frac{\tau^2}{\sigma_i^2 + \tau^2}. \tag{13}$$

The posterior distribution of DIF parameters in (12) is used as the basis for DIF inferences. (An alternative version of the EB DIF method allows estimation of the distribution of the item's DIF *statistic* in future administrations.) The posterior distribution can be used to probabilistically assign the item to the A, B, and C DIF categories described in an earlier section. In addition, the posterior mean serves as a point estimate of the DIF parameter for that item. The posterior mean can be regarded as a *shrinkage* estimator of Mantel-Haenszel DIF: The larger the MH standard error, σ_i^2 , the more the EB estimation procedure "shrinks" the observed MH_i value toward the prior mean (which is usually close to zero because MH statistics must sum to approximately zero in typical applications). On the other hand, as σ_i^2 approaches zero, the EB DIF estimate approaches MH_i.

In the initial phase of research (ZTL, 1997; 1999), the EB methods were extensively investigated through simulation study and were applied experimentally to data from paper-and-pencil tests, including the Graduate Record Examinations (GRE). Subsequent work involved an elaboration of the method which was based on the use of loss functions for DIF detection (ZTL, in press). The EB DIF methods are now being applied by the Defense Department to the CAT version of the Armed Services Vocational Aptitude Battery (CAT-ASVAB; see Krass

& Segall, 1998) and are slated to be used in the upcoming National Assessment of Adult Literacy. Also, Miller and Fan (1998) compared the ZTL approach to a method identical to the MH version of the ZTW procedure and concluded that the ZTL approach was more promising for the detection of DIF in high-dimensional CATs. The most recent EB DIF research (Zwick & Thayer, in press), sponsored by the Law School Admission Council (LSAC), was an investigation of the applicability of these methods to a large-scale computerized adaptive admissions test. Some findings from this study, which was part of an investigation (Pashley, 1997) of the feasibility of a computerized adaptive Law School Admission Test (LSAT), are described in the subsequent sections. The EB approach has been experimentally applied to data from NCLEX, the computerized adaptive licensure exam of the National Council of State Boards of Nursing.

2.5. LSAT SIMULATION STUDY

In developing a modification of the EB DIF methods for the LSAT CAT context, we needed to accommodate LSAC's interest in CATs that are adaptive on the testlet rather than the item level. To test the EB CAT procedure, therefore, we designed a simulation involving testlet-based CAT administration. The CAT pool consisted of 10 five-item testlets at each of three difficulty levels–a total of 150 items. The simulation included several conditions that varied in terms of focal group ability distribution (same as or different from reference group) and in terms of sample size (3,000 per group or 1,000 per group). In the large-n conditions, item-level sample sizes (within a group) ranged from 86 to 842; for the small-n conditions, the range was from 16 to 307. The data were generated using the 3PL model (equation 8). As in our previous simulation studies of the EB method, we defined true DIF as follows in the LSAC research:

$$True\ DIF\ =\ -2.35 \int \ln \left\{ \frac{p_{iR}(\theta)/q_{iR}(\theta)}{p_{iF}(\theta)/q_{iF}(\theta)} \right\} f_R(\theta)d\theta, \qquad (14)$$

where $p_{iG}(\theta)$ is the item response function for group G, given by equation 8, $q_{iG}(\theta) = 1 - p_{iG}(\theta)$ and $f_R(\theta)$ is the reference group ability distribution. Pommerich, Spray, and Parshall (1995) proposed similar indexes in other contexts. This quantity can be viewed as the true MH value, unaffected by sampling or measurement error (see ZTL, 1997).

We matched examinees for DIF analysis on the basis of the expected true score for the entire item pool, as in the ZTW (1994a; 1995) studies; this seemed most consistent with available LSAC scoring plans. Our procedures for estimating the parameters of the prior, which had been developed for nonadaptive tests, needed some modification for application to CATs (see Zwick & Thayer, in press). A major goal of the study was to determine whether the EB method, previously tested on samples no smaller than 200 examinees for the reference group and 50 for the focal group, could be applied successfully with even smaller samples.

2.6. PROPERTIES OF EB DIF ESTIMATES

How close were the EB DIF values to the target values given by equation 14, and how did their accuracy compare to that of the non-Bayesian version of the MH statistics? We compared these two types of DIF estimates using root mean square residuals ($RMSR$s), defined for each item as follows:

$$RMSR = \sqrt{\frac{1}{R} \sum_{r=1}^{R} (\hat{D}_r - True\ DIF)^2}, \tag{15}$$

where r indexes replications, R is the number of replications, $\hat{D}_r$ is either the $MH\ D\text{-}DIF$ statistic or the EB posterior mean from the rth replication, and $True\ DIF$ is the appropriate value from equation 14. The $RMSR$ represents the average departure, in the MH metric, of the DIF estimate from the $True\ DIF$ value. If these $True\ DIF$ values are regarded as the estimands for the DIF statistics, then these $RMSR$ values give estimates of the mean squared error (the average distance between the parameter estimate and the parameter) for the DIF statistics.

2.7. PROPERTIES OF EB POINT ESTIMATES IN THE NO-DIF CASE.

We first investigated the performance of the EB method when DIF was absent (i.e., the $True\ DIF$ value for each item was zero). Abilities for both examinee groups were drawn from a standard normal distribution. A large-sample case, with 3,000 examinees per group and 200 replications, and a small-sample case, with 1,000 examinees per group and 600 replications, were considered. (In this portion of the study, item sample sizes per group ranged from about 290 to 800 in the large-sample condition and from about 80 to 300 in the small-sample condition.) The top panel of Table 2 gives, for each of the two sample sizes, the 25th, 50th, and 75th percentiles of the distribution of $RMSR$ values across the 150 items. The difference in the performance of the EB DIF approach and that of the non-Bayesian MH statistic is quite striking: The median $RMSR$ for the MH method was roughly ten times the median

Table 2. RMSR results for EB and MH DIF statistics in LSAT simulation study (Zwick & Thayer, 1998)

	Initial Group $n = 1,000$		Initial Group $n = 3,000$	
	EB	MH	EB	MH
DIF Absent; Reference $N(0,1)$, Focal $N(0,1)$				
25th %ile	.068	.543	.031	.298
Median	.072	.684	.034	.365
75th %ile	.078	.769	.037	.417
DIF Present; Reference $N(0,1)$, Focal $N(0,1)$				
25th %ile	.460	.565	.284	.317
Median	.509	.713	.341	.390
75th %ile	.542	.787	.380	.444
DIF Present; Reference $N(0,1)$, Focal $N(-1,1)$				
25th %ile	.464	.585	.302	.322
Median	.517	.641	.361	.366
75th %ile	.560	1.190	.442	.594

Note. Each RMSR summarizes results across replications (600 in the small-n condition and 200 in the large-n condition). The results above are summaries over the 150 items.

$RMSR$ for the EB approach in both sample-size conditions. The EB DIF statistic departed from its target value of zero by an average of about .03 in the large-sample case and .07 in the small-sample case; the corresponding values for MH D-DIF were .37 and .68.

2.8. PROPERTIES OF EB POINT ESTIMATES IN THE DIF CASE.

In the conditions for which DIF was present, the *True DIF* values in this study (see equation 14) ranged from -2.3 to 2.9 in the MH metric, with a standard deviation of about one. Here, as in the no-DIF conditions, we compared the EB point estimates of DIF to the MH D-DIF statistics using root mean square residuals, defined in equation 15. The bottom two panels of Table 2 summarize the results for the 150 items in four simulation conditions. The table gives, for each condition, the 25th, 50th, and 75th percentiles of the distribution of $RMSR$ values across the 150 items. In the two small-n simulation conditions, the

RMSR tended to be substantially smaller for the EB estimate than for *MH D-DIF*. The difference in median *RMSR* values was larger in the case in which both reference and focal ability distributions were standard normal. In the large-n conditions, the advantage of the EB estimates was greatly reduced. The lesser difference between the two DIF estimates is to be expected, since the MH standard errors are small when samples are large, causing the EB DIF estimate to be close to the MH values.

The small-n results were also examined separately for easy, medium, and hard items. The smallest sample sizes occurred for the 50 hard items when the focal group ability distribution was $N(-1, 1)$, implying that it was centered more than two standard deviations lower than the mean difficulty of the items, which was 1.27. Here, reference group sample sizes ranged from 80 to 151, with a mean of 117; focal group sample sizes ranged from 16 to 67 with a mean of 40. These sample sizes are substantially smaller than is ordinarily considered acceptable for application of the MH procedure. Table 3 summarizes the *RMSR* results for these items, as well as the number of *RMSR* values exceeding 1 (about 1 SD unit in the *True DIF* metric). The range of item sample sizes across the 50 items and 600 replications was from 80 to 151, with a mean of 117 for the reference group and from 16 to 67, with a mean of 40 for the focal group. While only two of the 50 values exceeded one for the EB method, all 50 *RMSR*s for the MH procedure were greater than one. The median *RMSR* for the EB method for these items was .53, compared to 1.25 for the MH. It is interesting to note that, in a different subset of the results (not shown) for which the MH *RMSR* had a median of .53 (medium-difficulty items, $N(-1, 1)$ focal group distribution), the sample sizes averaged about 240 per group. Roughly, speaking, then, the EB procedure achieved the same stability for samples averaging 117 and 40 reference and focal group members, respectively, as did the MH for samples averaging 240 per group.

The generally smaller *RMSR* values for the EB estimates are consistent with theory. Such estimates have smaller mean squared error than their non-Bayesian counterparts. They are not, however, unbiased; in fact, the bias of these estimates is greatest for the extreme parameter values. Analyses of the bias of the EB DIF statistics are described in ZTL (1997; 1999; 2000) and in Zwick and Thayer (in press). Further investigation of the bias issue is planned, along with additional research on improving model estimation procedures.

Table 3. Distribution of RMSRs
for the 50 hard items in the
small-sample condition (Zwick
& Thayer, 1998)

	EB	MH
25th %ile	.514	1.190
Median	.532	1.252
75th %ile	.558	1.322
Number > 1	2	50

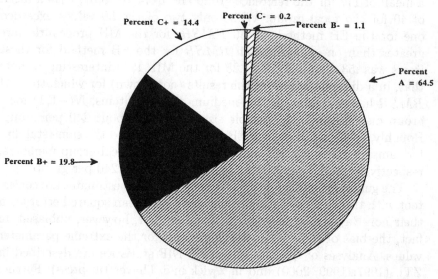

MH D-DIF = 4.71 , SE(MH D-DIF) = 2.22   ETS DIF Status C+
Posterior Mean (EB DIF Value) = 0.69 , Posterior Standard Deviation = 0.76

Estimate of True DIF Status

Percent C+ = 14.4   Percent C- = 0.2
                                                         Percent B- = 1.1

                                                              Percent
                                                              A = 64.5

Percent B+ = 19.8 ➤

Figure 1. EB DIF results for a CAT simulation item with true classifica-
tion "A". Reference group: $N(0,1), n = 101$, focal group: $N(-1,1), n = 23$. (Adapted from Zwick & Thayer, 1998).

2.9. PROBABILISTIC CLASSIFICATION OF DIF RESULTS

In addition to offering an alternative point estimate of DIF, the EB method provides a probabilistic version of the A, B, and C DIF classification system. Two related problems associated with the traditional classification approach are that (1) when sample sizes are small, the DIF category is unstable and may vary substantially from one test administration to another and (2) attaching an A, B, or C label to an item may convey the mistaken notion that an item's DIF category is deterministic. The EB approach yields an estimate of the *probability* that the true DIF for an item falls into the A, B, and C categories, based on an estimate of the posterior distribution of DIF parameters (see ZTL, 1997; 1999 for details). The estimated A, B, and C probabilities can be regarded as representing our state of knowledge about the true DIF category for the item.

A possible advantage of the EB method of probabilistic DIF classification is that it may convey information about the sampling variability of DIF results in a more comprehensible way than do the current procedures. This alternative way of representing the variability of DIF findings lends itself well to graphical display. Pie charts can be used effectively to represent the posterior probabilities associated with the A, B, and C categories, as shown in Figure 1. The displayed item, which had an actual status of A (*True DIF* near zero), was incorrectly classified as a C+ using the standard ETS procedure. According to the EB approach however, the estimated probability of A status was .65. The EB methods can be modified easily if the current rules used to assign items to categories are adjusted (e.g., see Miller & Fan, 1998) or, if other hypothesis-testing approaches are substituted, for the Mantel-Haenszel procedure.

2.10. CATSIB: A MODIFICATION OF THE SIBTEST METHOD OF SHEALY AND STOUT

In CATSIB, the modification of SIBTEST for CAT (Roussos, 1996; Nandakumar & Roussos, in press; Roussos & Nandakumar, 1998), examinees are matched on a regression-corrected version of an IRT-based ability estimate (grouped into intervals). So far, CATSIB has been applied only to simulated "pretest" items that have been "administered" nonadaptively to at least 250 members in each examinee group; that is, CAT administration affected only the items (25 per examinee, out of a pool of 1,000) used to match examinees for DIF analysis, not the 16 suspect items themselves, all of which were administered to each examinee (Nandakumar & Roussos, in press). In this respect, the study

Table 4. Summary of two-tailed rejection rates from CATSIB simulation study (adapted from Nandakumar & Roussos, in press.)

	No DIF	DIF = .05	DIF = .10
Ref. & Focal means same			
$n_R = 250, n_F = 250$	.051	.275	.728
$n_R = 500, n_F = 250$	.044	.336	.840
$n_R = 500, n_F = 500$	.057	.478	.940
Focal θ mean .5 SD lower			
$n_R = 250, n_F = 250$	.046	.260	.715
$n_R = 500, n_F = 250$	.045	.328	.811
$n_R = 500, n_F = 500$	.051	.462	.935
Focal θ mean 1 SD lower			
$n_R = 250, n_F = 250$	.050	.235	.638
$n_R = 500, n_F = 250$	.057	.294	.719
$n_R = 500, n_F = 500$	.058	.397	.894

resembles the pretest DIF study of ZTW, 1994b. The CATSIB simulation findings on DIF parameter estimation, Type I error, and power were favorable; a partial summary of the Type I error and power results from Nandakumar and Roussos (in press) is given in Table 4 [2] (for $\alpha =$ .05, with combined reference and focal sample sizes ranging from 500 to 1,000). When DIF was equal to .05 in the SIBTEST metric (i.e., probabilities of correct response for matched reference and focal group examinees differed by an average of .05), CATSIB's power ranged from .24 to .48; when DIF was equal to .10, the power ranged from .64 to 94. Power decreased as the difference between reference and focal group ability distributions increased. Type I error was quite well controlled overall.

It is difficult to compare the CATSIB results to the ZTW and ZTL results for several reasons. First, the item sample sizes and administration mode (adaptive versus nonadaptive) are different. Also, Nandakumar and Roussos (in press) included estimation of Type I error rates and power, while the ZTW and ZTL research focused on parameter estimation and probabilistic DIF classification. Yet another factor that

---

[2] The nominal Type I error rate was .05. Results are averaged across six items in the No-DIF condition and across five items in each of the two DIF conditions. In all conditions, results are also averaged across 400 replications. The DIF metric here represents the average difference between the probabilities of correct response for matched reference and focal group examinees.

makes CATSIB and the other CAT DIF procedures hard to compare is a particular feature of the simulation procedures used in the CAT-SIB studies. Most aspects of the Nandakumar and Roussos (in press) simulation were carefully designed to be realistic. For example, great care was taken in choosing the properties of the generating item parameters so as to produce data resembling actual test results. Another strong feature of the simulation is that it involved administration of 25 CAT items from a pool of 1000. This ratio of pool size to test length is more realistic than that used in ZTW and ZTL; the exposure control features implemented in the Nandakumar and Roussos simulation were also more elaborate. However, as the authors themselves mentioned, the simulated CAT administration and DIF analyses departed in one major way from actual practice and from the ZTW and ZTL studies: The true item parameters–those used in data generation–were used in all computations. In the CATSIB context, this means that true, rather than estimated parameters were involved in three major aspects of the simulation and analysis: the assignment of items to examinees via the CAT algorithm, the computation of the regression correction (discussed below), and the calculation of examinee ability estimates, which are used for DIF matching. Nandakumar and Roussos noted that their future research will use item parameter estimates in applying the CATSIB method.

In future CATSIB research, it would also seem worthwhile to reconsider the formulation of CATSIB's "correction" for measurement error. The correction is achieved in much the same way as in the original version of SIBTEST: A model analogous to that of classical test theory is assumed for the relation of $\hat{\theta}$, the ability estimate, and θ, the true ability. One concern about the correction described by Nandakumar and Roussos (in press) is that, in estimating the slope of the regression of θ on $\hat{\theta}$, the variance of $\hat{\theta}$ is estimated by the inverse of the test information function, despite the fact that the $\hat{\theta}$ values in the study are not maximum likelihood estimates (for which this procedure for variance estimation would be appropriate), but Bayes modal estimates. Perhaps a reformulation of the correction could improve CATSIB's already favorable performance. (Roussos [personal communication, November 17, 1998] indicated that, in further applications, the variance computation will be appropriately modified.) In summary, the CATSIB procedure seems promising, and clearly warrants further study.

3. Future Research

A number of important questions remain to be addressed in future CAT DIF research:

1. How does the performance of existing CAT DIF methods compare when simulation design and analysis features are held constant? As noted earlier, the existing studies used differing simulation approaches and assessed different aspects of the DIF procedures. It would be useful to create a common simulation data set on which all existing methods could be applied. A common set of criteria could then be used to evaluate the results, including the performance of the DIF methods in terms of parameter estimation, Type I error rate, power, and DIF classification.

2. Can automated test assembly (ATA) procedures be used effectively to reduce DIF? Some work has been conducted which is relevant to this question. The Defense Department, in the early stages of the development of its ATA algorithms for paper-and-pencil tests, considered using these algorithms to regulate the amount of DIF. The focus, however, was on balancing DIF across forms, rather than reducing the presence of DIF (Gary Thomasson, personal communication, September 4, 1998). Stocking, Jirele, Lewis, and Swanson (1998) explored the feasibility of using an ATA algorithm to reduce score differences between African-American and White examinees, and between male and female examinees, on the SAT I Mathematical Reasoning test. The goal of impact reduction was incorporated into a previously developed ATA algorithm, which was designed to "select items from a pool...in such a way as to minimize the weighted sum of deviations from constraints reflecting desirable test properties..." (Stocking et al., p. 203). If DIF information from a pretest were available, a similar approach could be used to minimize DIF, rather than impact. Furthermore, although the SAT application involved a paper-and-pencil test, the DIF reduction feature could be incorporated into a CAT algorithm. Exploration of ATA-based approaches to DIF minimization would be fruitful.

3. What sources of DIF are of particular concern in CATs, and how can they can be reduced? An entirely different, but extremely important type of investigation which needs to be undertaken is field research to study the sources of DIF (as well as other threats to validity) which are of particular concern in CATs, and to determine how they can be reduced. It is not difficult to imagine situations in which CAT administration could introduce DIF into an item that was DIF-free in its paper-and-pencil incarnation. Suppose, for example, that for most items on a math test, computer experience has little effect on the probability of correct response, but that, on complex figural response items

that require examinees to use a mouse to point to a graphical display, examinees who are computer-savvy have an advantage. Now suppose that computer familiarity (given a particular level of math ability) is more likely to occur in certain demographic groups, a conjecture that appears quite plausible (e.g., see Legg & Buhr, 1992; Wenglinsky, 1998). This phenomenon would create DIF on the figural response items. Another interesting hypothesis of this kind is the following: Suppose that non-native speakers of English rely on the ability to make notes directly on the test booklet, perhaps consisting of a partial translation of the item. If this type of note-taking were particularly important on certain types of items, computer administration could result in DIF.

In summary, a research effort that includes both simulation-based technical investigations and a program of field studies is needed to further our understanding of DIF in CATs and, more generally, to help us evaluate the fairness of computerized adaptive tests.

References

Agresti, A. (1990). *Categorical data analysis*. New York: Wiley.

Camilli, G., & Shepard, L. A. (1994). *Methods for identifying biased test items*. Thousand Oaks, CA: Sage.

Donoghue, J. R., Holland, P. W., & Thayer, D. T. (1993). A Monte Carlo study of factors that affect the Mantel-Haenszel and standardization measures of differential item functioning. In P. W. Holland and H. Wainer (Eds.), *Differential Item Functioning*. Hillsdale, NJ: Erlbaum.

Dorans, N. J., & Holland, P. W. (1993). DIF detection and description: Mantel-Haenszel and standardization. In P. W. Holland and H. Wainer (Eds.), *Differential item functioning*, (pp. 35-66). Hillsdale, NJ: Erlbaum.

Dorans, N. J., & Kulick, E. (1986). Demonstrating the utility of the standardization approach to assessing unexpected differential item performance on the Scholastic Aptitude Test. *Journal of Educational Measurement*, 23, 355-368.

Fischer, G.H. (1995). Some neglected problems in IRT. *Psychometrika*, 60, 459-487.

Gelman, A., Carlin, J. B., Stern, H. S., & Rubin, D. B. (1995). *Bayesian data analysis*. London: Chapman & Hall.

Holland, P.W., & Thayer, D.T.(1985) *An alternative definition of the ETS delta scale of item difficulty*. (ETS Research Report No. 85-43). Princeton, NJ: Educational Testing Service.

Holland, P.W., & Thayer, D.T. (1988) Differential item performance and the Mantel-Haenszel procedure. In H. Wainer & H. I. Braun (Eds.), *Test validity*, (pp. 129-145). Hillsdale, NJ: Erlbaum.

Holland, P.W., & Zwick, R. (1991). *A simulation study of some simple approaches to the study of DIF for CAT's*. (Internal memorandum). Princeton, NJ: Educational Testing Service.

Holland, P.W., & Wainer, H. (Eds.) *Differential item functioning*. Hillsdale, NJ: Erlbaum

Kelley, T. L. (1923). *Statistical methods*. New York: Macmillan.

Krass, I., & Segall, D. (1998). *Differential item functioning and on-line item calibration.* (Draft report). Monterey, CA: Defense Manpower Data Center.

Legg, S.M., & Buhr, D.C. (1992). Computerized adaptive testing with different groups. *Educational Measurement: Issues and Practice, 11,* 23-27.

Mantel, N. & Haenszel, W. (1959). Statistical aspects of the analysis of data from retrospective studies of disease. *Journal of the National Cancer Institute, 22,* 719-748.

Miller, T.R. (1992, April). *Practical considerations for conducting studies of differential item functioning in a CAT environment.* Paper presented at the annual meeting of the American Educational Research Association, San Francisco.

Miller, T.R., & Fan, M. (1998, April). *Assessing DIF in high dimensional CATs.* Paper presented at the annual meeting of the National Council on Measurement in Education, San Diego.

Nandakumar, R. & Roussos, L. (in press). *CATSIB: A modified SIBTEST procedure to detect differential item functioning in computerized adaptive tests.* (Research report) Newtown, PA: Law School Admission Council.

Pashley, P. J. (1997). *Computerized LSAT research agenda: Spring 1997 update.* (LSAC report). Newtown, PA: Law School Admission Council.

Phillips, A. & Holland, P.W. (1987). Estimation of the variance of the Mantel-Haenszel log-odds-ratio estimate. *Biometrics, 43,* 425-431.

Pommerich, M., Spray, J.A., & Parshall, C.G. (1995). *An analytical evaluation of two common-odds ratios as population indicators of DIF.* (ACT Report 95-1). Iowa City: American College Testing Program.

Powers, D. E., & O'Neill, K. (1993). Inexperienced and anxious computer users: Coping with a computer-administered test of academic skills. *Educational Assessment, 1,* 153-173.

Robins, J., Breslow, N., & Greenland, S. (1986). Estimators of the Mantel-Haenszel variance consistent in both sparse data and large-strata limiting models. *Biometrics, 42,* 311-323.

Roussos, L. (1996, June). *A type I error rate study of a modified SIBTEST DIF procedure with potential application to computerized-adaptive tests.* Paper presented at the annual meeting of the Psychometric Society, Banff, Alberta, Canada.

Roussos, L., & Nandakumar, R. (1998, June). *Kernel-smoothed CATSIB.* Paper presented at the annual meeting of the Psychometric Society, Urbana-Champaign, IL.

Roussos, L., & Stout, W.F. (1996). Simulation studies of effects of small sample size and studied item parameters on SIBTEST and Mantel-Haenszel Type I error performance. *Journal of Educational Measurement, 33,* 215-230.

Schaeffer, G., Reese, C., Steffen, M., McKinley, R. L., & Mills, C. N. (1993). *Field test of a computer-based GRE general test.* (ETS Research Report No. RR 93-07). Princeton, NJ: Educational Testing Service.

Shealy, R., & Stout, W.F. (1993a). A model-based standardization approach that separates true bias/DIF from group ability differences and detects test bias/DTF as well as item bias/DIF. *Psychometrika, 58,* 159-194.

Shealy, R., & Stout, W.F. (1993b). An item response theory model for test bias and differential test functioning. In P. W. Holland & H. Wainer (Eds.), *Differential item functioning* (pp. 197-239). Hillsdale, NJ: Erlbaum.

Steinberg, L., Thissen, D, & Wainer, H. (1990). Validity. In H. Wainer (Ed.), *Computerized adaptive testing: A primer* (pp. 187-231). Hillsdale, NJ: Erlbaum.

Stocking, M.L., Jirele, T., Lewis, C., & Swanson, L. (1998). Moderating possibly irrelevant multiple mean score differences on a test of mathematical reasoning. *Journal of Educational Measurement, 35,* 199-222.

Thissen, D., Steinberg, L., & Wainer, H. (1993). Detection of differential item functioning using the parameters of item response models. In P. W. Holland & H. Wainer (Eds.), *Differential item functioning* (pp. 67-113). Hillsdale, NJ: Erlbaum.

Way, W. D. (1994). *A simulation study of the Mantel-Haenszel procedure for detecting DIF for the NCLEX using CAT.* (Internal technical report). Princeton, NJ: Educational Testing Service.

Wenglinsky, H. (1998). *Does it compute? The relationship between educational technology and student achievement in mathematics.* (ETS Policy Information Center report) Princeton, NJ: Educational Testing Service.

Wingersky, M. S., Patrick, R., & Lord, F. M. (1988). *LOGIST user's guide: LOGIST Version 6.00.* Princeton, NJ: Educational Testing Service.

Zieky, M. (1993). Practical questions in the use of DIF statistics in test development. In P. W. Holland & H. Wainer (Eds.), *Differential item functioning* (pp. 337-347). Hillsdale, NJ: Erlbaum.

Zwick, R. (1990). When do item response function and Mantel-Haenszel definitions of differential item functioning coincide? *Journal of Educational Statistics, 15,* 185-197.

Zwick, R. (1992). *Application of Mantel's chi-square test to the analysis of differential item functioning for functioning for ordinal items.* (Technical memorandum). Princeton, NJ: Educational Testing Service.

Zwick, R. (1997). The effect of adaptive administration on the variability of the Mantel-Haenszel measure of differential item functioning. *Educational and Psychological Measurement, 57,* 412-421.

Zwick, R., & Thayer, D. T. (1996). Evaluating the magnitude of differential item functioning in polytomous items. *Journal of Educational and Behavioral Statistics, 21,* 187-201.

Zwick, R., & Thayer, D. T. (in press). Application of an empirical Bayes enhancement of Mantel-Haenszel DIF analysis to computer-adaptive tests. *Applied Psychological Measurement.*

Zwick, R., Thayer, D. T., & Lewis, C. (1997) *An Investigation of the Validity of an Empirical Bayes Approach to Mantel-Haenszel DIF Analysis.* (ETS Research Report No. 97-21). Princeton, NJ: Educational Testing Service.

Zwick, R., Thayer, D. T., & Lewis, C. (1999). An empirical Bayes approach to Mantel-Haenszel DIF analysis. *Journal of Educational Measurement, 36,* 1-28.

Zwick, R., Thayer, D.T., & Lewis, C. (2000). Using loss functions for DIF detection: An empirical Bayes approach. *Journal of Educational and Behavioral Statistics, 25,* 225-247.

Zwick, R., Thayer, D.T., & Mazzeo, J. (1997). Descriptive and inferential procedures for assessing DIF in polytomous items. *Applied Measurement in Education, 10,* 321-344.

Zwick, R., Thayer, D.T., & Wingersky, M. (1993). *A simulation study of methods for assessing differential item functioning in computer-adaptive tests.* (ETS Research Report 93-11). Princeton, NJ: Educationl Testing Service.

Zwick, R., Thayer, D. T., & Wingersky, M. (1994a) A simulation study of methods for assessing differential item functioning in computerized adaptive tests. *Applied Psychological Measurement, 18,* 121-140.

Zwick, R., Thayer, D.T., & Wingersky, M. (1995). Effect of Rasch calibration on ability and DIF estimation in computer-adaptive tests. *Journal of Educational Measurement, 32*, 341-363.

Chapter 13

Testlet Response Theory: An Analog for the 3PL Model Useful in Testlet-Based Adaptive Testing

Howard Wainer, Eric T. Bradlow, & Zuru Du
Educational Testing Service, USA
The Wharton School of the University of Pennsylvania, USA
Professional Examination Service, USA*

1. Introduction

The invention of short multiple choice test items provided an enormous technical and practical advantage for test developers; certainly the items could be scored easily, but that was just one of the reasons for their popular adoption in the early part of the 20th century. A more important reason was the increase in validity offered because of the speed with which such items could be answered. This meant that a broad range of content specifications could be addressed, and hence an examinee need no longer be penalized because of an unfortunate choice of constructed response (e.g., essay) question. These advantages, as well as many others (see Anastasi, 1976, 415-417) led the multiple choice format to become, by far, the dominant form used in large-scale standardized mental testing throughout this century. Nevertheless, this breakthrough in test construction, dominant at least since the days of Army α, is currently being reconsidered.

Critics of tests that are made up of large numbers of short questions suggest that decontextualized items yield a task that is abstracted too far from the domain of inference for many potential uses. For several reasons, only one of them as a response to this criticism, variations in test theory were considered that would allow the retention of the short-answer format while at the same time eliminating the shortcomings expressed by those critics. One of these variations was the development of item response theory (IRT), an analytic breakthrough in test scoring. A key feature of IRT is that examinee responses are conceived of as reflecting evidence of a particular location on a single underlying latent

---

* This research was supported by the research division of Educational Testing Service, the Graduate Record Board and the TOEFL Policy Council. We are pleased to be able to acknowledge their help. We also thank Jane Rodgers and Andrew Gelman for help of various sorts during the development of the procedure.

W.J. van der Linden and C.A.W. Glas (eds.),
Computerized Adaptive Testing: Theory and Practice, 245-269.
© 2000 *Kluwer Academic Publishers. Printed in the Netherlands.*

trait. Thus the inferences that are made are about that trait, and not constrained to what some thought of as atomistic responses to specific individual items. A second approach, with a long and honored history, was the coagulation of items into coherent groups that can be scored as a whole. Such groups of items are often referred to as testlets (Wainer & Kiely, 1987), since they are longer than a single item (providing greater context) yet shorter than an entire test. This chapter provides a history and motivation for our approach to the inclusion of testlets into a coherent psychometric model for test calibration and scoring.

We discuss three reasons for using testlets; there are others. One reason is to reduce concerns about the atomistic nature of single independent small items. A second is to reduce the effects of context in adaptive testing; this is accomplished because the testlet, in a very real sense, carries its context with it (except for the items at the boundaries). The third reason is to improve the efficiency (information per unit time) of testing when there is an extended stimulus (e.g., a reading passage, a diagram, a table) associated with an item. A substantial part of an examinee's time is spent processing the information contained in the stimulus material, thus greater efficiency is obtained if more than a single bit of information is collected from each stimulus. Obviously diminishing returns eventually sets in; one cannot ask 200 independent questions about a 250 word passage. Usually the depths of meaning carried by such a passage are considered plumbed by four to six questions. More recent trends have yielded longer passages, often with as many as ten to twelve items per passage.

Traditionally the items that made up testlets were scored as if they were independent units just like all other items. Research has shown that such scoring, ignoring local dependence, tends to overestimate the precision of measurement obtained from the testlet (Sireci, Wainer & Thissen, 1991; Wainer & Thissen, 1996; Yen, 1993), although with testlets of modest length (e.g., a four-item testlet and a 250-word passage) the effect may be minimal. A scoring method that accounts for the item information more accurately treats the testlet as a single item and scores it polytomously. This approach has worked well (Wainer, 1995) in a broad array of situations and is a good practical approach. But there are two circumstances where something more is needed.

Circumstance 1: If you need to extract more information from the testlet. When a testlet is fit by a polytomous IRT model the testlet score is represented by the number correct. As evidenced by the number of successful applications of the Rasch model, such a representation is often good enough. Yet there is some information in the exact pattern of correct scores. To extract this information requires a more complex model.

Circumstance 2: Ad hoc testlet construction within a CAT. A common current practice is to build testlets on the fly within the context of a computerized adaptive test (CAT). For example in the CAT's item pool there might be an item stimulus (e.g., reading passage) and 15 associated items. There was never any intent that all 15 items would be presented to any specific examinee; indeed, quite often those items were written with the express notion that some of the items are incompatible with one another (i.e. one might give the answer to another, or one might be an easier version of another). The item selection algorithm chooses the testlet stimulus (e.g., the passage) and then picks one item on the basis of item content, the psychometric properties of that item, and the previous examinee responses. After the examinee responds to that item the examinee's proficiency estimate is updated and a second item from those associated with that stimulus is chosen. This process is continued until a prespecified number of items have been presented and the item selection algorithm moves on. At the present time, the algorithm behaves as if all of the items it chooses within the testlet are conditionally independent.

These two testing circumstances have been occurring more frequently of late due to the increasing popularity of adaptive testing and the strong desire for test construction units (the testlets) to be substantial and coherent. If test formats continue into the future as they have existed in the past with testlets being of modest length (4-6 items/testlet), we do not expect that the unmodeled local dependence will prove to be a serious obstacle in the path of efficient testing or accurate scoring. But as testlets grow longer explicit modeling of their effects will be needed. Our approach is to develop a parametric framework that encompasses as many alternative test formats as possible. To put it another way, we need a psychometric theory of testlets — what we shall term Testlet Response Theory (TRT).

In the balance of this chapter we describe a very general approach to modeling test (independent and testlet) items. The formal structure of the model as well as the motivation for its development is in Section 2. In Section 3 we discuss estimation issues and describe our solution. Section 4 contains some simulation results that can clarify intuition with regards to the functioning of the model. Section 5 applies this model to two interesting data sets drawn from the SAT and the GRE; and in Section 6 we summarize this work and place it into a broader context.

2. A Bayesian Random Effects Model for Testlets

Item response theory (Lord & Novick, 1968) is one of today's most popular test scoring models and is currently applied in the development and scoring of many standardized educational tests. One important assumption made in item response theory is that of local independence. The assumption of local independence is that the examinee's trait (ability or proficiency) value, denoted θ, provides all the necessary information about the examinee's performance, and once trait level is taken into account, all other factors affecting examinee performance are random. This implies that for a given examinee, there is no relationship between the examinees' responses to any pair of items given their ability. A violation of this assumption occurs when examinees' responses to test items are conditionally correlated, either positively or negatively. Responses to two items are conditionally positively correlated when examinees of given ability perform either higher or lower than their expected performance level on both items. Negative conditional correlation of item responses occurs when a higher-than-expected performance on one item is consistently related to a lower-than-expected performance on the other item. In most testlet situations, violations of local independence are due to a positive conditional correlation among item scores. Although not a focus of this chapter, nonparametric procedures to detect conditional item dependence are considered in Stout (1987, 1990), Zhang and Stout (1999).

As motivated in Section 1, many current standard educational tests have sections that are composed of coherent groups of multiple choice items based on a common stimulus — testlets. Research has shown that testlet items often violate local independence (Rosenbaum, 1988). Some of the plausible causes might be an unusual level of background knowledge necessary to understand the reading passage, misreading of a data display for a group of math questions (Yen, 1993), or general frustration with the stimulus. By applying standard IRT models which assume local independence, overestimation of the precision of proficiency estimates as well as a bias in item difficulty and discrimination parameter estimates result (Sireci, Wainer & Thissen, 1991; Wainer 1995; Wainer & Thissen, 1996).

To model examinations that are composed of a mixture of independent items and testlets, Bradlow, Wainer and Wang (1999), referred to hereafter as BWW, proposed a modification to the standard two-parameter logistic (2PL) IRT model to include an additional random effect for items nested within the same testlet. To facilitate model parameter estimation they embedded this model into a larger Bayesian hierarchical framework. The computation of the posterior distributions of model parameters was completed via a data augmented Gibbs sampler (Tanner & Wong, 1987; Albert, 1992; Albert & Chib, 1993).

BWW showed, within a simulation, that the 2PL testlet model was able to recover the parameters better than the standard model, and that, unsurprisingly, the size of estimation bias resulting from the use of the standard 2PL model is a function of the magnitude (the variance of the distribution for the testlet effects) of the testlet effects.

Compared to the approach of treating testlets as units of analysis and fitting a multiple-categorical-response model, the approach taken by BWW has two advantages. First, the units of analysis are still test items, not testlets, thus information contained in the response patterns within testlets is not lost. Second, the familiar concepts of item parameters such as an item's discriminating power and difficulty are still valid and operational. This second feature facilitates an easy transfer from using the standard 2PL IRT model to using the 2PL testlet model. It also allows CAT item selection algorithms to build tests according to their usual pattern but still calculate both examinee ability and its standard error accurately, since it can take the excess local dependence into account. However, two aspects of BWW required extending and are the focus of this chapter.

First, BWW chose initially to modify the 2PL IRT model because the modified model was analytically tractable. However, since multiple choice items allow an examinee to answer an item correctly by chance, a more general model that addresses guessing would have greater practical value. The extension to the 3PL requires more complex computation. The Gaussian-based 2PL version derived by BWW uses a conjugate hierarchical structure that facilitates straightforward and fast computation that is unavailable for the 3PL testlet model. Therefore, one aspect of our extension is to develop efficient computational algorithms for this extension.

A second aspect of BWW's 2PL testlet model that we generalize is its assumption that the variance of the testlet effect is constant across all testlets within an examination. This assumption is unlikely to be satisfied in real test data as some passages exhibit more context effects, etc., than others.

What follows is a description of a modified model for tests, now allowing for guessing (3PL) and testlet effects to vary by testlet.

The three-parameter logistic model as proposed by Birnbaum (1968) has a mathematical form given by

$$p(y_{ij} = 1) = c_j + (1 - c_j)\frac{exp(a_j(\theta_i - b_j))}{1 + exp(a_j(\theta_i - b_j))} \tag{1}$$

where y_{ij} is the score for item j received by examinee i, $p(y_{ij} = 1)$ is the probability that examinee i answers item j correctly, θ_i is the ability of examinee i, b_j is the difficulty of test item j, a_j denotes the discrimination (slope) of the item j, and c_j denotes the pseudo-chance level parameter of item j. For a full explication of this model see Hambleton & Swaminathan, 1985; Lord, 1980; Wainer & Mislevy, 1990. One concern with this model is that the conditional correlation, $cor(y_{ij}, y_{ij'} | \theta_i, a_j, a_{j'}, b_j, b_{j'}, c_j, c_{j'})$ is zero for all item pairs $j \neq j'$ (whether or not items j, j' are in the same testlet). Our modification of (1) presented next addresses this concern.

We extend the model in (1) to allow for testlets, by retaining the 3PL structure yet modifying the logit's linear predictor to include a random effect, $\gamma_{id(j)}$, for person i on testlet $d(j)$. Specifically, we posit the 3PL testlet model given by

$$p(y_{ij} = 1) = c_j + (1 - c_j) \frac{exp(a_j(\theta_i - b_j - \gamma_{id(j)}))}{1 + exp(a_j(\theta_i - b_j - \gamma_{id(j)}))}. \tag{2}$$

Now, two items j and j' which are both in testlet $d(j)$, have predicted probability correct which share the common testlet effect $\gamma_{id(j)}$ and therefore higher marginal correlation than items where $d(j) \neq d(j')$. By definition, $\gamma_{id(j)} = 0$ if item j is an independent item, and the testlets are mutually exclusive and exhaustive with respect to the test items.

Two interesting properties of the 3PL testlet model given in (2) are of note. First, the expected Fisher information for θ_i, $I(\theta_i)$, for a single item response is given by

$$
\begin{aligned}
I(\theta_i) &= E[-\partial^2/\partial\theta_i^2 \log(p_{ij}^{y_{ij}}(1 - p_{ij})^{1-y_{ij}})] \tag{3} \\
&= E[a_j^2(\frac{exp(t_{ij})}{(1 + exp(t_{ij}))^2} - y_{ij}c_j \frac{exp(t_{ij})}{(c_j + exp(t_{ij}))^2})] \\
&= a_j^2(\frac{exp(t_{ij})}{1 + exp(t_{ij})})^2 \frac{1 - c_j}{c_j + exp(t_{ij})},
\end{aligned}
$$

where E is the expectation operation, $exp(x) = e^x$, $t_{ij} = a_j(\theta_i - b_j - \gamma_{id(j)})$, the logit linear predictor, and p_{ij} is the probability of correct response for person i on item j for the 3PL model given in (2). From (3), we observe the well-established properties that: (a) $I(\theta_i) \geq 0$, (b) $I_{c_j=0}(\theta_i) > I_{c_j \neq 0}(\theta_i)$, that is the information for θ_i is greater when there is no guessing, (c) for the 2PL model $I(\theta_i) = a_j^2 p_{ij}(1 - p_{ij})$ and maximized at $p_{ij} = 0.5$ (that is $\theta_i = b_j$, and hence the design of CATs based on the 2PL), and (d) for the 3PL model $I(\theta_i)$ is maximized when

$p_{ij} = exp(t_{ij})/2$ and thus for c_j around 0.2, $p_{ij} = 0.6$. Not surprisingly, results (c) and (d) mirror current practice in large scale standardized CATs.

One additional and intuitive property of (2) is given by

$$\frac{\partial p_{ij}}{\partial \gamma_{id(j)}} = \frac{a_j exp(t_{ij})(c_j - 1)}{(1 + exp(t_{ij}))^2} < 0. \qquad (4)$$

That is an increasing testlet effect makes an item appear more difficult as $b_j^* = b_j + \gamma_{id(j)}$, and hence yields a lower value of p_{ij}.

As in BWW, we embed the model in (2) in a larger hierarchical Bayesian framework thereby allowing for sharing of information across units where commonalities are likely to occur. This is done by specifying prior distributions for the model parameters. Our choice of priors was chosen out of convention and is given by $\theta_i \sim N(0, 1)$, $a_j \sim N(\mu_a, \sigma_a^2)$, $b_j \sim N(\mu_b, \sigma_b^2)$, $\gamma_{id(j)} \sim N(0, \sigma_{\gamma d(j)}^2)$ where $N(\mu, \sigma^2)$ denotes a normal distribution with mean μ and variance σ^2. To better approximate normality, c_j was reparameterized to $q_j = log(c_j/(1 - c_j))$ and a prior $q_j \sim N(\mu_q, \sigma_q^2)$ was chosen. The mean and variance of the distribution for θ_i and the mean of the distribution for $\gamma_{id(j)}$ are fixed for identifiability. If we set $c_j = 0$ (the 2PL model) and $\sigma_{\gamma d(j)}^2 = \sigma_\gamma^2$ (the same variance for all testlets), this is the model given in BWW. It is important to note that $\sigma_{\gamma d(j)}^2$ indicates the strength of the testlet effect for testlet $d(j)$, that is the larger $\sigma_{\gamma d(j)}^2$ the greater the proportion of total variance in test scores that is attributable to the given testlet. Obtaining inferences for the model parameters is not straightforward as the model given in (2) and the Gaussian prior distributions are non-conjugate. We draw inferences from the model parameters by choosing samples from their marginal posterior distributions using Markov Chain Monte Carlo (MCMC) simulation. This approach is briefly described next.

3. Parameter Estimation

Estimates of the unknown model parameters are obtained by drawing samples from their marginal posterior distributions using a form of MCMC simulation, the Gibbs sampler (Gelfand & Smith 1990). To implement the MCMC simulation, we draw in turn from the set of conditional distributions of each unknown parameter, conditional on the previously drawn values of all other parameters and the data. After a determined number of complete cycles, the draws may be considered from the target distribution of interest. Inferences from the parameters are then derived by computing any quantity of inference from the obtained samples. A difficulty with this model is that it is not straightforward to sample from the conditional distribution of the model parameters, even after conditioning on all other parameters and the data. Specifically, the conditional distribution for each parameter δ in the set $\Delta = (\theta_1, ..., \theta_I, a_1, ..., a_J, b_1, ..., b_j, q_1, ..., q_J, \gamma_{11}, ..., \gamma_{Id(J)})$ given all other parameters and the data is given by

$$[\delta_i | \Delta_{-\delta_i}, Y] \propto \prod_{i=1}^{I} \prod_{j=1}^{J} p_{ij}^{y_{ij}} (1 - p_{ij})^{1-y_{ij}} N(\mu_{\delta_i}, \sigma_{\delta_i}^2), \qquad (5)$$

where $[X|Y]$ denotes the conditional density of random variable X given Y, $N(\mu_{\delta_i}, \sigma_{\delta_i}^2)$ is a normal density with mean μ_{δ_i} and variance $\sigma_{\delta_i}^2$, $\Delta_{-\delta_i}$ is the set of all parameters excluding δ_i, and Y is the observed $I \times J$-dimensional binary outcome matrix. The conditional distributions given by (5) are the product of a product Bernoulli and normal density, known to not permit a closed-form solution. The approach we use is briefly described next but details are available in Du (1998).

For these conditional distributions, which do not permit straightforward sampling, we implemented a Metropolis-Hastings algorithm (Hastings 1970). The basic idea is that to sample from target distribution f (which is hard to sample from), for arbitrary parameter ω, you first draw a single candidate from approximating distribution g (easy to sample from, e.g., normal, or t-distribution), call this draw ω^*. Then at iteration $t+1$ of the MCMC simulation you accept draw ω^*, i.e. $\omega^{t+1} = \omega^*$, with probability equal to $\min[1, (f(\omega^*)/g(\omega^*))/(f(\omega^{(t)})/g(\omega^{(t)}))]$, and set $\omega^{(t+1)} = \omega^{(t)}$ otherwise. The efficiency of the Metropolis-Hastings approach is determined by the "closeness" of target distribution f and approximating distribution g. We selected normal distributions for all of the approximating distributions, with mean equal to the previously drawn parameter value and variance set so as to yield an optimal acceptance rate.

The conditional distributions of the variance components, σ_a^2, σ_b^2, σ_q^2, $\sigma_{\gamma d(j)}^2$, however do permit straightforward sampling from inverse χ^2 distributions given the assumed conjugate structure described in Section 2.

Diagnostics performed on the sampler, and alternative approaches tried but rejected, made us confident that our sampler was obtaining draws from the true posterior, and further that our approach was computationally fast and efficient. We describe simulation findings next that confirmed this.

4. Simulation Results

The simulation was designed to study the performance of model (2) and its implementation. We tested four different models:

(i) MML - the standard 3PL model fit using maximum marginal likelihood as implemented in the computer program BILOG (Mislevy & Bock, 1983)

(ii) Gibbs - the standard 3PL model (without testlet effects) fit using the MCMC simulation.

(iii) Gibbs (= γ) - a testlet based version of the 3PL with a single value of the parameter variance σ_γ^2 for all testlets, a 3PL version of the BWW model.

(iv) Gibbs ($\neq \gamma$) - a testlet based version of the 3PL with possibly different values of the parameter variance $\sigma_{\gamma d(j)}^2$ for each testlet.

These four models were fit to three different sets of test data, which were simulated to match closely the marginal distribution observed for a typical form of the Scholastic Assessment Test (SAT). Each data set consists of binary scores of 1,000 simulees responding to a test of 70 multiple-choice items. Each of the three 70-item tests is composed of 30 independent items and four testlets containing 10 items each. The data sets varied in the amount of local dependence simulated as represented by the size of $\sigma_{\gamma d(j)}^2$ used. The other parameters for the simulation were drawn from Gaussian distributions as follows: $\theta_i \sim N(0,1)$, $a_j \sim N(0.8, 0.2^2)$, $b_j \sim N(0,1)$, $c_j \sim N(0.2, 0.03^2)$. The character of the three data sets is shown in Table 1.

Table 1. Simulation design showing the variance of γ

Data Set	Testlet 1	Testlet 2	Testlet 3	Testlet 4
1 (no effect)	0.00	0.00	0.00	0.00
2 (equal effects)	0.80	0.80	0.80	0.80
3 (unequal effects)	0.25	0.50	1.00	2.00

In the first dataset the size of the testlet effect is set to zero for all four testlets ($\sigma_{\gamma k}^2 = 0$, $k = 1, 2, 3, 4$). We expected that under this condition all the estimation procedures would perform equally well in recovering the true model parameters. In the second data set where $\sigma^2_{\gamma k}$ was set to a modest 0.8 but equal size for all of the testlets, we expected that the models that had the capacity to characterize such excess dependence (models (iii) and (iv)) would do better at recovering the true underlying parameters. In the last dataset, we expected that model (iv) would do the best job.

We anticipated that the size of the advantage of the more general models would be modest in our simulation because (a) a large proportion of the test is comprised of independent items, and (b) the values chosen for the $\sigma^2_{\gamma k}$ were not over a particularly large range. We chose this particular simulation structure, despite its lack of dramatic impact, because it reflects the structure of the SAT. As we shall see in Section 5, a test that is built with greater local dependence of its items within testlets will yield a considerably more dramatic effect.

There are many plausible summary statistics that we can use to convey the results of the simulations. We present four different ones here. In Table 2 are shown the correlations between the estimated parameters and the true ones. These indicate the sorts of modest advantages of the new model over older approaches that were anticipated. Note that our results match previous simulation results (Lord, 1980; Thissen & Wainer, 1982; Wingersky & Lord, 1984) in that the location parameters θ and b appear to be easy to estimate well, whereas the slope (a) and guessing parameter (c) are more difficult. The new model does considerably better on these, more difficult to estimate, parameters.

We next examine the mean absolute error associated with each estimator. The results of this, shown in Table 3, mirror the results seen in the correlations of Table 2; there is a modest reduction in the size of the mean absolute errors for the most general model. But note the efficiencies (the ratio of test information of other estimators to the MML approach in turn). When it's generality is required (unequal γs) the model that fits such excessive local dependence is 28% more efficient than the commonly used MML estimator. This increase in efficiency is principally due to the bias in the estimates obtained from the less general model. Note further that in the most general condition – larger, unequal testlet effects – the size of the mean absolute error of b and c is considerably worse (almost twice the size) for the BILOG estimates than for the Gibbs($\neq \gamma$) model.

Another measure of the success of the model fitting procedure is the extent to which the model-based proficiency estimates match the true values. To some extent this was examined in the correlations shown in Table 2, but another measure that might be of use is counting how

Table 2. Correlations of estimated parameters with true values

		θ	a	b	c	Mean
No effects						
	MML	0.96	0.84	0.98	0.56	0.84
	Gibbs	0.96	0.93	0.99	0.64	0.88
	Gibbs $(= \gamma)$	0.96	0.93	0.99	0.63	0.88
	Gibbs $(\neq \gamma)$	0.96	0.93	0.99	0.66	0.89
Equal effects						
	MML	0.92	0.77	0.98	0.64	0.83
	Gibbs	0.92	0.80	0.98	0.65	0.84
	Gibbs $(= \gamma)$	0.93	0.87	0.99	0.69	0.87
	Gibbs $(\neq \gamma)$	0.93	0.86	0.99	0.70	0.87
Unequal effects						
	MML	0.91	0.76	0.97	0.57	0.80
	Gibbs	0.91	0.80	0.97	0.60	0.82
	Gibbs $(= \gamma)$	0.92	0.75	0.98	0.68	0.83
	Gibbs $(\neq \gamma)$	0.93	0.83	0.99	0.68	0.86
	Mean	0.93	0.84	0.98	0.64	0.85

many of the top 100 candidates were correctly identified by the model. The results of this are shown in Table 4.

One advantage of the Bayesian approach espoused here is an increased capacity for model checking that is not available in more traditional approaches (Gelman, Carlin, Stern & Rubin, 1995). Such checking is possible since the Gibbs sampler produces a sample from the entire posterior distribution of each parameter. Thus one can calculate whatever characteristic of the posterior that is of particular interest. For example, to check on the validity of the estimated confidence intervals, which rely heavily on a Gaussian assumption, one can compute a 95% coverage region. We expect that for a correctly specified model, the true value falls within the nominal 95% confidence region at least 95% of the time. In Table 5 we see that when the model suits the data simulated, the coverage probability is correct, but as the model's assumptions depart from the character of the simulated data the estimated bounds become overly optimistic (too narrow). MML interval estimates are omitted as a complete discussion and comparison to the MCMC approach presented here are given in the next chapter. A sum-

Table 3. Mean absolute error and relative efficiency of each estimator

		θ	a	b	c	Relative Efficiency
No effects						
	MML	0.22	0.19	0.15	0.04	1.00
	Gibbs	0.22	0.14	0.11	0.03	1.00
	Gibbs $(= \gamma)$	0.22	0.15	0.12	0.03	1.00
	Gibbs $(\neq \gamma)$	0.22	0.14	0.11	0.03	1.00
Equal effects						
	MML	0.31	0.22	0.16	0.04	1.00
	Gibbs	0.30	0.23	0.17	0.05	1.00
	Gibbs $(= \gamma)$	0.28	0.14	0.13	0.03	1.08
	Gibbs $(\neq \gamma)$	0.28	0.14	0.13	0.03	1.16
Unequal effects						
	MML	0.34	0.18	0.24	0.07	1.00
	Gibbs	0.33	0.16	0.20	0.05	1.06
	Gibbs $(= \gamma)$	0.31	0.18	0.15	0.04	1.20
	Gibbs $(\neq \gamma)$	0.30	0.14	0.13	0.04	1.28
	Mean	0.28	0.17	0.15	0.04	1.07

Note: The relative efficiency compares each method with MML as baseline

mary of those results suggest overly narrow MML interval estimates by about 9%.

The analyses described here were run on a desktop minicomputer with a Pentium 233 mghz processor. Although we typically ran each analysis for something in excess of 4,000 iterations, we found that convergence generally (always) occurred within 200-300 iterations. One iteration took less than 2 seconds. Thus most analyses (e.g., 80 items and 1,000 people) were satisfactorily completed within 15 minutes. We view this as acceptable, even (especially) for operational tests. Since future computers will be faster and cheaper it is certain that MCMC procedures within IRT are a viable practical alternative to all other methodologies.

To provide a better intuitive picture of the performance of the new testlet model, we plotted the parameter estimates obtained from the simulation against their residual from the true value. To use these plots one estimates the parameter of interest and locates the value of the

Table 4. Percentage of top 100 examinees correctly identified

	No effects	= Effects	≠ Effects	Mean
MML	83	73	69	75
Gibbs	83	73	69	75
Gibbs $(= \gamma)$	83	76	69	76
Gibbs $(\neq \gamma)$	83	76	72	77
Mean	83	75	70	76

Table 5. 95% Coverage probability

	θ	a	b	c	Mean
No effects					
Gibbs	0.94	0.96	0.92	0.92	0.94
Gibbs $(= \gamma)$	0.95	0.95	0.93	0.98	0.95
Gibbs $(\neq \gamma)$	0.96	0.96	0.96	0.95	0.96
Equal effects					
Gibbs	0.89	0.66	0.94	0.93	0.86
Gibbs $(= \gamma)$	0.95	0.96	0.96	0.98	0.96
Gibbs $(\neq \gamma)$	0.95	0.96	0.96	1.00	0.97
Unequal effects					
Gibbs	0.87	0.88	0.90	0.97	0.91
Gibbs $(= \gamma)$	0.94	0.85	0.87	0.95	0.90
Gibbs $(\neq \gamma)$	0.94	0.97	0.97	0.95	0.96
Mean	0.93	0.91	0.93	0.96	

parameter on the horizontal axis, then reads the accuracy of that parameter on the vertical axis. Of course different circumstances (tests of different lengths, examinee samples of different sizes, different testlet structures) will yield different results, but these provide a plausible idea for the current circumstances, which are reasonably common.

In Figure 1, Panels a, b and c, are the values obtained when there is no testlet effect ($\gamma = 0$). From Panel a we see that 55 of the 70 items' slopes are estimated within 0.2 units and only 3 items have errors larger than 0.4; Panel b shows us that difficulties are estimated at about the same level of accuracy. Panel c shows that the guessing parameter is

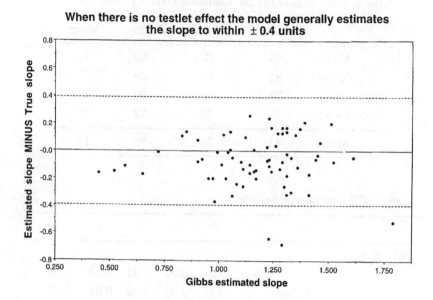

Figure 1. Panel a. Slope parameter *a*.

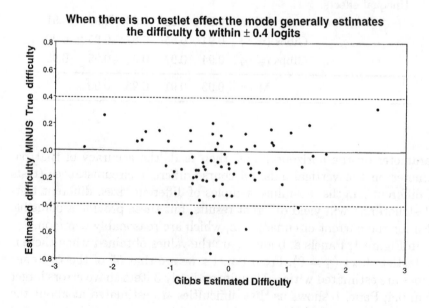

Figure 1. Panel b. Difficulty parameter *b*.

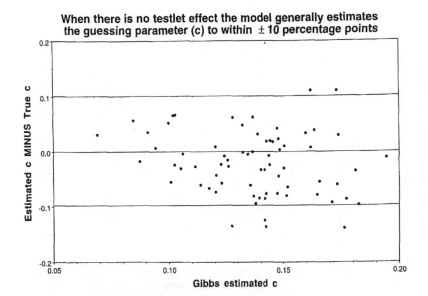

Figure 1. Panel c. Guessing parameter *c.*

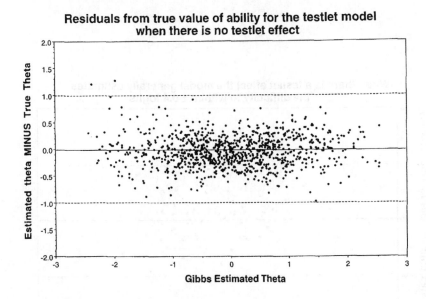

Figure 1. Panel d. Proficiency parameter θ.

Figure 1. Model parameters estimated by Gibbs($\neq \gamma$), for 70 simulated items, and 1,000 simulees, when there are no testlet effects $(\gamma = 0)$, plotted against their residuals from their true value.

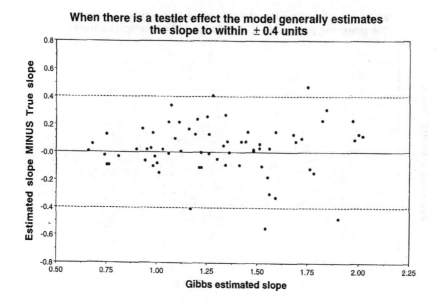

Figure 2. Panel a. Slope parameter *a.*

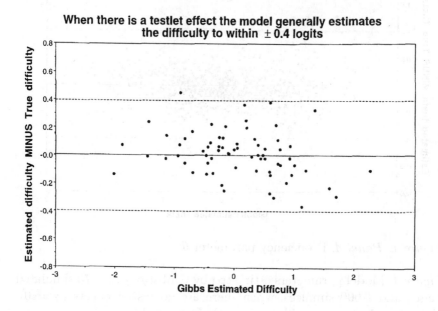

Figure 2. Panel b. Difficulty parameter *b.*

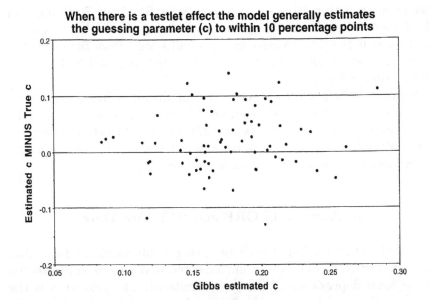

Figure 2. Panel c. Guessing parameter *c*.

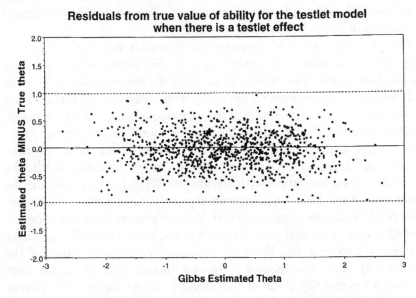

Figure 2. Panel d. Proficiency parameter θ.

Figure 2. A plot of the model parameters estimated by Gibbs($\neq \gamma$) for 70 simulated items and 1,000 simulees when there are testlet effects ($\gamma \neq 0$) plotted against their residuals from their true value.

estimated to within 10 percentage points for 64 of the 70 items, and all of them with 15 percentage points. Panel d shows that examinee proficiency is tightly clustered within a half a logit from the estimated value for most simulees, and that only two (out of 1,000) had an error greater than 1 logit.

Figure 2 is identical to Figure 1 except that it is based on the situation in which 40 items are contained in one of four testlets and the variances of γ are 0.25, 0.50, 1.00 and 2.00 in each of the testlets respectively. Visually it is hard to detect any difference in fit for the model when there is a testlet effect from when there is not.

5. Analysis of GRE and SAT Test Data

The results obtained from the simulations provide ample evidence that the new testlet model can be implemented and that it captures the excess local dependence of the type simulated. The next step is the fitting of the model to real test data and the examination of the outcome. We fit data from two well known operational large scale tests; the verbal section of one administered form of the Scholastic Assessment Test (SAT-V) and the verbal section of an administered form of the Graduate Record Examination (GRE-V)[1]

Table 6 contains a brief summary of the testlet structure of the two tests; both contained four testlets as well as a substantial number of independent items. The results from the SAT were obtained from an analysis of the test responses from 4,000 examinees; the GRE results are the average of the results obtained from two samples of 5,000 examinees. A detailed description of these analyses is found in Du(1998). These results contain much that was expected: testlet effects exist and they are unequal across testlets. There was also one unexpected result; the size of the effect was larger with the shorter GRE testlets than for the longer ones of the SAT. Of course we expected the opposite, since more items on the same set of material should show greater local dependence, ceteris paribus. But obviously all else was not equal; the GRE passages apparently were not like the SAT passages. A detailed study of the reasons why local dependence is greater in one kind of testlet than another is beyond the reach of this chapter. Nevertheless, it is worthy of note that GRE testlet 1 contains a 400 word long reading passage, whereas testlet 2's passage is only 150 words. Relating characteristics

---

[1] The SAT form analyzed was operationally administered in 1996 and the GRE form was administered in 1997. Both are released forms of the test. We are grateful to the College Board and the Graduate Record Research Board for their generous permission to use their data.

of the passage (e.g., length, type of material, etc.) with the size of its testlet effect is certainly an area for future research.

Table 6. The structure of two tests and estimated test-let effects

Test	Testlet	Number of items	$Var(\gamma)$
	independent	38	
	1	12	0.13
SAT-Verbal	2	6	0.35
	3	10	0.33
	4	12	0.11
	TOTAL	78	
	independent	54	
	1	7	0.61
GRE-Verbal	2	4	0.96
	3	7	0.60
	4	4	0.63
	TOTAL	76	

With real data we do not have the benefit of knowing what are the right answers, and so calculating the correlation of estimates with the correct answers (as in Table 2) is not possible. We can however calculate the correlation of the estimates obtained in two different ways. These are shown in Table 7. We see that for the SAT the estimates obtained using BILOG are almost perfectly correlated with those using the most general testlet model. This is especially true for the ability parameter (θ) and for the difficulty (b) and is not unexpected since the size of σ_γ^2 for the SAT testlets is modest. The GRE results are somewhat different because the size σ_γ^2 is larger. Again we find that there is almost perfect agreement for ability and difficulty, but considerably less agreement for slope (a) and lower asymptote (c). We conjecture that because these latter two parameters are much more difficult to estimate in general, when extra error is in the mix (as is the case when there is unmodeled local dependence) the parameters' estimates are perturbed.

Table 7 represents correlations between two <u>different</u> models for the <u>same</u> set of data. The next question we will examine is the correlation of parameter estimates for the <u>same</u> model on <u>different</u> data samples.

Table 7. Correlation between parameter estimates obtained from BILOG and Gibbs ($\neq \gamma$)

Test	a	b	c	θ
SAT-Verbal	0.97	0.99	0.96	0.998
GRE-Verbal	0.86	0.98	0.75	0.995

The results of this study are shown in Table 8. These were obtained just from the GRE data (we had over 100,000 examinees and so drawing two large independent data samples from it was very easy). We chose two independent samples of 5,000 examinees each and fit these samples with both BILOG and Gibbs ($\neq \gamma$). There is ample empirical evidence suggesting that statistically stable parameter estimates would be obtained with a sample size this large (Lord, 1983). We found that indeed this was true for the difficulty parameter, but not necessarily for the other two; at least when the other two are estimated using maximum marginal likelihood. In Table 8 we show not only the correlation between the estimates of the same parameters in different samples (analogous to the reliability of the parameter estimates), but also partition this correlation into two pieces. One piece is the same correlation, but only between the 54 independent items, and the second piece is calculated only for the items embedded in testlets. Interpretation is aided by keeping in mind that the standard error for the correlations associated with the independent items is about 0.1, whereas for the testlet items is about 0.2; differences of 0.05 should not be regarded with undue seriousness. The messages that we can derive from this analysis are clear:

(i)   estimating item difficulty (b) is easy, its estimates are not seriously affected by excess local dependence.

(ii) estimating item discrimination (a) is harder, at least for BILOG. The testlet model seems to do a very good job and does it equally well whether or not the item is embedded in a testlet. BILOG's estimate is clearly adversely affected by the excess local dependence.

(iii) estimating the lower asymptote (c) is the toughest of all, but local dependence doesn't seem to have much of an effect.

Thus we conclude that the principal area in which local dependence has its effect is, as expected, the slope of the item characteristic curve. This finding matches that of BWW who also found that when there is excess local dependence the item discriminations are systematically underestimated by any model that does not account for it. The apparent reason for this is that when there is unmodeled local dependence the model looks upon the resulting nonfit as noise and so the regression of the item on the underlying trait is, in the face of this noise, more gradual. When the local dependence is modeled, the 'noise' caused by it is gone and so the regression on this more cleanly estimated trait is steeper.

Table 8. Correlation between parameter estimates on the GRE-V obtained by BILOG and Gibbs($\neq \gamma$) on independent samples

Model	a	b	c	number of items
BILOG	0.79	0.98	0.71	76
Independent	0.78	0.98	0.68	54
Testlet	0.57	0.97	0.77	22
Gibbs($\neq \gamma$)	0.95	0.98	0.78	76
Independent	0.94	0.98	0.75	54
Testlet	0.94	0.99	0.88	22

This effect is clearly seen when one examines paired plots of the values of the slope parameter. In Figure 3 is a plot of the two values of a obtained from the two samples obtained from BILOG. Note that the slopes for the testlet-based items are less reliably estimated (their regression line has a more gradual incline) than the independent items. Figure 4 is a plot of the same thing estimated by the Gibbs($\neq \gamma$) model. Note that the two regression lines are essentially identical, implying that the slopes of all items are estimated with equal reliability.

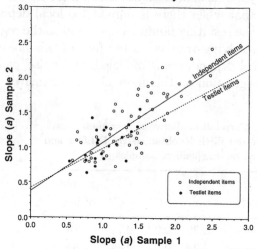

Figure 3. A plot of the slopes (*a*'s) estimated by BILOG for 76 GRE verbal items. The filled in plotting points represent the testlet-based items; the open circles are independent items. The independent items are more reliably estimated.

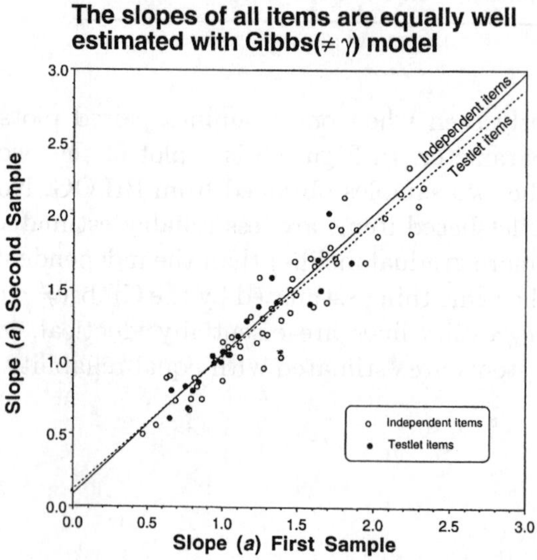

The slopes of all items are equally well estimated with Gibbs(≠ γ) model

Figure 4. A plot of the slopes (*a*'s) estimated by Gibbs(≠ γ) for 76 GRE verbal items. The filled in plotting points represent the testlet-based items; the open circles are independent items. All item types are estimated equally reliably.

6. Discussion and Conclusions

In this chapter we have presented motivation for testlet based exams, and provided a general psychometric model for scoring them. We have spent considerable effort trying to demonstrate that our generalization of the traditional 3PL IRT model both fits the local dependence that is the hallmark of items-nested-within-testlets, as well as being practical for operational tests. We would like to emphasize how the estimation of the parameters of this model was neatly accomplished within the context of a Bayesian paradigm implemented by Markov Chain Monte Carlo procedures.

The simulations we report are a small fraction of the analyses done. Our analyses reported here of the data from the SAT and GRE exams were meant to illustrate how this model can be applied to operational test data and show what benefits accrue from its use. The simulations show that the estimation of item difficulty and examinee proficiency are well done without recourse to our, more complex, parameterization. This finding should be well expected — how else could one explain the success of the Rasch model in so many circumstances where it is clearly inappropriate? It is in the estimation of item discriminations, and other important statistics like item information which are functions of the discriminations, that the power of this new model is demonstrated. This power shows itself in the analysis of the GRE data where we found the same reliability of measurement of the discrimination parameter as we found for the difficulty. Also, and in sharp contrast to BILOG, we found essentially the same reliability for testlet-based items as for independent items.

At this point it seems worthwhile to become a little more discursive about the plausible uses and usefulness of this model. To begin this discussion we need to emphasize an important distinction about multidimensionality in tests (a distinction previously made by Wainer & Thissen (1996)). There are two ways a test can contain multidimensionality: fixed multidimensionality and random multidimensionality.

(a) Fixed multidimensionality is predetermined by the test builder and appears in much the same way in all forms of the test. An example of fixed multidimensionality might be a test like the GRE, which could be made up of three components (verbal, quantitative and analytic reasoning) that are combined in equal quantities to make up a composite score. None would deny that this is a multidimensional test, but its multidimensionality is of fixed and predetermined character.

(b) Random multidimensionality occurs without design and is not consistent across different forms of the test. An example of random multidimensionality might be the multidimensional structure induced by the character of a reading passage. One goal of a reading comprehension item is to test how well an examinee can read and understand descriptive prose; the passage's content is a source of uncontrolled variation. Thus a passage about baseball might introduce some extra knowledge dimensions that are not present in a passage of similar difficulty about rituals in Samoa, which has, in its turn, some of its own, unique, content dimensions. These extra dimensions are a nuisance when it comes to estimating overall proficiency, because they are idiosyncratic to a particular form and hence make form to form comparisons difficult. Any sensible scoring method ought to minimize the impact of random multidimensionality on itself.

The model for testlets that we have proposed is essentially one for fitting random multidimensionality, and thereby controlling its influence. The model can also be used as a confirmatory method to check fixed multidimensionality akin to a confirmatory factor analysis applied to grouped questionnaire items. By so doing it allows us an unobscured view of the other parameters. We have seen that what this seems to mean in practice are better estimates of everything, but much better estimates of item discrimination. In the end the practical value of this means that we can have better estimates of precision. Thus, if we choose to stop an individualized testlet built test on the basis of the standard error of proficiency, we can do so and have reasonable assurance that the model's estimate of standard error is closer to being correct.

References

Albert, J. H. (1992). Bayesian estimation of normal ogive item response curves using Gibbs sampling. *Journal of Educational Statistics*, 17, 251-269.

Albert, J. H., & Chib, S. (1993). Bayesian analysis of binary and polychotomous response data. *Journal of the American Statistical Association*, 88, 669-679.

Anastasi, A. (1976). *Psychological testing* (4th edition). New York: Macmillan.

Birnbaum, A. (1968). Some latent trait models and their use in inferring an examinee's ability. In F.M. Lord and M.R. Novick, *Statistical theories of mental test scores* (chapters 17-20). Reading, MA: Addison-Wesley.

Bradlow, E. T., Wainer, H., & Wang, X. (1999). A Bayesian random effects model for testlets. *Psychometrika*, 64, 153-168.

Du, Z. (1998). Modeling conditional item dependencies with a three-parameter logistic testlet model. Unpublished doctoral dissertation. Columbia University, New York, NY.

Gelfand, A. E. & Smith, A. F. M. (1990). Sampling-based approaches to calculating marginal density. *Journal of the American Statistical Association*, 85, 398-409.

Gelman, A., Carlin, J. B., Stern, H. S. & Rubin, D. B. (1995). *Bayesian data analysis*. London: Chapman & Hall.

Hambleton, R.K., & Swaminathan, H. (1985). *Item response theory: Principles and applications*. Boston: Kluwer.

Hastings, W. K. (1970). Monte Carlo sampling methods using Markov chains and their applications. *Biometrika*, 54, 93-108.

Lord, F. M. (1980). *Applications of item response theory to practical testing problems*. Hillsdale, NJ: Lawrence Erlbaum Associates.

Lord, F. M. (1983). Small N justifies the Rasch model. In D. J. Weiss (Ed.) *New horizons in testing: Latent trait test theory and computerized adaptive testing* (pp. 51-62). New York: Academic Press.

Lord, F.M., & Novick, M.R. (1968). *Statistical theories of mental test scores*. Reading, MA: Addison-Wesley.

Mislevy, R.J., & Bock, R.D. (1983). *BILOG: Item analysis and test scoring with binary logistic models* [computer program]. Mooresville, IN: Scientific Software.

Rosenbaum, P.R. (1988). Item bundles. *Psychometrika*, 53(3), 349-359.

Sireci, S.G., Wainer, H., & Thissen, D. (1991). On the reliability of testlet-based tests. *Journal of Educational Measurement*, 28, 237-247.

Stout, W. F. (1987). A nonparametric approach for assessing latent trait dimensionality. *Psychometrika*, 52, 589-617.

Stout, W. F., Habing, B., Douglas, J., Kim, H. R., Russos, L., & Zhang, J. (1996). Conditional covariance-based nonparametric multidimensional assessment. *Applied Psychological Measurement*, 20, 331-354.

Tanner, M. A. & Wong, W. H. (1987). The calculation of posterior distributions by data augmentation (with discussion). *Journal of the American Statistical Association*, 82, 528-550.

Thissen, D. M. & Wainer, H. (1982). Some standard errors in item response theory. *Psychometrika*, 47, 397-412.

Wainer, H. (1995). Precision and differential item functioning on a testlet-based test: The 1991 Law School Admissions Test as an example. *Applied Measurement in Education*, 8, 157-187.

Wainer, H., & Kiely, G. (1987). Item clusters and computerized adaptive testing: A case for testlets. *Journal of Educational Measurement*, 24, 185-202.

Wainer, H., & Mislevy, R. J. (1990). Item response theory, item calibration and proficiency estimation. In H. Wainer, D.J. Dorans, R. Flaugher, B.F. Green, R.J. Mislevy, L. Steinberg, & D. Thissen, (1990). *Computerized adaptive testing: A Primer*. (pp. 65-102). Hillsdale, NJ: Lawrence Erlbaum Associ-ates.

Wainer, H. & Thissen, D. (1996). How is reliability related to the quality of test scores? What is the effect of local dependence on reliability? *Educational Measurement: Issues and Practice*, 15(1), 22-29.

Wingersky, M. S. & Lord, F. M. (1984). An investigation of methods for reducing sampling error in certain IRT procedures. *Applied Psychological Measurement*, 8, 347-364.

Yen, W. (1993). Scaling performance assessments: Strategies for managing local item dependence. *Journal of Educational Measurement*, 30, 187-213.

Zhang, J. & Stout, W. F. (1999). The theoretical DETECT index of dimensionality and its application to approximate simple structure. *Psychometrika*, 64, 213-250.

Chapter 14
MML and EAP Estimation in Testlet-based Adaptive Testing

Cees A.W. Glas, Howard Wainer, & Eric T. Bradlow
University of Twente, The Netherlands
Educational Testing Service, USA
University of Pennsylvania, USA*

1. Introduction

In the previous chapter, Wainer, Bradlow and Du (this volume) presented a generalization of the three-parameter item response model in which the dependencies generated by a testlet structure are explicitly taken into account. That chapter is an extension of their prior work that developed the generalization for the two-parameter model (Bradlow, Wainer, & Wang, 1999). Their approach is to use a fully Bayesian formulation of the problem, coupled with a Markov Chain Monte Carlo (MCMC) procedure to estimate the posterior distribution of the model parameters. They then use the estimated posterior distribution to compute interval and point estimates. In this chapter, we derive estimates for the parameters of the Wainer, Bradlow, and Du testlet model using more traditional estimation methodology; maximum marginal likelihood (MML) and expected a posteriori (EAP) estimates. We also show how the model might be used within a CAT environment.

After deriving the estimation equations, we compare the results of MCMC and MML estimation procedures and we examine the extent to which ignoring the testlet structure affects the precision of item calibration and the estimation of ability within a CAT procedure.

2. MML Estimation

In Glas (this volume), a concise introduction to maximum marginal likelihood (MML, Bock & Aitkin, 1982; Mislevy, 1986) was presented.

* The work of Cees Glas was supported by the Law School Admission Council (LSAC); the work of Eric Bradlow and Howard Wainer on this research was supported by the Graduate Record Examination Research Board and the TOEFL Research Committee.

W.J. van der Linden and C.A.W. Glas (eds.),
Computerized Adaptive Testing: Theory and Practice, 271–287.
© 2000 *Kluwer Academic Publishers. Printed in the Netherlands.*

It was shown that ability parameters could be viewed as unobserved data. Then the complete data likelihood could be integrated over these unobserved variables to produce a so-called marginal likelihood. Finally, it was shown that the likelihood equations are easily derived by applying an identity derived by Louis (1982; also see, Glas, 1992, 1998, 1999) in the framework of the EM-algorithm, which is an algorithm for finding the maximum of a likelihood marginalized over unobserved data. In the present chapter, this framework will be adopted to the testlet model. In our approach, both the ability parameters θ_j and testlet parameters γ_{jd} will be treated as unobserved data.

Consider a test consisting of D testlets, which will be labeled $d = 1, ..., D$. Let $\mathbf{u}_d$ be the response pattern on testlet d, and let $\mathbf{u}$ be the response pattern on the complete test, that is, $\mathbf{u} = (\mathbf{u}_1, \mathbf{u}_2, ..., \mathbf{u}_d, ..., \mathbf{u}_D)$. It is assumed that the person parameters θ_j have a standard normal distribution with a density denoted by $g(\theta_j)$. For every testlet, every person j independently draws γ_{jd} from a normal distribution with zero mean and variance $\sigma^2_{\gamma d}$. The density of γ_{jd} will be denoted $h(\gamma_{jd} \mid \sigma^2_{\gamma d})$. Let $p(\mathbf{u}_{jd} \mid \theta_j, \gamma_{jd}, \boldsymbol{\beta}_d)$ be the probability of observing response pattern $\mathbf{u}_{jd}$ as a function of the ability parameter θ_j, the parameters a_i, b_i and c_i of the items in testlet d, which are stacked in a vector $\boldsymbol{\beta}_d$, and the testlet parameter γ_{jd}. Responses are given according to the testlet model defined by (2) in the previous chapter. Let $\boldsymbol{\eta}$ be the vector of all parameters to be estimated, so $\boldsymbol{\eta}$ contains the item parameters $\boldsymbol{\beta}_d$, and the variance parameters $\sigma^2_{\gamma d}$, $d = 1, ..., D$. The marginal probability of observing response pattern $\mathbf{u}_j$ is given by

$$p(\mathbf{u}_j; \boldsymbol{\eta}) = \int \left[\prod_d \int p(\mathbf{u}_{jd} \mid \theta_j, \gamma_{jd}, \boldsymbol{\beta}_d) h(\gamma_{jd} \mid \sigma^2_{\gamma d}) d\gamma_{jd} \right] g(\theta_j) d\theta_j.$$

$$(1)$$

Notice that the marginal likelihood entails a multiple integral over θ_j and γ_{jd}, $d = 1, ..., D$. Fortunately, the integrals for the testlets factor, hence there is no need to compute integrals of dimension $D + 1$; computing D two-dimensional integrals suffices. In this respect the approach to model estimation presented here is related to the bi-factor full-information factor analysis model by Gibbons and Hedeker (1992).

Using the framework of Glas (this volume), the likelihood equations are given by

$$\frac{\partial}{\partial \boldsymbol{\eta}} \log L(\boldsymbol{\eta}; \mathbf{U}) = \sum_j E(\boldsymbol{\omega}_j(\boldsymbol{\eta}) \mid \mathbf{u}_j, \boldsymbol{\eta}) = \mathbf{0},$$

$$(2)$$

with $\boldsymbol{\omega}_j(\boldsymbol{\eta})$ the first-order derivatives of the complete data log-likelihood for person j. In the present case, the "complete data" are $\mathbf{u}_j, \boldsymbol{\gamma}_j =$

$(\gamma_{j1}, ..., \gamma_{jd}, ..., \gamma_{jD})$ and θ_j, so the derivative of the complete data log-likelihood with respect to η is

$$\omega_j(\eta) = \frac{\partial}{\partial \eta} \log p(\mathbf{u}_j, \gamma_j, \theta_j; \eta). \tag{3}$$

Notice that the first-order derivatives in (2) are expectations with respect to the posterior density

$$p(\gamma_j, \theta_j \mid \mathbf{u}_j, \eta) \propto \prod_d p(\mathbf{u}_{jd} \mid \theta_j, \gamma_{jd}, \beta_d) h(\gamma_{jd} \mid \sigma_{\gamma d}^2) g(\theta_j). \tag{4}$$

Further, the loglikelihood in the right-hand side of (3) can be written as

$$\log p(\mathbf{u}_j, \gamma_j, \theta_j; \beta) =$$

$$\sum_{d,i} \log P(u_{ij}; \theta_j, \gamma_{jd(i)}, \beta_{di}) + \sum_d \log h(\gamma_{jd} \mid \sigma_{\gamma d}^2) + \log g(\theta_j),$$

where β_{di} is the vector of the three item parameters of item i in testlet d. Therefore, the likelihood equation for $\sigma_{\gamma d}^2$ can be written as

$$\sum_j E(\partial \log h(\gamma_{jd} \mid \sigma_{\gamma d}^2)/\partial \sigma_{\gamma d} \mid \mathbf{u}_j, \eta) . \tag{5}$$

Setting (5) equal to zero results in

$$\sigma_{\gamma d}^2 = \frac{1}{N} \sum_j E(\gamma_{jd}^2 \mid \mathbf{u}_j, \eta). \tag{6}$$

So estimation boils down to setting $\sigma_{\gamma d}^2$ equal to the posterior expectation of γ_{jd}^2. The equations for the item parameters a_i, b_i, and c_i are analogous to the equivalent equations for the 3PL model, the difference being that the first-order derivatives of the complete data log-likelihood function are now posterior expectations with respect to (4). So the expressions for the first order derivatives of the item parameters, say $\omega_j(a_i)$, $\omega_j(b_i)$ and $\omega_j(c_i)$ are given by (4), (5) and (6) in Glas (this volume). Inserting these expressions for $\omega_j(a_i)$, $\omega_j(b_i)$ and $\omega_j(c_i)$ as elements of the vector $\omega_j(\eta)$ into (2) gives the likelihood equations for the item parameters. Finally, the computation of the standard errors of the parameters estimates is a straightforward generalization of the method for the 3PL model, presented in Glas (this volume) so the standard errors can be found upon inverting an approximate information matrix

$$\mathbf{H}(\eta, \eta) \approx \sum_j E(\omega_j(\eta) \mid \mathbf{u}_j, \eta) E(\omega_j(\eta) \mid \mathbf{u}_j, \eta)'.$$

Like the 3PL model, the MML estimation equations can be solved either by an EM or a Newton-Raphson algorithm.

In Glas (this volume), several generalizations of the basic MML framework were outlined. These generalizations also pertain to the MML estimation procedure for the testlet model. So data can be collected in an incomplete test administration design, and the testlet model can be enhanced by introducing multiple populations with unique normal ability distributions. Also the description of the Bayes modal estimation procedure for the 3PL model in Glas (this volume) directly applies the testlet model.

3. Alternative Model Formulations

There are at least three mathematically isomorphic ways to include the testlet parameter $\gamma_{jd(i)}$ in the model. While they all yield identical fit to the data, each supports a different interpretation for the testlet parameter.

Alternative 1. $\gamma_{jd(i)}$ as part of ability

In the current model the logit is configured as $a_i(\theta_j - b_i + \gamma_{jd(i)})$. If we think of γ as being part of ability this can be regrouped as $a_i[(\theta_j + \gamma_{jd(i)}) - b_i]$. An interpretation of this approach is that on the specific material of this testlet this particular examinee is either especially good ($\gamma_{jd(i)} > 0$) or especially poor ($\gamma_{jd(i)} < 0$). The value of this approach is that it leaves the item parameters untouched, which hence maintain the same value for all examinees. This feature conforms well with the notions of what standardized testing means. Interestingly, this interpretation of the model also supports an hierarchical interpretation. Consider the reparametrization $\xi_{jd} = \theta_j + \gamma_{jd}$. Since θ_j and γ_{jd} have independent normal distributions with zero means and variances equal to unity and $\sigma^2_{\gamma d}$, respectively, ξ_d given θ has a normal distribution with mean θ and variance $\sigma^2_{\gamma d}$. This result supports the interpretation that for every testlet the respondent independently draws an ability parameter ξ_{jd} from a distribution with mean θ and within-person ability variance $\sigma^2_{\gamma d}$ and then proceeds to respond to testlet item i with ability equal to $\xi_{jd(i)}$.

Alternative 2. $\gamma_{jd(i)}$ as part of difficulty

Just as well, one could group $\gamma_{jd(i)}$ as part of item difficulty $a_i[\theta_j - (b_i - \gamma_{jd(i)})]$ from which we could interpret the result of all the items in a particular testlet being particularly easy ($\gamma_{jd(i)} > 0$) or difficult ($\gamma_{jd(i)} < 0$) for this examinee. Such an approach might be sensible if the estimation of ability is the primary objective.

Alternative 3. $\gamma_{jd(i)}$ as an independent entity

A third way to characterize the logit is to separate $\gamma_{jd(i)}$ from both ability and difficulty, as $a_i(\theta_j - b_i) + \gamma_{jd(i)}$. In this formulation, the scale of $\gamma_{jd(i)}$ is unaffected by the slope of the item characteristic function and hence may be a useful in comparing the extent of local dependence from one testlet to another. Since the choice of formulation is not aided by any sort of fit statistic, it must be guided by the most likely question that the analysis was meant to answer. We believe that for the most common psychometric uses, the first two formulations are likely to be most helpful. But, there are common situations in which the last formulation is better suited. For example, a current study of oral comprehension testlets on the Test of English as a Foreign Language (TOEFL) discovers substantial variation in $\gamma_{jd(i)}$ across testlets. One important question that emerged is: "What are the characteristics of the testlets that yield low $\gamma_{jd(i)}$?" For such a question we would like to have estimates of $\gamma_{jd(i)}$ that are not intertwined with the value of the slopes of the items within each testlet.

4. Testlet-Based Calibration

In this section, two studies on item calibration will be presented: the first one on the extent to which MCMC and MML calibrations produce comparable results, and the second one on the effects of ignoring the testlet structure on the precision of the calibration.

The first topic was studied using the sample of 5,000 respondents to the verbal section of the Graduate Record Examination (GRE-V) also analyzed in Wainer, Bradlow and Du (this volume). In Table 1, the correlations between the item parameter estimates of the MCMC and BILOG procedures reported in Wainer, Bradlow and Du (this volume), and the parameter estimates of the MML estimation procedure of the testlet model are shown.

It can be seen that the MML and MCMC item parameter estimates are highly correlated; the correlations between the MML estimates of the 3PL model and the 3PL testlet model are somewhat lower.

Although the estimates of the $\sigma_{\gamma d}^2$ were quite close for the two different estimation methods, the width of the interval estimates on the proficiency parameter θ differed, on average, by about 30%. In Figure

Table 1. Correlation between item parameter estimates

Parameter	MCMC	TESTLET	MML	BILOG
a				
	1.00	.95		.86
	.95	1.00		.82
	.86	.82		1.00
b				
	1.00	.95		.98
	.95	1.00		.93
	.98	.93		1.00
c				
	1.00	.87		.75
	.87	1.00		.85
	.75	.85		1.00

1, the confidence regions generated for the four testlet variances by the two estimation methods are shown. The region derived from the MCMC procedure is the 90% highest posterior density region (HPD-region); it is depicted as a solid line in the figure. The 90% confidence interval shown for the MML procedure was computed from the usual assumptions of asymptotic normality of the estimates. These intervals are depicted with dotted lines.

We note that while MML always underestimates the width of the interval, it does so most profoundly for Testlets 2 and 4. If we express the size of the underestimation of interval width as a percentage of the MML interval, we find ranges from 6% (Testlet 3) to 81% (Testlet 4) with an average, over the four testlets, of 30%. The posterior distribution and the normal approximation generated by the MCMC and MML procedure, respectively, are displayed in Figure 2.

It can be seen that for three of the four testlets the posterior distribution is skewed to the right. This result is not surprising since the distribution concerns a variance parameter and variances cannot be negative. The explanation for the smaller MML interval estimates is that they are based on the assumption of asymptotic normality, which, in turn, is based on a Taylor-expansion of the likelihood which terms of order higher two ignored. So a sample size of 5,000 respondents was not enough to eliminate the skewness of the distribution of the parameter estimates.

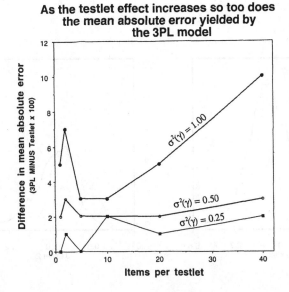

Figure 1. MML estimates of posterior density tending to be too narrow

Next, we assess the effect of ignoring the testlet structure on the precision of item calibration through a simple simulation study. We assume one population of respondents, with every simulee responding to the same 40 items. In addition, a simplified version of the testlet model is utilized in which $\sigma^2_\gamma = \sigma^2_{\gamma d}$ for all testlets d. The data were generated according to the 3PL model from the following structure: $a_i \sim U(0.8, 1.2)$, $b_i \sim U(-1, 1)$, $c_i = 0.25$, and $\theta \sim N(0, 1)$.

The fixing of $c_i = 0.25$ for all items was sufficient for convergence of the MML estimation procedure without any priors on the parameter distributions. There were two values for each of the three independent variables in the design,

(i)   number of testlets (4 or 8) and hence 10 and 5 items per testlet,

(ii)   number of simulees (2,000 or 5,000), and

(iii)   testlet effect size ($\sigma^2_\gamma = 0.25$ or 1.00).

The first 10 or 5 items formed the first testlet, the following 10 or 5 items the second testlet, etcetera. Because items were ordered in ascending difficulty, the mean difficulty of the testlets varied substantially. The estimated and true values of the testlet effects and their standard errors are shown in Table 2 (averaged over 10 replications).

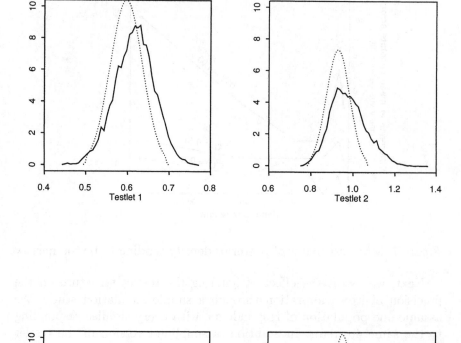

Figure 2. The posterior and the MML normal-based distribution of the variance parameters

Table 2. Estimation of within-person variance $\sigma^2_{\gamma d}$

Number of Testlets	Number of Items per Testlet	N	$\sigma^2_{\gamma d}$	$\hat{\sigma}^2_{\gamma d}$	$SE(\sigma^2_{\gamma d})$
4	10	2000	.25	.256	.040
			1.00	.996	.036
		5000	.25	.250	.026
			1.00	1.016	.022
8	5	2000	.25	.251	.059
			1.00	1.004	.033
		5000	.25	.249	.035
			1.00	.991	.020

The estimates are well within their confidence regions, so the algorithm did not produce any surprises.

The standard errors of the testlet effect for the design with 4 testlets of 10 items are smaller than the standard errors with 8 testlets of 5 items. So the fact that the first design produces more distinct values of $\gamma_{jd(i)}$ per simulee does not result in a more precise estimate of $\sigma^2_{\gamma d}$, probably because the shorter testlet length decreases the precision of the individual values for $\gamma_{jd(i)}$. Therefore, a simple relation between the testlet design and the precision of the estimate of $\sigma^2_{\gamma d}$ cannot be expected to hold. Exploring the relationship between the testlet design and its standard error more fully is an interesting area of future research.

In Table 3, the effect of ignoring the testlet structure is assessed. The eight rows of the table relate to the same studies as in Table 2. In the two columns labeled with MAE(a) and MAE(b), the mean absolute error of the estimates of the discrimination and difficulty parameters computed are presented using conventional MML estimation for the 3 PL ignoring the testlet structure as well as for the proper model. The difference between the parameter estimates is negligible for the cases where $\sigma^2_{\gamma} = 0.25$, while moderate effects appear for the, somewhat more substantial, within-persons standard deviation $\sigma^2_{\gamma} = 1.00$. As described in Wainer, Bradlow and Du (this volume), values of σ^2_{γ} near or greater than 1 are often observed in real data sets and hence ignoring testlet induced biases can have significant effects.

Table 3. Mean absolute error of item parameter estimates and likelihood ratio statistic for 3PL versus 3PL testlet model

$\sigma^2_{\gamma d}$	Number Items Testlet	N	LR	3PL Model $MAE(a)$	$MAE(b)$	3PL Testlet Model $MAE(a)$	$MAE(b)$
0.25	10	2000	21	.083	.070	.083	.071
		5000	25	.046	.046	.046	.046
	5	2000	8	.083	.068	.081	.068
		5000	8	.046	.036	.048	.037
1.00	10	2000	1181	.092	.064	.072	.062
		5000	3233	.052	.060	.038	.035
	5	2000	778	.112	.082	.072	.066
		5000	1848	.087	.075	.039	.041

The bias in the parameter estimates caused by ignoring the testlet structure is modest, at least for the value of σ^2_γ used in these studies. Nevertheless, the two models are easily distinguished, as can be seen from the one-degree-of-freedom likelihood-ratio statistics displayed in the fourth column of Table 3.

5. Testlet-Based Computer Adaptive Testing

In this section, we study the impact of a testlet structure on CAT. We consider the situation in which one is interested in estimating an examinee's ability parameter θ and the within-person ability variance $\sigma^2_{\gamma d}$ is considered as error variance. It is assumed that the CAT procedure consists of administering testlets from a large testlet bank. The most obvious way to do so is to administer the complete testlet as it has been pretested in a calibration phase. However, the same approach is also suitable when only parts of testlets are administered. The main psychometric procedures of CAT are item or testlet selection and estimation of ability. Van der Linden (1998; see also van der Linden & Pashley, this volume) gives an overview of item selection criteria; the most commonly used criterion is the maximum information criterion. Alternative criteria are usually also based on the information function, but attempt to take the uncertainty about the ability parameter into account. Examples are likelihood weighted information and posterior expected information (van der Linden, 1998; van der Linden & Pashley, this volume; Veerkamp & Berger, 1998).

In the framework of testlet-based CAT, the maximum information criterion needs no fundamental adjustment since the within-person ability variance $\sigma^2_{\gamma d}$ is considered as unsystematic error variance. Therefore, item information can be evaluated using the current estimate of θ, item information can be summed to testlet information, and the testlet with the maximum information is the best candidate for the next testlet to be administered. Bayesian selection criteria, however, must explicitly take the testlet structure into account. This also holds for Bayesian ability and precision estimates, which are usually based on the posterior distribution of the ability parameter θ. Let $f(\theta)$ be any function of θ. Then the posterior expectation of this function, say the EAP estimate of θ or the posterior variance, is computed as

$$E(\,f(\theta_j)\mid \mathbf{u}_j, \boldsymbol{\eta}) = \int \cdots \int f(\theta_j)\, p(\boldsymbol{\gamma}_j, \theta_j \mid \mathbf{u}_j,\ \boldsymbol{\eta})\, d\boldsymbol{\gamma}_j\, d\theta_j, \qquad (7)$$

with $p(\boldsymbol{\gamma}_j, \theta_j \mid \mathbf{u}_j)$ the posterior density of the ability parameters defined by (4). The question addressed in the last part of this chapter concerns the impact of disregarding the testlet structure, in particular, the impact of using

$$E(\,f(\theta_j)\mid \mathbf{u}_j, \boldsymbol{\eta}) = \int f(\theta_j)\, p(\theta_j \mid \mathbf{u}_j, \boldsymbol{\eta})\, d\theta_j, \qquad (8)$$

rather than (7) for computing the EAP estimate of ability.

To answer this question a number of simulation studies were carried out. All studies concerned an adaptive test of 40 items. The variables for the first set of simulations were the number of testlets, the within-person variance $\sigma^2_{\gamma d}$, and the method of computing the EAP. The setup of the study can be seen in Table 4. In the first column, the values of $\sigma^2_{\gamma d}$ are given, the second and third columns give the number of testlets and the number of items within a testlet. The ability parameters θ were drawn from a standard normal distribution. The testlets were chosen from a bank of 25 testlets generated as follows: First, 1,000 items were generated by drawing the item difficulty from a standard normal distribution and drawing the discrimination parameter from the log-normal distribution with mean zero and variance 0.25. The guessing parameter was equal to 0.25 for all items. Subsequently, these 1,000 items were ordered with respect to the point on the ability continuum where maximum information was attained. The first items comprised the first testlet in the bank, the following items comprised the second testlet, etcetera. In this way, 25 testlets were created that were homogeneous in difficulty and attained their maximum information at distinct points

Table 4. Mean absolute error for 3PL and 3PL testlet model

$\sigma_{\gamma d}^2$	Number of Testlets	Number Items Testlet	3PL Model	Testlet Model
.00	1	40	.31	
.00	2	20	.28	
.00	4	10	.26	
.00	8	5	.26	
.00	20	2	.26	
.00	40	1	.25	
.25	1	40	.35	.33
.25	2	20	.31	.30
.25	4	10	.30	.28
.25	8	5	.27	.27
.25	20	2	.26	.25
.25	40	1	.26	.26
.50	1	40	.47	.44
.50	2	20	.38	.36
.50	4	10	.34	.32
.50	8	5	.31	.29
.50	20	2	.29	.26
.50	40	1	.28	.26
1.000	1	40	.71	.61
1.000	2	20	.57	.52
1.000	4	10	.47	.44
1.000	8	5	.40	.37
1.000	20	2	.37	.30
1.000	40	1	.34	.29

of the latent ability scale. Next 1,000 replications of a CAT with maximum information testlet selection were generated for the four values of
σ_{γ}^2, six testlet sizes, and the two algorithms for computing the EAP.

Table 5. Differences in mean absolute error between 3PL and 3PL testlet model

$\sigma^2_{\gamma d}$	Number of Testlets	Number Items Testlet	θ -2.0	-1.0	0.0	1.0	2.0
0.25	1	40	-0.07	0.04	0.06	0.04	-0.06
0.25	2	20	-0.02	0.03	0.04	0.03	-0.02
0.25	4	10	0.00	0.02	0.03	0.02	0.01
0.25	8	5	0.00	0.02	0.02	0.02	0.01
0.25	20	2	0.01	0.01	0.01	0.01	0.01
0.25	40	1	0.00	0.01	0.01	0.01	0.00
0.50	1	40	-0.22	0.05	0.15	0.08	-0.24
0.50	2	20	-0.10	0.05	0.10	0.09	-0.06
0.50	4	10	-0.03	0.05	0.10	0.08	0.03
0.50	8	5	0.02	0.04	0.06	0.06	0.04
0.50	20	2	0.03	0.04	0.04	0.05	0.05
0.50	40	1	0.03	0.02	0.03	0.04	0.06
1.00	1	40	-0.52	0.08	0.38	0.12	-0.59
1.00	2	20	-0.28	0.06	0.16	0.12	-0.24
1.00	4	10	-0.10	0.05	0.12	0.12	-0.03
1.00	8	5	0.01	0.06	0.10	0.11	0.10
1.00	20	2	0.09	0.08	0.10	0.14	0.24
1.00	40	1	0.12	0.06	0.08	0.14	0.22

The results can be seen in the last two columns of Table 4, where the mean absolute errors of the final estimates of ability are displayed. The six rows labeled $\sigma^2_{\gamma d} = 0.00$ relate to the cases where the within-person ability variance was zero, so only the standard 3PL model is used for computing the EAP's. Notice that the first row concerns a test of 40 items. Since the first testlet is always chosen to be optimal for an ability $\theta = 0.0$, this row relates to a fixed test that has maximum information for the average examinee.

In the same manner, the second row relates to a two-stage testing design were all examinees get the same first 20-item routing test, followed by a test that is maximally informative at their EAP ability estimate. Finally, the sixth row relates to a classical CAT where the items are administered one at a time and the EAP is computed after every response. Notice that increasing the number of testlets invariably

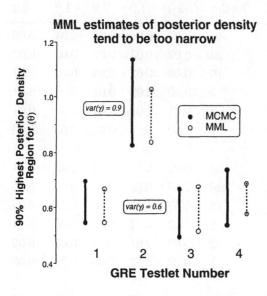

Figure 3. As the testlet effect increases, so does the mean absolute error yielded by the 3PL model

improves the precision of the ability estimates but that the greatest gain is made moving from one to two testlets. If $\sigma^2_{\gamma d}$ is increased, the uncertainty about θ grows. It can be seen in Table 4 that this results in increasing mean absolute errors. Comparing the last two columns, it can be seen that ignoring $\sigma^2_{\gamma d}$ when computing EAP's invariably leads to a decrease in precision. So for a random sample of examinees, (7) is definitely preferable to (8) for administering a CAT. These results are shown graphically in Figure 3.

Do these conclusions also hold locally on the θ-scale? To answer this question, a comparable set of simulation studies was done, the only difference from the previous studies being that ability was fixed for all 1,000 replications. The results are summarized in Table 5. Consider the first row. It relates to 10 simulation studies of 1,000 replications each. In the first two studies, ability was fixed at -2.0; in the second two it was fixed at -1.0; etcetera. In the last two studies in the first row, ability was fixed at 2.0. Then, for each value of θ, the first study used the EAP's computed using (8) whereas in the second study the EAP's were computed using (7).

In both studies, the mean absolute error was computed. The differences are shown in the last five columns of Table 5. So the difference .064 in the cell related to the studies with one testlet of 40 items, $\sigma^2_{\gamma d} = .250$, $\theta = 0.0$ means that using an EAP estimator that takes the testlet structure into account, that is, an estimator computed using (7), resulted in the smallest mean absolute error. It can be seen that this result generally holds, except for the extreme ability values $\theta = -2.0$ and $\theta = 2.0$ in combination with four or fewer testlets. For a standard normal ability distribution, such extreme abilities are rare.

6. Discussion

In this chapter, it was shown how the testlet model can be used in a CAT administration by using the EAP estimator to estimate the main ability dimension θ and by treating the within-person variance $\sigma^2_{\gamma d}$ as random error. It was shown that ignoring the testlet effect, that is, setting $\sigma^2_{\gamma d} = 0$, resulted in a reduction of measurement precision. This result also held for the calibration phase: setting $\sigma^2_{\gamma d} = 0$ leads to imprecise parameter estimates. However, the correlations between the MML estimates for the testlet model and the estimates obtained using BILOG were such that the latter estimates can be called quite close. Also, the MML and MCMC estimates for the testlet model were highly correlated. However, on average, MML underestimated the width of the 90% confidence region for $\sigma^2_{\gamma d}$ by about 30%. This result was obtained because the asymptotic normal approximation was unable to account for the right skewness of the posterior distribution.

Of course, the MML procedure should not be blamed for this underestimation; the fault lies instead with the normal approximation of the posterior. If we had evaluated the posterior directly, perhaps at a set of quadrature points, it is likely that the two results would have been in much closer. Such a calculation is not standard and requires additional programming. This feature reflects one of the key advantages of the MCMC procedure: Because it generates the entire posterior distribution pro forma, it allows for the easy post hoc calculation of a myriad of measures.

Let us next consider the relative merits of the two procedures. A fully Bayesian procedure may produce better interval estimates. However, this advantage may not be the main reason for the recent interest in MCMC procedures for estimating the parameters of IRT models (Albert, 1992, Patz & Junker, 1997, Baker, 1998). For the present model, inspection of (1) shows that the computations for the MML method entail evaluation of a number of two-dimensional integrals over the domain of (θ, γ_p). A possible generalization of the hierarchical approach of modeling testlet response behavior to multidimensional IRT models (see, for instance, McDonald, 1997, or Reckase, 1997) would entail evaluation of integrals of much higher dimensionality, and this may be quite difficult. In a fully Bayesian approach, on the other hand, marginalization is not an issue. In the MCMC process, the posterior distribution of a parameter need only be evaluated conditionally on draws of the other parameters. Therefore, the fully Bayesian framework seems especially promising for evaluation of complex IRT models, such as the generalization of the testlet response model to a multidimensional framework.

Finally, from a purely practical point of view, the MCMC procedure is much easier to program than MML: A new model can be fit in hours not days or weeks. The costs of this ease of implementation, however, is substantially greater computing time; fitting with MML may take a few minutes whereas MCMC could take hours or days. Moreover, it is usually good practice with MCMC to fit the same data several times with different starting points to assure convergence and to obtain good estimates of standard error (although such a conservative practice is probably a good idea with MML as well to avoid having settled into a local extremum). Fortunately, computing keeps getting faster. We began this project with an 80 MHz machine and are writing this paper on one with a 400 MHz capacity. Last week's newspaper offered a gigahertz chip. MCMC is now a practical alternative for very large operational problems.

References

Albert, J.H. (1992). Bayesian estimation of normal ogive item response functions using Gibbs sampling. *Journal of Educational Statistics, 17,* 251-269.

Baker, F.B. (1998). An investigation of item parameter recovery characteristics of a Gibbs sampling procedure. *Applied Psychological Measurement 22,* 153-169.

Bock, R.D., & Aitkin, M. (1981). Marginal maximum likelihood estimation of item parameters: An application of an EM-algorithm. *Psychometrika, 46,* 443-459.

Bradlow, E.T., Wainer, H., and Wang, X. (1999). A Bayesian random effects model for testlets. *Psychometrika, 64,* 153-168.

Gibbons, R.D. & Hedeker, D.R. (1992). Full-information bi-factor analysis. *Psychometrika, 57,* 423-436.

Glas, C.A.W. (1992). A Rasch model with a multivariate distribution of ability. In M. Wilson, (Ed.), *Objective measurement: Theory into practice* (Vol. 1) (pp. 236-258). New Jersey: Ablex Publishing Corporation.

Glas, C.A.W. (1998) Detection of differential item functioning using Lagrange multiplier tests. *Statistica Sinica, 8.* 647-667.

Glas, C.A.W. (1999). Modification indices for the 2PL and the nominal response model. *Psychometrika, 64,* 273-294.

Louis, T.A. (1982). Finding the observed information matrix when using the EM algorithm. *Journal of the Royal Statistical Society, Series B, 44,* 226-233.

McDonald, R.P. (1997). Normal-ogive multidimensional model. In W.J. van der Linden and R.K. Hambleton (Eds.). *Handbook of modern item response theory* (pp. 257-269). New York: Springer.

Mislevy, R.J. (1986). Bayes modal estimation in item response models. *Psychometrika, 51,* 177-195.

Patz, R.J. & Junker, B.W. (1997). *Applications and extensions of MCMC in IRT: Multiple item types, missing data, and rated responses.* (Technical Report No.670). Pittsburgh: Carnegie Mellon University, Department of Statistics.

Reckase, M.D. (1997). A linear logistic multidimensional model for dichotomous item response data. In W.J. van der Linden and R.K. Hambleton (Eds.). *Handbook of modern item response theory.* (pp. 271-286). New York: Springer.

van der Linden, W.J. (1998). Bayesian item selection criteria for adaptive testing. *Psychometrika, 63,* 201-216.

Veerkamp, W.J.J. & Berger, M.P.F. (1997). Some new item selection criteria for adaptive testing. *Journal of Educational and Behavioral Statistics, 221,* 203-220.

Chapter 15
Testlet-Based Adaptive Mastery Testing

Hans J. Vos & Cees A.W. Glas*
University of Twente, The Netherlands

1. Introduction

In mastery testing, the problem is to decide whether a test taker must be classified as a master or a non-master. The decision is based on the test taker's observed test score. Well-known examples of mastery testing include testing for pass/fail decisions, licensure, and certification. A mastery test can have both a fixed-length and variable-length form. In a fixed-length mastery test, the performance on a fixed number of items is used for deciding on mastery or non-mastery. Over the last few decades, the fixed-length mastery problem has been studied extensively by many researchers (e.g., De Gruijter & Hambleton, 1984; van der Linden, 1990). Most of these authors derived, analytically or numerically, optimal rules by applying (empirical) Bayesian decision theory (e.g. DeGroot, 1970; Lehmann, 1986) to this problem. In the variable-length form, in addition to the action of declaring mastery or non-mastery, also the action of continuing to administer items is available (e.g., Kingsbury & Weiss, 1983; Lewis & Sheehan, 1990; Sheehan & Lewis, 1992; Spray & Reckase, 1996).

In either case, items may be administered one at a time or in batches of more than one item. If it is plausible that the responses to items within a batch are more strongly related than the responses to items of different batches, these batches are usually referred to as testlets. The main advantage of variable-length mastery tests as compared to fixed-length mastery tests is that the former offer the possibility of providing shorter tests for those test takers who have clearly attained a certain level of mastery (or non-mastery) and longer tests for those for whom the mastery decision is not as clear-cut (Lewis & Sheehan, 1990). For instance, Lewis and Sheehan (1990) showed in a simulation study that average test lengths could be reduced by half without sacrificing classification accuracy.

* This study received funding from the Law School Admission Council (LSAC). The opinions and conclusions contained in this paper are those of the author and do not necessarily reflect the position or policy of LSAC. The authors thank Charles Lewis for his valuable comments.

W.J. van der Linden and C.A.W. Glas (eds.),
Computerized Adaptive Testing: Theory and Practice, 289–309.
© 2000 *Kluwer Academic Publishers. Printed in the Netherlands.*

Two approaches to variable-length mastery testing can be distinguished. The first approach involves specification of the costs of missclassifications and the cost of test administration, with the decision
to continue testing being guided by expected losses. Item and testlet
selection is not an issue in this approach; items (or testlets) are randomly selected and administered. In this case, the stopping rule (i.e.,
termination criterion) is adaptive but the item selection procedure is
not. This type of variable-length mastery testing is known as sequential mastery testing, and, in the sequel, it will be referred to as SMT.
The procedure is usually modeled using sequential Bayesian decision
theory. In the second approach, item (or testlet) selection is tailored to
the test taker's estimated proficiency level. Kingsbury and Weiss (1983)
denote this type of variable-length mastery testing as adaptive mastery
testing (AMT). This approach, however, does not involve specification
of the cost of missclassifications and the cost of test administration.
The approach is usually modelled using an IRT model and the process
of testing continues until the difference between the test taker's proficiency θ and a cut-off point θ_c on the latent continuum can be estimated
with a certain precision.

The purpose of the present chapter is to present a combination of
sequential and adaptive mastery testing which will be referred to as
ASMT (adaptive sequential mastery testing). The strong points of both
approaches are combined; that is, the selection as well as the stopping
rule are adaptive and the cost per observation is explicitly taken into
account. To support decision making and adaptive item (or testlet)
selection, the 1PL model (Rasch, 1960) and the 3PL model (Birnbaum,
1968) will be used to describe the relation between proficiency level and
observed responses.

The chapter is organized as follows. First, a concise review of the existing literature and earlier approaches to the variable-length mastery
problem is presented. Second, the combined approach of sequential and
adaptive mastery testing will be described. Then this general approach
will be applied to the 1PL and 3PL models, and to the 3PL testlet
model discussed by Wainer, Bradlow and Du (this volume) and Glas,
Wainer and Bradlow (this volume). Following this, a number of simulation studies will be presented that focus on the gain of a sequential
procedure over a fixed-length test and the gain of an adaptive sequential test over a classical sequential test. Gain will be defined in terms
of the average loss, the average number of items administered, and the
percentage of correct decisions. Further, as in the previous chapter, the
impact of ignoring the testlet structure on the performance of the procedure will be studied. The chapter concludes with some new lines of
research.

2. Earlier Approaches to the Variable-Length Mastery Problem

In this section, earlier approaches to the variable-length mastery problem will be briefly reviewed. First, the application of the sequential probability ratio test (SPRT, Wald, 1947)) to SMT is considered. Next, IRT-based adaptive mastery testing strategies will be reviewed. Finally, contributions of Bayesian decision theory to sequential mastery testing will be presented.

2.1. CONTRIBUTIONS OF SPRT TO VARIABLE-LENGTH MASTERY TESTING

The application of Wald's SPRT, originally developed as a statistical quality control test in a manufacturing setting, to SMT dates back to Ferguson (1969). In this approach, a test taker's responses to items were assumed to be binomially distributed. Reckase (1983) proposed alternative sequential procedures within an SPRT-framework, which, in contrast to Ferguson's approach, did not assume that items have equal difficulty but allowed them to vary in difficulty and discrimination by using an IRT model instead of a binomial distribution (also see, Spray & Reckase, 1996).

2.2. IRT-BASED ITEM SELECTION STRATEGIES APPLIED TO ADAPTIVE MASTERY TESTING

Two IRT-based item selection strategies have been primarily used in implementing AMT. In the first approach, Kingsbury and Weiss (1983) proposed selecting the item that maximizes the amount of information at the test taker's last proficiency estimate. In the second approach, a Bayesian item selection strategy, the item to be administered next is the one that minimizes the posterior variance of the test taker's last proficiency estimate. A prior distribution for the test taker's proficiency level must be specified in this approach before the administration of the first item. As pointed out by Chang and Stout (1993), it may be noted that the posterior variance converges to the reciprocal of the test information when the number of items goes to infinity. Therefore, the two methods of IRT-based item selection strategies should yield similar results when the number of administered items is large.

2.3. SEQUENTIAL MASTERY TESTING BASED ON BAYESIAN DECISION THEORY

As mentioned before, most researchers in this area have applied (empirical) Bayesian decision theory to the fixed-length mastery problem. Within a Bayesian decision-theoretic framework, the following two basic elements must be specified: A psychometric model for the probability of answering an item correctly given a test taker's proficiency level (i.e., the item response function), and a loss structure evaluating the total costs and benefits of all possible decision outcomes. These costs may reflect all relevant psychological, social, and economic aspects involved in the decision. The Bayesian approach allows the decision maker to incorporate into the decision process the costs of missclassifications. Furthermore, a prior distribution must be specified representing prior knowledge of test taker's proficiency level. Finally, a cut-off point on the latent proficiency scale separating masters and non-masters must be specified in advance by the decision maker using a method of standard setting (e.g., Angoff, 1971). In a Bayesian approach, optimal rules are obtained by minimizing the posterior expected loss associated with each possible decision outcome.

Lewis and Sheehan (1990), Sheehan and Lewis (1992), and Smith and Lewis (1995) have recently applied Bayesian sequential decision theory to the variable-length mastery problem. In addition to the elements needed in the previous Bayesian decision-theoretic approach, the cost per observation is explicitly specified in this framework. The cost of administering one additional item (or testlet) can be considered as an extension of the loss structure for the fixed-length mastery problem to the variable-length mastery problem. Posterior expected losses associated with non-mastery and mastery decisions can now be calculated at each stage of testing. The posterior expected loss associated with continuing to test is determined by averaging the posterior expected loss associated with each of the possible future decision outcomes relative to the probability of observing those outcomes (i.e., the posterior predictive probability). Analogous to the fixed-length mastery problem, the optimal sequential rule is found by selecting the action (i.e., mastery, non-mastery, or to continue testing) that minimizes posterior expected loss at each stage of sampling using techniques of dynamic programming (i.e., backward induction). This technique makes use of the principle that at each stage of an optimal procedure, the remaining portion of the procedure is optimal when considered in its own right. As indicated by Lewis and Sheehan (1990), the action selected at each stage of testing is optimal with respect to the entire sequential mastery testing procedure.

Vos (1999) also applied Bayesian sequential decision theory to SMT. Like Smith and Lewis (1995), he assumed three classification categories (i.e., non-mastery, partial mastery, and mastery). However, as in Ferguson's (1969) SPRT-approach, for the conditional probability of a correct response given the test taker's proficiency level, the binomial distribution instead of an IRT model is considered. This modeling of response behavior corresponds to the assumption that all items have equal difficulty or are sampled at random from a large (real or hypothetical) pool of items. Assuming that prior knowledge about the test taker's proficiency can be represented by a beta prior $B(\alpha, \beta)$ (i.e., its natural conjugate), it is shown that the number-correct score is sufficient to calculate the posterior expected losses at future stages of the mastery test.

3. Bayesian Sequential Decision Theory Applied to Adaptive Mastery Testing

In this section, the approach of applying Bayesian sequential decision theory to ASMT will be described. Before doing so, some necessary notation will be introduced and the general variable-length mastery problem will be formalized. Then, the loss function assumed will be discussed. Next, it will be shown how IRT models can be incorporated into ASMT.

3.1. FORMALIZATION OF THE VARIABLE-LENGTH MASTERY PROBLEM

In the following, it will be assumed that the variable-length mastery problem consists of S $(S \geq 1)$ stages labeled $s = 1, ..., S$ and that at each stage one of the available testlets can be given. At each stage, one or more items labeled i are administered and the observed item response will be denoted by a discrete random variable U_i, with realization u_i. It is assumed that U_i takes the values 1 and 0 for a correct and incorrect response, respectively. Let $\mathbf{u}_s$ be the response to the s-th testlet. For $s = 1, ..., S$ the decisions will be based on a statistic $\mathbf{w}_s$ which is a function of the response patterns $\mathbf{u}_s$, that is, $\mathbf{w}_s = f(\mathbf{u}_1, ..., \mathbf{u}_s)$. In many cases, $\mathbf{w}_s$ will be the response pattern $\mathbf{u}_1, ..., \mathbf{u}_s$ itself. However, below it will become clear that some computations are only feasible if the information of the complete response pattern is aggregated. At each stage of sampling s $(s = 1, ..., S - 1)$ a decision rule $d(\mathbf{w}_s)$ can be defined as

$$d(\mathbf{w}_s) = \begin{cases} m & \text{test taker is judged a master, sampling stops,} \\ n & \text{test taker is judged a non-master, sampling stops,} \\ c & \text{sampling is continued.} \end{cases}$$

At the final stage of sampling only the two mastery classification decisions m and n are available. Mastery will be defined in terms of the latent proficiency continuum of the IRT model. Therefore, let θ and θ_c denote test taker's proficiency level and some prespecified cut-off point on the latent continuum, respectively. Examinees with proficiency θ below this cut-off point are considered non-masters, while test takers with proficiency θ above this cut-off point are considered masters.

3.2. LINEAR LOSS

As noted before, a loss function evaluates the total costs and benefits for each possible combination of action and test taker's proficiency θ. Unlike Lewis and Sheehan (1990), Sheehan and Lewis (1992), and Smith and Lewis (1995), threshold loss will not be adopted here. The reason is that this loss function, although frequently used in the literature, may be less realistic in some applications. An obvious disadvantage of threshold loss is that it does not depend on the distance between θ and θ_c. It seems more realistic to assume that loss is an increasing function of θ for non-masters and a decreasing function of θ for masters. Moreover, the threshold loss function is discontinuous; at the cut-off point θ_c this function "jumps" from one constant value to another. This sudden change seems unrealistic in many decision making situations. In the neighborhood of θ_c, the losses for correct and incorrect decisions should change smoothly rather than abruptly (van der Linden, 1981).

To overcome these shortcomings, van der Linden and Mellenbergh (1977) proposed a continuous loss function for the fixed-length mastery problem that is a linear function of test taker's proficiency level θ (see also Huynh, 1980; van der Linden & Vos, 1996; Vos, 1997a, 1997b, 1999).

For the variable-length mastery problem, the piecewise linear loss functions for the master and non-master decision can be restated at each stage of sampling as

$$L(m, \theta) = max\{sC, sC + A(\theta - \theta_c)\} \tag{1}$$

with $A < 0$ and

$$L(n, \theta) = max\{sC, sC + B(\theta - \theta_c)\}, \tag{2}$$

with $B > 0$; C is the cost of delivering one testlet, sC is the cost of delivering s testlets. For the sake of simplicity, following Lewis and Sheehan (1990), these costs are assumed to be equal for each decision outcome as well as for each sample. The above defined function consists of a constant term and a term proportional to the difference between test taker's proficiency level θ and the specified cut-off point θ_c. The condition $A < 0$ and $B > 0$ is equivalent to the statement that for action m the loss is a decreasing function of the latent proficiency θ, whereas the loss for action n is assumed to be increasing in θ. The definitions (1) and (2) guarantee that the losses are at least sC. Unlike the specification of loss in van der Linden and Mellenbergh (1977), this specific formulation of linear loss is chosen because in many problems it has been convenient to work with nonnegative loss functions (see, for instance, DeGroot, 1970, p.125).

The loss parameters A, B and C have to be either theoretically or empirically assessed. For assessing loss functions empirically, most texts on decision theory propose lottery methods (e.g., Luce & Raiffa, 1957, chap. 2). In general, these methods use the notion of desirability of outcomes to scale the consequences of each pair of actions and the test taker's proficiency level.

At stage s, the decision as to whether the test taker is a master or a non-master, or whether another testlet will be administered, is based on the expected losses of the three possible decisions given the observation $\mathbf{w}_s$. The expected losses of the first two decisions are computed as

$$E(L(m,\theta) \mid \mathbf{w}_s) = sC + A \int_{-\infty}^{\theta_c} (\theta - \theta_c)p(\theta \mid \mathbf{w}_s)d\theta \qquad (3)$$

and

$$E(L(n,\theta) \mid \mathbf{w}_s) = sC + B \int_{\theta_c}^{\infty} (\theta - \theta_c)p(\theta \mid \mathbf{w}_s)d\theta, \qquad (4)$$

where $p(\theta \mid \mathbf{w}_s)$ is the posterior density of θ given $\mathbf{w}_s$. The expected loss of the third possible decision is computed as the expected risk of continuing to test. If the expected risk of continuing to test is smaller than the expected loss of a master or a non-master decision, testing will be continued. The expected risk of continuing to test is defined as follows. Let $\{\mathbf{w}_{s+1}|\mathbf{w}_s\}$ be the range of $\mathbf{w}_{s+1}$ given $\mathbf{w}_s$. Then, for $s = 1, ..., S - 1$, the expected risk of continuing to test is defined as

$$E(R(\mathbf{w}_{s+1}) \mid \mathbf{w}_s) = \sum_{\{\mathbf{w}_{s+1}|\mathbf{w}_s\}} p(\mathbf{w}_{s+1} \mid \mathbf{w}_s)R(\mathbf{w}_{s+1}), \qquad (5)$$

where the posterior predictive distribution $p(\mathbf{w}_{s+1} \mid \mathbf{w}_s)$ is given by

$$p(\mathbf{w}_{s+1} \mid \mathbf{w}_s) = \int p(\mathbf{w}_{s+1} \mid \theta, \mathbf{w}_s)p(\theta \mid \mathbf{w}_s)d\theta, \qquad (6)$$

and risk is defined as

$$R(\mathbf{w}_{s+1}) = min\{E(L(m,\theta) \mid \mathbf{w}_{s+1}),$$

$$E(L(n,\theta) \mid \mathbf{w}_{s+1}), E(R(\mathbf{w}_{s+2}) \mid \mathbf{w}_{s+1})\}. \tag{7}$$

The risk associated with the last testlet is defined as

$$R(\mathbf{w}_S) = min\{E(L(m,\theta) \mid \mathbf{w}_S), E(L(n,\theta) \mid \mathbf{w}_S)\}. \tag{8}$$

So, given an observation $\mathbf{w}_s$, the expected distribution of $\mathbf{w}_{s+1}, \mathbf{w}_{s+2}, ...,$ $\mathbf{w}_S$ is generated and an inference about future decisions is made. Based on these inferences, the expected risk of continuation in (5) is computed and compared with the expected losses of a mastery or non-mastery decision. If the risk of continuation is less than these two expected losses, testing is continued. If this is not the case the classification decision with the lowest expected loss is made.

Notice that the definitions (5) through (8) imply a recursive definition of the expected risk of continuing to test. In practice, the computation of the expected risk of continuing to test can be done by backward induction as follows. First, the risk of the last testlet is computed for all possible values of $\mathbf{w}_S$. Then the posterior predictive distribution $p(\mathbf{w}_S \mid \mathbf{w}_{S-1})$ is computed using (6), followed by the expected risk $E(R(\mathbf{w}_S) \mid \mathbf{w}_{S-1})$ defined in (5). This, in turn, can be used for computing the risk $R(\mathbf{w}_{S-1})$ for all $\mathbf{w}_{S-1}$ using (7). The iterative process continues until s is reached and the decision can be made to administer testlet $s+1$ or to decide on mastery or non-mastery.

3.3. THE RASCH MODEL

In the Rasch model, the probability of a response pattern $\mathbf{u}$ on a test of K items is given by

$$p(\mathbf{u} \mid \theta, \mathbf{b}) = \prod_{i=1}^{K} \frac{\exp(u_i(\theta - b_i))}{1 + \exp(\theta - b_i)}$$

$$= \exp(t\theta)\exp(-\mathbf{u}'\mathbf{b})P_0(\theta), \tag{9}$$

where $\mathbf{b} = (b_1, ..., b_K)$ is a vector of item parameters, $\mathbf{u}'\mathbf{b}$ is the inner product of $\mathbf{u}$ and $\mathbf{b}$, t is the sum score $t = \sum_i u_i$, and

$$P_0(\theta) = \prod_{i=1}^{K} (1 + \exp(\theta - b_i))^{-1}. \tag{10}$$

Notice that t is the minimal sufficient statistic for θ. Further, it is easily verified that $P_0(\theta)$ is the probability, given θ, of a response pattern with all item responses equal to zero. The probability of observing t given θ is given by

$$
\begin{aligned}
p(t \mid \theta) &= \sum_{\{u\mid t\}} p(\mathbf{u} \mid \theta) \\
&= \sum_{\{u\mid t\}} \exp(t\theta - \mathbf{u}'\mathbf{b})P_0(\theta) \\
&= \gamma_t(\mathbf{b})\exp(t\theta)P_0(\theta),
\end{aligned}
$$

with $\gamma_t(\mathbf{b})$ an elementary symmetric function defined by $\gamma_t(\mathbf{b}) = \sum_{\{u\mid t\}} \exp(-\mathbf{u}'\mathbf{b})$, and where $\{\mathbf{u} \mid t\}$ stands for the set of all possible response patterns resulting in a sum score t. Let $g(\theta)$ be the prior density of θ. Usually the prior is taken to be standard normal. An important feature is that the posterior distributions of θ given $\mathbf{u}$ and t are the same, that is

$$
\begin{aligned}
p(\theta \mid \mathbf{u}) &= \frac{\exp(t\theta - \mathbf{u}'\mathbf{b})P_0(\theta)g(\theta)}{\int \exp(t\theta - \mathbf{u}'\mathbf{b})P_0(\theta)g(\theta)d\theta} \\
&= \frac{\exp(t\theta)P_0(\theta)g(\theta)}{\int \exp(t\theta)P_0(\theta)g(\theta)d\theta} \\
&= \frac{\gamma_t(\mathbf{b})\exp(t\theta)P_0(\theta)g(\theta)}{\int \gamma_t(\mathbf{b})\exp(t\theta)P_0(\theta)g(\theta)d\theta} = p(\theta \mid t).
\end{aligned}
$$

At this point, an assumption will be introduced that may not be completely realistic. It will be assumed that local independence simultaneously holds within and between testlets, that is, all item responses are independent given θ. So at this point, no attempt is made here to model a possible dependence structure of testlet responses. This point will be addressed in the following section.

Applying the general framework of the previous section to the Rasch model entails choosing the minimal sufficient statistics for θ, that is, the unweighted sum scores, for the statistics $\mathbf{w}_s$. Let $\mathbf{t}_s, \mathbf{t}_s = (t_1, ..., t_s)$ be the score pattern on the first s testlets, and define $r_s = \sum_{d=1}^{s} t_d$. Let $p(\theta \mid r_s)$ stand for the posterior density of proficiency given r_s. Then the expected losses (3) and (4) and the expected risk (5) can be written as $E(L(m,\theta) \mid r_s)$, $E(L(n,\theta) \mid r_s)$ and $E(R(r_{s+1}) \mid r_s)$. More specifically, the expected risk is given by

$$
E(R(r_{s+1}) \mid r_s) = \sum_{r_{s+1}\mid r_s} p(r_{s+1} \mid r_s)R(r_{s+1}), \tag{11}
$$

and (6) specializes to

$$p(r_{s+1} \mid r_s) \quad = \quad \int p(r_{s+1} \mid \theta, r_s) p(\theta \mid r_s) d\theta$$

$$= \quad \int \gamma_{t_{s+1}} \, \exp(t_{s+1}\theta) P_{0(s+1)}(\theta) \, p(\theta \mid r_s) d\theta, \quad (12)$$

where $t_{s+1} = r_{s+1} - r_s$, $\gamma_{t_{s+1}}$ is a shorthand notation for the elementary symmetric function of the item parameters of testlet $s+1$ and $P_{0(s+1)}(\theta)$ is equal to (10) evaluated using the item parameters of testlet $s+1$. That is, $P_{0(s+1)}(\theta)$ is equal to the probability of a zero response pattern on testlet $s+1$, given θ. Since elementary functions can be computed very quickly and with a high degree of precision (Verhelst, Glas & van der Sluis, 1984), the risk functions can be explicitly computed.

3.4. THE 3PL MODEL AND THE 3PL TESTLET MODEL

Unlike the Rasch model, the 3PL model has no minimal sufficient statistic for θ. Therefore, one approach of applying the general framework for sequential testing to the 3PL model would be to substitute the complete response pattern $(\mathbf{u}_1, ..., \mathbf{u}_s)$ for $\mathbf{w}_s$. For the testlets where the responses are already known, say the testlets $1, ..., s^*$, this presents no problem. But for evaluation of $E(R(\mathbf{w}_{s+1}) \mid \mathbf{w}_s)$, $s \geq s^*$, however, this entails a summation over the set of all possible response patterns on the future testlets, and exact computation of this expected risk generally presents a major problem. One of the approaches to this problem is approximating (5) using Monte Carlo simulation techniques, that is, simulating a large number of draws from $p(\mathbf{w}_{s+1} \mid \mathbf{w}_s)$ to compute the mean of $R(\mathbf{w}_{s+1})$ over these draws. However, this approach proves quite time consuming and is beyond the scope of the present chapter. The approach adopted here assumes that the unweighted sum score contains much of the relevant information provided by the testlets $s + 1, ..., S$ with respect to θ. This is motivated by the fact that the expected number right score $\sum_i P_i(\theta)$ is monotonically increasing in θ (see, for instance, Lord, 1980, pp. 46-49).Therefore, the following procedure is used.

Suppose that s^* testlets have been administered. Then $\mathbf{w}_{s^*,s}$ will be the observed response patterns on the s^* testlets and the, as yet, unobserved sum score on the testlets $d = s^* + 1, ..., s$. So let t_d be the sum score on testlet d, that is, $t_d = \sum_i u_{di}$, and let $r_{s^*,s}$ be the sum over the scores of the testlets $d = s^* + 1, ..., s$, that is, $r_{s^*,s} = \sum_{d=s^*+1}^{s} t_d$. Then $\mathbf{w}_s$ is defined as $\mathbf{w}_s = (\mathbf{u}_1, ..., \mathbf{u}_{s^*}, r_{s^*,s})$. Using these definitions, the formulas (3) to (8) are evaluated with response patterns to support the computation of the posterior proficiency distribution given the observations, and sum scores as summary statistics for future response behavior.

Testlet response behavior will be modeled by the testlet response model by Wainer, Bradlow and Du (this volume). In Glas, Wainer and Bradlow (this volume), it was shown that one of the interpretations of the model is that for every testlet, the respondent independently draws a proficiency parameter ξ_s from a normal distribution with mean θ and within-person proficiency variance, say $\sigma_{\xi s}^2$. As the items parameters, also the parameters $\sigma_{\xi s}^2$ are assumed known.

The probability of response pattern $\mathbf{u}_s$ can be written as

$$p(\mathbf{u}_s \mid \xi_s) = \prod_i P_{is}(\xi_s)^{u_{is}} (1 - P_{is}(\xi_s))^{1-u_{is}}.$$

The probability of a sum score t_s is computed by summing over the set of all possible response patterns $\mathbf{u}_s$ resulting in a sum score t_s, denoted by $\{\mathbf{u}_s \mid t_s\}$, that is,

$$p(t_s \mid \xi_s) = \sum_{\{\mathbf{u}_s \mid t_s\}} p(\mathbf{u}_s \mid \xi_s).$$

The recursion formulas needed for the computation of these probabilities can, for instance, be found in Kolen and Brennan (1995, pp. 181-183). The probability of t_s conditional on θ is given by

$$p(t_s \mid \theta; \sigma_{\xi s}) = \int p(t_s \mid \xi_s) h(\xi_s \mid \theta; \sigma_{\xi s}) d\xi_s.$$

Finally, let $\{t_{s^*+1,...,}t_s \mid r_{s^*,s}\}$ be the set of all testlet scores $t_{s^*+1,...,}t_s$ compatible with a total score $r_{s^*,s}$. Then

$$p(r_{s^*,s} \mid \theta; \sigma_{\xi s}) = \sum_{\{t_{s^*+1,...,}t_s \mid r_{s^*,s}\}} p(t_s \mid \theta; \sigma_{\xi s})$$

and the posterior distribution of θ given $\mathbf{w}_s$ is given by $p(\theta \mid \mathbf{w}_s; \sigma_{\xi s}) \propto p(\mathbf{u}_1, ..., \mathbf{u}_{s^*} \mid \theta; \sigma_{\xi s}) p(r_{s^*,s} \mid \theta; \sigma_{\xi s}) g(\theta)$, with

$$p(\mathbf{u}_1, ..., \mathbf{u}_{s^*} \mid \theta; \sigma_{\xi s}) = \prod_{d=1}^{s^*} \int p(\mathbf{u}_s \mid \xi_s) h(\xi_s \mid \theta; \sigma_{\xi s}) d\xi_s.$$

Inserting these definitions into (3) to (8) defines the sequential mastery testing procedure for the 3PL testlet model.

3.5. ADAPTIVE SEQUENTIAL MASTERY TESTING

One of the topics addressed in this chapter is how the sequential testing procedure can be optimized when a large testlet bank is available.
The question is which testlets must be administered next upon observing $\mathbf{w}_s$. Three approaches will be considered. The first two are taken
directly from the framework of non-Bayesian adaptive mastery testing (see, for instance, Kingsbury & Weiss, 1983, Weiss & Kingsbury,
1984). Both are based on the maximum information criterion; the first
approach entails choosing items or testlets with maximum information
at θ_c, and the second one chooses items or testlets with maximum
information at $\hat{\theta}_s$, which is an estimate of θ at stage s. The third approach relates to a distinct difference between the non-Bayesian and
Bayesian approach. In the former approach, one is interested in a point
estimate of θ or in whether θ is below or above some cut-off point. In
the latter approach, however, one is primarily interested in minimizing
possible losses due to missclassifications and the costs of testing. This
can be directly translated into a selection criterion for the next testlet. In a Bayesian framework for traditional computer adaptive testing,
one might be interested in the posterior expectation of θ. One of the selection criteria suited for optimizing testlet administration is choosing
the testlet with the minimal expected posterior variance. If $\mathbf{w}_s$ is some
function of the observed response pattern, and $\{\mathbf{w}_{s+1}|\mathbf{w}_s\}$ is the set of
all possible values $\mathbf{w}_{s+1}$ given $\mathbf{w}_s$, one may select the testlet where

$$\sum_{\{\mathbf{w}_{s+1}|\mathbf{w}_s\}} var(\theta \mid \mathbf{w}_{s+1})p(\mathbf{w}_{s+1} \mid \mathbf{w}_s)$$

is minimal (see, for instance, van der Linden, 1998). In a sequential
mastery testing framework, however, one is interested in minimizing
possible losses, so as a criterion for selection of the next testlet the
minimization of

$$\sum_{\{\mathbf{w}_{s+1}|\mathbf{w}_s\}} var(L(m,\theta) - L(n,\theta) \mid \mathbf{w}_{s+1})p(\mathbf{w}_{s+1} \mid \mathbf{w}_s) \qquad (13)$$

will be considered. That is, a testlet is chosen such that the expected
reduction in the variance of the difference between the losses of the
mastery and non-mastery decision is maximal. This criterion focuses
on the posterior variance of the difference between the losses $L(m,\theta)$
and $L(n,\theta)$ given $\mathbf{w}_{s+1}$, and the criterion entails that the sum over all
possible response patterns $\mathbf{w}_{s+1}$ of this posterior variance weighted by
its posterior predictive probability $p(\mathbf{w}_{s+1} \mid \mathbf{w}_s)$ is minimal. In the case
of the Rasch model, (13) is relatively easy to compute because in that
case sum scores can be substituted for $\mathbf{w}_{s+1}$ and $\mathbf{w}_s$.

4. Performance of Sequential and Adaptive Sequential Mastery Testing

4.1. THE 1PL MODEL

The main research questions addressed in this section will be whether, and under what circumstances, sequential testing improves upon a fixed test, and whether, and under what circumstances, adaptive sequential testing improves upon sequential testing. The design of the studies will be explained using the results of the first study, reported in Table 1.

The study concerns the 1PL model, 40 items and a cut-off point θ_c equal to 1.00. The 13 rows of the table represent 13 simulation studies of 2000 replications each. For every replication a true θ was drawn from a standard normal distribution. In the first simulation study, every simulee was presented one test with a fixed length of 40 items. For every simulee, the item parameters were drawn from a standard normal distribution. Also the prior distribution of θ was standard normal. The remaining 12 rows relate to a two-factor design, the first factor being the test administration design, the second the selection method. The test administration design is displayed in the first two columns of the table. The three designs used were 4 testlets of 10 items, 10 testlets of 4 items and 40 testlets of one item each. Four selection methods were studied. In the studies labeled "Sequential" the Bayesian SMT procedure was used. The studies labeled "Cut-off Point", "Eap Estimate" and "Min Variance" used ASMT procedures. The label "Cut-off Point" refers to studies where testlets were selected with maximum information at the cut-off point, "Eap Estimate" refers to studies with a selection procedure based on maximum information at the EAP estimate of proficiency, and "Min Variance" refers to studies with adaptive testlet selection using the Bayesian criterion defined by (13). For all simulations, the parameters of the loss functions (1) and (2) were equal to $A = -1.00$, $B = 1.00$ and $C = 0.01k_t$, where k_t stands for the number of items in a testlet. The motivation for this choice of C is keeping the total cost of administering 40 items constant.

For the SMT condition, the item parameters of the first testlet were all equal to zero and the item parameters of all other testlets were randomly drawn from a standard normal distribution. In the ASMT conditions, it was also the case that the first testlet had all item parameters equal to zero. The reason for starting both the SMT and ASMT procedures with testlets with similar item parameters was to create comparable conditions in the initial phase of the procedures. The following testlets were chosen from a bank of 50 testlets that was generated as follows.

Table 1. Relation between selection method and loss in the 1PL model

Maximum Number Testlets	Number Items Testlet	Selection Method	Proportion Testlets Given	Proportion Correct Decisions	Mean Loss
1	40	Fixed Test	1.00	.94	.4171
4	10	Sequential	.28	.89	.1622
4	10	Cut-off Point	.27	.92	.1555
4	10	Eap Estimate	.27	.89	.1630
4	10	Min Variance	.29	.90	.1623
10	4	Sequential	.17	.91	.1094
10	4	Cut-off Point	.19	.91	.1211
10	4	Eap Estimate	.17	.90	.1151
10	4	Min Variance	.16	.91	.1068
40	1	Sequential	.09	.89	.0996
40	1	Cut-off Point	.10	.90	.0899
40	1	Eap Estimate	.10	.90	.0997
40	1	Min Variance	.10	.89	.1028

For every simulee, $50k_t$ item parameters were drawn from the standard normal distribution first. Then, these $50k_t$ item parameters were ordered in magnitude from low to high. The first k_t items comprised the first testlet in the bank, the second k_t items comprised the second testlet, etcetera. In this way, 50 testlets were created that were homogeneous in difficulty and attained their maximum information at distinct points of the latent proficiency scale. In the "Eap Estimate" condition, at stage s, $s = 1, ..., S - 1$, an expected a posteriori estimate of proficiency was computed and the expected risk of a "Continue Sampling" decision was computed using the $S - s$ testlets with highest information at this estimate. If a "Continue Sampling" decision was made, the next testlet administered was the most informative testlet of the $S - s$ testlets initially selected. The procedure in the "min variance" condition was roughly similar, only here the minimum variance criterion defined by (13) was used. Finally, in the "Cut-off Point" condition, testlets were selected from the testlet bank described above that were most informative at the cut-off point θ_c. The last three columns of Table 1 give the average proportion of testlets administered, the proportion of correct decisions and the mean loss over 2000 replications for each of the thirteen conditions, where the loss in every replication was computed using (1) or (2) evaluated at the true value of θ, with s the number of testlets actually administered.

The study described in the previous paragraph was carried out for three total test lengths, $K = 10$, $K = 20$ and $K = 40$, and two choices of cut-off points, $\theta_c = 1.00$ and $\theta_c = 0.10$. The results for the combination $K = 40$ and $\theta_c = 1.00$ are given in Table 1. Notice that the mean loss in the SMT and ASMT conditions was much lower than in the fixed test condition, and mean loss decreased as a function of the number of testlets. Further, it can be seen that the decrease of mean loss was mainly due to a dramatic reduction in the proportion of testlets given. The number of correct classifications remained stable. Finally, it can be seen that there were no systematic or pronounced differences between SMT and ASMT. This picture also emerged in the $K = 10$ and $K = 20$ studies. Choosing a cut-off point $\theta_c = 0.10$ resulted in increased mean loss. For instance, for $K = 40$ mean loss rose from 0.4171 to 0.4299 for the fixed test condition and from 0.0980 to 0.1541 for the 40 testlets condition, with mean loss averaged over the 4 selection methods. The reason for this increase is that moving the cut-off point closer to the mean of the proficiency distribution increases the number of test takers near the cut-off point. Summing up, it was found that sequential mastery testing did indeed lead to a considerable decrease of mean loss, mainly due to a significant decrease of testlets administered. Across studies, ASMT did only fractionally better than SMT, and again across studies, the minimum variance criterion (13) and selection of testlets with maximum information near the cut-off point θ_c produce the best results, but the difference between them and the maximum information criterion is very small.

4.2. THE 3PL MODEL AND THE 3PL TESTLET MODEL

This section focuses on the question whether the picture that emerged in the previous section for the 1PL model also holds for the 3PL model. In addition, the impact of the testlet structure will be studied. The simulation studies generally have the same set-up as the studies of the previous section. The testlet bank was generated as above, with the difference that besides drawing the item difficulties from a standard normal distribution, item discrimination parameter values were drawn from a log-normal distribution with mean zero and variance 0.25. The guessing parameter value was equal to 0.25 for all items. As above, the testlets were composed in such a way that they attained their maximum information at distinct points on the latent proficiency scale. Since the differences among the three selection procedures for ASMT in the

Table 2. Relation between selection method and loss in the 3PL model

Maximum Number Testlets	Number Items Testlet	Selection Method	Proportion Testlets Given	Proportion Correct Decisions	Mean Loss
1	40	Fixed Test	1.00	.93	.4278
4	10	Sequential	.32	.91	.1699
4	10	Cut-off Point	.36	.93	.1730
4	10	Eap eatimate	.36	.92	.1748
10	4	Sequential	.29	.91	.1526
10	4	Cut-off Point	.26	.93	.1324
10	4	Eap eatimate	.25	.92	.1322
40	1	Sequential	.39	.91	.1922
40	1	Cut-off Point	.11	.90	.0990
40	1	Eap eatimate	.13	.96	.0645

Table 3. Relation between selection method and loss in the 3PL testlet model

Maximum Number Testlets	Number Items Testlet	Selection Method	Proportion Testlets Given	Proportion Correct Decisions	Mean Loss
1	40	Fixed Test	1.00	.88	.4795
4	10	Sequential	.36	.88	.2106
4	10	Cut-off Point	.36	.88	.2089
4	10	Eap eatimate	.35	.87	.2233
10	4	Sequential	.19	.89	.1442
10	4	Cut-off Point	.17	.88	.1339
10	4	Eap eatimate	.17	.89	.1381
40	1	Sequential	.20	.90	.1332
40	1	Cut-off Point	.16	.90	.1080
40	1	Eap eatimate	.18	.92	.1100

Table 4. Relation between selection method and loss in the 3PL model when ignoring the testlet structure

Maximum Number Testlets	Number Items Testlet	Selection Method	Proportion Testlets Given	Proportion Correct Decisions	Mean Loss
1	40	Fixed Test	1.00	.82	.5564
4	10	Sequential	.31	.85	.2209
4	10	Cut-off Point	.36	.88	.2202
4	10	Eap eatimate	.37	.88	.2199
10	4	Sequential	.28	.88	.1773
10	4	Cut-off Point	.25	.90	.1477
10	4	Eap eatimate	.23	.89	.1506
40	1	Sequential	.37	.90	.1955
40	1	Cut-off Point	.12	.87	.1386
40	1	Eap eatimate	.15	.92	.1120

1PL studies were very small, and the minimum variance criterion is quite time consuming to compute, the latter selection criterion was not included in these studies. The results for a study with $K = 40$, $\theta_c = 1.00$ and no within-person variation of proficiency, that is, with $\sigma_{\xi s} = 0.00$, are shown in Table 2. It can be seen that the overall conclusion from the 1PL model studies still holds: there is a considerable decrease of mean loss as the number of testlets increases and the decrease is not bought at the expense of an increased proportion of incorrect decisions. However, contrary to the results of Table 1, it can be observed that these studies showed a clear effect of adaptive testlet selection in terms of a decrease of mean loss. The magnitude of this decrease was positively related to the maximum number of testlets given.

In Table 3, analogous results are given for a situation where $\sigma_{\xi s} = 1.00$, and this within-person variance is explicitly taken into account in the testlet selection and decision procedure. Comparing the results to the results of Table 2, it can be seen that increasing the within-person variance resulted in an increase of mean loss. This increase is due to the addition of within-person variance that acts as a random error component. However, the positive effects of increasing the number of testlets and adaptive testlet selection remained evident.

Finally, in Table 4, results are given for a setup where the responses follow a testlet model with $\sigma_{\xi s} = 1.00$ but decisions and testlet selection were governed by the standard 3PL model, that is, a model with $\sigma_{\xi s} = 0.00$. In other words, for the computation of losses and making decisions, the testlet structure was not taken into account. It can be seen that mean loss is further inflated in all conditions. However, the advantage of SMT over using a fixed test and the advantage of ASMT over SMT were still apparent.

5. Discussion

In this paper, a general theoretical framework for adaptive sequential mastery testing (ASMT) based on a combination of Bayesian sequential decision theory and item response theory was presented. It was pointed out how IRT-based sequential mastery testing (SMT) could be generalized to adaptive item and testlet selection rules, that is, to the case where the choice of the next item or testlet to be administered is optimized using the information from previous responses. The impact of IRT-based sequential and adaptive sequential mastery testing on average loss, proportion correct decisions, and proportion testlets given was investigated in a number of simulations using the 1PL as well as the 3PL model. Two different dependence structures of testlet responses were introduced for the 3PL testlet model. In the first approach, it was assumed that local independence simultaneously holds within and among testlets, that is, all item responses are independent given the test taker's proficiency level. In the second approach, a hierarchical IRT model was used to describe a greater similarity of responses to items within than between testlets.

As far as the 1PL model is concerned, the results of the simulation studies indicated that the average loss in the SMT and ASMT conditions decreased considerably compared to the fixed test condition, while the proportion of correct decisions hardly changed. This result could mainly be ascribed to a significant decrease in the number of testlets administered. With the 3PL model, ASMT produced considerably better results than SMT, while with the 1PL model the results of ASMT were only fractionally better. When testlet response behavior was simulated by a hierarchical IRT model with within-person proficiency variance, average loss increased. Ignoring the within-person variance in the decision procedure resulted in a further inflation of losses.

In summary, the conclusion is that the combination of Bayesian sequential decision theory and modeling response behavior by an IRT model provides a sound framework for adaptive sequential mastery testing where both the cost of test administration and the distance between the test takers' proficiency and the cut-off point are taken into account.

The general approach sketched here can be applied to several other IRT models, for instance, to multidimensional IRT models (see, for instance, McDonald, 1997, or Reckase, 1997). The loss structure involved must allow for both conjunctive and compensatory testing strategies in this case. In decision theory, much work has already been done in this area under the name of "multiple-objective decision making" (Keeney & Raiffa, 1976). It still needs to be examined how the results reported there could be applied to the problems of ASMT in the case of multidimensional IRT models.

Another point of further study is the adoption of minimax sequential decision theory instead of Bayesian sequential decision theory (e.g., DeGroot, 1970; Lehmann, 1986). Optimal rules are found in this approach by minimizing the maximum expected losses associated with all possible decision rules. As pointed out by van der Linden (1981), the minimax principle assumes that it is best to prepare for the worst and establish the maximum expected loss for each possible decision rule. Minimax rules, therefore, can be characterized as either conservative or pessimistic (Coombs, Dawes, & Tversky, 1970). Analogous to Bayesian sequential decision theory, the cost of test administration is also explicitly taken into account in this approach.

References

Angoff, W.H. (1971). Scales, norms, and equivalent scores. In R.L.Thorndike (ed.), *Educational measurement* (2nd ed., pp. 508-600). Washington DC: American Council of Education.

Birnbaum, A. (1968). Some latent trait models. In F.M. Lord & M.R. Novick (Eds.), *Statistical theories of mental test scores.* Addison-Wesley: Reading (Mass.).

Chang, H.-H., & Stout, W.F. (1993). The asymptotic posterior normality of the latent trait in an IRT model. *Psychometrika, 58,* 37-52.

Coombs, C.H., Dawes, R.M., Tversky, A. (1970). *Mathematical psychology: An elementary introduction.* Englewood Cliffs, New Jersey: Prentice-Hall Inc.

DeGroot, M.H. (1970). *Optimal statistical decisions.* New York: McGraw-Hill.

De Gruijter, D.N.M., & Hambleton, R.K. (1984). On problems encountered using decision theory to set cutoff scores. *Applied Psychological Measurement, 8,* 1-8.

Ferguson, R.L. (1969). *The development, implementation, and evaluation of a computer-assisted branched test for a program of individually prescribed instruction.* Unpublished doctoral dissertation, University of Pittsburgh, Pittsburgh PA.

Huynh, H. (1980). A nonrandomized minimax solution for passing scores in the binomial error model. *Psychometrika, 45,* 167-182.

Keeney, D., & Raiffa, H. (1976). *Decisions with multiple objectives: Preferences and value trade-offs.* New York: John Wiley and Sons.

Kingsbury, G.G., & Weiss, D.J. (1983). A comparison of IRT-based adaptive mastery testing and a sequential mastery testing procedure. In D.J. Weiss (Ed.): *New horizons in testing: Latent trait test theory and computerized adaptive testing* (pp. 257-283). New York: Academic Press.

Kolen, M.J., & Brennan, R.L. (1995). *Test equating.* New York: Springer.

Lehmann, E.L. (1986). *Testing statistical hypothesis. (second edition).* New York: Wiley.

Lewis, C., & Sheehan, K. (1990). Using Bayesian decision theory to design a computerized mastery test. *Applied Psychological Measurement, 14,* 367-386.

Lord, F.M. (1980). *Applications of item response theory to practical testing problems.* Hillsdale, N.J., Erlbaum.

Luce, R.D., & Raiffa, H. (1957). *Games and decisions.* New York: John Wiley and Sons.

McDonald, R.P. (1997). Normal-ogive multidimensional model. In W.J. van der Linden and R.K. Hambleton (Eds.). *Handbook of modern item response theory.* (pp. 257-269). New York: Springer.

Rasch, G. (1960). *Probabilistic models for some intelligence and attainment tests.* Copenhagen: Danish Institute for Educational Research.

Reckase, M.D. (1983). A procedure for decision making using tailored testing. In D.J. Weiss (Ed.): *New horizons in testing: Latent trait test theory and computerized adaptive testing* (pp. 237-255). New York: Academic Press.

Reckase, M.D. (1997). A linear logistic multidimensional model for dichotomous item response data. In W.J. van der Linden and R.K. Hambleton (Eds.). *Handbook of modern item response theory.* (pp. 271-286). New York: Springer.

Sheehan, K., & Lewis, C. (1992). Computerized mastery testing with non-equivalent testlets. *Applied Psychological Measurement, 16,* 65-76.

Smith, R.L., & Lewis, C. (1995). A Bayesian computerized mastery model with multiple cut scores. Paper presented at the annual meeting of the National Council on Measurement in Education, San Francisco, CA, April.

Spray, J.A., & Reckase, M.D. (1996). Comparison of SPRT and sequential Bayes procedures for classifying examinees into two categories using a computerized test. *Journal of Educational and Behavioral Statistics, 21,* 405-414.

van der Linden, W.J. (1981). Decision models for use with criterion-referenced tests. *Applied Psychological Measurement, 4,* 469-492.

van der Linden, W.J. (1990). Applications of decision theory to test-based decision making. In R.K. Hambleton & J.N. Zaal (Eds.): *New developments in testing: Theory and applications,* 129-155. Boston: Kluwer.

van der Linden, W.J. (1998). Bayesian item selection criteria for adaptive testing. *Psychometrika, 63,* 201-216.

van der Linden, W.J., & Mellenbergh, G.J. (1977). Optimal cutting scores using a linear loss function. *Applied Psychological Measurement, 1,* 593-599.

van der Linden, W.J., & Vos, H.J. (1996). A compensatory approach to optimal selection with mastery scores. *Psychometrika, 61,* 155-172.

Verhelst, N.D., Glas, C.A.W. & van der Sluis, A. (1984). Estimation problems in the Rasch model: The basic symmetric functions. *Computational Statistics Quarterly, 1,* 245-262.

Vos, H.J. (1997a). Simultaneous optimization of quota-restricted selection decisions with mastery scores. *British Journal of Mathematical and Statistical Psychology, 50,* 105-125.

Vos, H.J. (1997b). A simultaneous approach to optimizing treatment assignments with mastery scores. *Multivariate Behavioral Research, 32,* 403-433.

Vos, H.J. (1999). Applications of Bayesian decision theory to sequential mastery testing. *Journal of Educational and Behavioral Statistics, 24,* 271-292.

Wald, A. (1947). *Sequential analysis.* New York: Wiley.

Weiss, D.J., & Kingsbury, G.G. (1984). Application of computerized adaptive testing to educational problems. *Journal of Educational Measurement, 21,* 361-375.

Author Index

A

Ackerman, T.A. 185, 198
Adema, J.J. 30, 50, 51, 118, 122, 124, 127
Agresti, A. 230, 241
Aitchison, J. 189, 190, 198
Aitkin, M. 184, 198, 271, 286
Albert, J.H. 249, 268, 285, 286
Anastasi, A. 245, 268
Andersen, E.B. 5, 24
Anderson, G.S. 84, 99, 151, 162
Anderson, T.W. 63, 72
Angoff, W.H. 292
Armstrong, R.D. 34, 50

B

Baecker, R.M. 147
Baker, E.L. 135, 141, 142, 146,
Baker, F.B. 212, 218, 285, 286
Balizet, S. 137, 148
Baranowski, R.A. 138, 148
Bashook, P.G. 148
Baxter, G.P. 139, 140, 142, 148
Bejar, I.I. 135, 143, 146
Bennett, R.E. 132, 133, 135, 136, 137, 138, 139, 140, 143, 144, 146, 147
Berger K. 137, 148

Berger, M.P.F. 14, 21, 25, 281, 287
Bernardo, J.M. 61, 72
Bernt, F.M. 136, 146
Bethke, A.D. 138, 148
Binet, A. iii, viii
Birnbaum, A. iii, viii , 9, 12, 24, 27, 50, 58, 72, 121, 127, 249, 268, 290
Bleiler, T. 137, 148
Bloxom, B.M. 54, 55, 72
Bock, R.D. 6, 7, 24, 175, 181, 184, 185, 188, 198, 199, 253, 269, 271, 286
Boekkooi-Timminga, E. 52, 151, 161
Booth, J. 146
Bosman, F. 138, 146
Bradlow, E.T. vii, 2, 30, 245, 248, 268, 271, 275, 279, 286, 290, 299
Brasswell, J. 20, 25
Braun, H.I. 141, 144, 146, 241
Breland, H.M. 134, 140, 142, 146
Brennan, R.L. 299, 308
Breslow, N. 224, 242
Bugbee, A.C. Jr. 136, 146
Buhr, D.C. 221, 241, 242
Burstein, J. 143, 144, 147
Buxton, W.A.S. 135, 147

C

Camilli, G. 221, 241
Carlin, J.B. 8, 24, 231, 241,
    255, 269
Case, S.M. 117, 124, 127
Casella, G. 13, 24
Chang, H.-H. 10, 13, 15, 19, 20,
    21, 24, 34, 50, 52, 190, 198,
    199, 291
Chang, L. 147
Chang, S.W. 172, 181
Chib, S. 249, 268
Chodorow, M. 143, 144, 147
Clauser, B.E. 130, 141, 143, 147
Clyman, S.G. 141, 143, 147
Coffman, D. 137, 148
Coombs, C.H. 307
Cordova, M.J. 34, 50
Cronbach, L.J. iv, viii, 118, 127

D

Davey, T. vi, 129, 132, 134, 140,
    142, 147, 149, 161, 164, 170,
    171, 172, 180, 181, 182
Dawes, R.M. 307
De Groot, M.H. 289, 295, 307
De Gruijter, D.N.M. 289, 307
Dempster, A.P. 186, 198
Dodd, B.G. 143, 147
Donoghue, J.R. 227, 241
Donovan, M.A. 138, 140, 148

Dorans, N.J. 223, 224, 225, 241,
    269
Douglas, J. 269
Drasgow, F. v, viii, 73, 98, 138,
    140, 141, 147, 148, 203, 206,
    218, 219
Du vii, 30, 245, 252, 262, 268,
    271, 275, 279, 290, 299
Durso, R. 77, 98

E

Efron, B. 186, 198
Eggen, T.J.H.M. 106, 116
Eignor, D.R. 174, 181
El-Bayoumi, G. 147
Englehard, G. 128

F

Fan, M. 232, 237, 242
Ferguson, T.S. 33, 51, 291, 293,
    307
Fischer, G.H. 4, 27, 116, 186,
    198, 227, 241
Fitch, W.T. 138, 147
Fitzpatrick, S.J. 144, 147
Flaugher, R. 150, 162, 269
Folk, V. 131, 132, 133, 140, 142,
    148
Fremer, J.J. 51
French, A. 133, 136, 139, 147
Frye, D. 144, 146

G

Gaver, W.W. 138, 147
Gelfand, A.E. 251, 269
Gelman, A. 8, 24, 231, 241, 255, 269
Gessaroli, M.E. 117, 128
Gibbons, R.D. 185, 198, 199, 272, 286
Glas, C.A.W. vi, vii, 2, 13, 20, 25, 30, 102, 103, 116, 183, 184, 186, 190, 198, 206, 216, 218, 271, 272, 273, 274, 286, 287, 289, 290, 299, 308
Gleser, G.C. iv, viii, 118, 127
Godwin, J. 132, 133, 134, 136, 137, 139, 140, 142, 147
Golub-Smith, M.L. 77, 98
Goodman, M. 135, 137, 138, 139, 146
Green, B.F. 20, 25, 175, 181, 269
Greenland, 224, 242
Gruber, J.S. 135, 147
Gulliksen, H. 1, 24

H

Habing, B. 269
Hadidi, A. 117, 118, 120, 122, 124, 128
Haenszel, W. 223, 224, 242
Hambleton, R.K. 198, 199, 202, 218, 250, 269, 287, 289, 307, 308
Hanson, B.A. 21, 23, 25

Hastings, W.K. 252, 269
Hedeker, D.R. 272, 286
Hessinger, J. 135, 146
Hetter, R.D. vi, 29, 45, 46, 51, 149, 162, 166, 168, 169, 170, 171, 172, 180, 182, 184, 199
Hirsch, T.M. 34, 51
Hoijtink, H. 204, 212, 219
Holland, P.W. 221, 222, 223, 224, 225, 227, 241, 242, 243
Holmes, R.M. 161, 162
Holtzman, W.H. viii
Hoogenboom, J. 138, 146
Humphreys, L.G. 175, 181
Huynh, H. 294, 307

J

Jacquemin, D. 135, 143, 146
Jannarone, R.J. 216, 218
Jirele, T. 240, 242
Johnson, C. 18, 19, 25
Jones, D.H. 34, 50
Junker, B.W. 285, 287

K

Kahn, H. 135, 146
Kaplan, B. 20, 25
Keenan, P.A. 138, 140, 148
Keeney, D. 307, 308
Kelley, T.L. 226, 241
Kiefer, J. 186, 199
Kiely, G.L. 30, 52, 128, 246, 269
Kim, H.R. 269

Kingsbury, G.G. 29, 51, 140, 166, 181, 183, 199, 289, 290, 291, 300, 308, 309

Klauer, K.C. 206, 215, 218

Koch, D.A. 130, 131, 147

Kolen, M.J. 299, 308

Kramer, G. 138, 147

Krass, I. 231, 241

Kukich, K. 143, 144, 147

Kulick, E. 223, 224, 241

L

Langdong, L.O. 138, 148

Larkin, K.C. iv, viii

Lau, C.-M.A. 21, 23, 25

Legg, S.M. 221, 241, 242

Lehmann, E.L. 13, 24, 289, 307, 308

Levine, M.V. 203, 206, 218, 219

Lewis, C. vi, 18, 20, 25, 40, 45, 46, 47, 51, 92, 98, 118, 123, 128, 149, 162, 163, 168, 171, 172, 180, 182, 207, 219, 223, 230, 240, 242, 243, 289, 292, 293, 294, 295, 308

Liard, N.M. 186, 198

Ligget, J. 135, 146

Linn, R.L. 175, 181

Lord, F.M. iv, viii, 1, 6, 8, 11, 24, 30, 50, 51, 59, 72, 101, 116, 118, 127, 165, 181, 228, 243, 248, 250, 254, 264, 268, 269, 298, 308

Louis, T.A. 186, 199, 272, 287

Lu, C. 143, 144, 147

Luce, R.D. 295, 308

Luecht, R.M. vi, 30, 34, 47, 51, 52, 55, 72, 117, 118, 120, 121, 122, 124, 125, 126, 127, 128, 130, 147

M

Malakoff, G.L. 147

Mancall, E.L. 148

Mantel, N. 223, 224, 242

Margolis, M.J. 141, 143, 147

Marshall, G. 135, 146

Martin, J.T. 165, 182

Martinez, M.E. 133, 139, 144, 147

Mazzeo, J. 225, 243

McBride, J.R. iv, viii, 21, 25, 51, 148, 165, 182

McDonald, R.P. 286, 287, 307, 308

McKinley, R.L. 77, 98, 221, 242

McLeod, L.D. 207, 219

Mead, A.D. 138, 140, 148

Meijer, R.R. vii, 201, 204, 205, 206, 207, 216, 218, 219

Mellenbergh, G.J. 294, 295, 308

Miller, T.R. 223, 232, 237, 242

Mills, C.N. v, vi, viii, 51, 75, 77, 78, 79, 98, 99, 221, 242

Mislevy, R.J. 6, 7, 10, 17, 24, 25, 147, 182, 184, 185, 188, 189, 190, 199, 250, 253, 269, 271, 287

Mittelholtz, D. 132, 134, 140, 142, 147

Moberg, P.J. 138, 140, 141, 147, 148

Molenaar, I.W. 116, 204, 207, 212, 219
Montgomery, D.C., 208, 219
Moreno, K.E. 55, 73, 136, 148
Morley, M. 133, 135, 136, 140, 143, 146
Muraki, E. 185, 198, 199

N

Nandakumar, R. 223, 237, 238, 239, 242
Nering, M.L. 149, 161, 171, 172, 180, 182, 205, 206, 219
Neyman, J. 186, 199
Nissan, S. 137, 148
Norcini, J.J. 138, 148
Novick, M.R. viii, 1, 11, 24, 50, 59, 72, 127, 248, 268, 269
Nungester, R.J. vi, 30, 51, 117, 118, 120, 121, 122, 124, 128, 147

O

O'Hagan, A. 62, 73
O'Niell K. 131, 132, 133, 140, 142, 148, 221, 242
O'Niell, H.F. Jr. 135, 142, 146
Olson-Buchanan, J.B. v, viii, 73, 98, 138, 140, 141, 147, 148
Owen, R.J. iv, viii, 3, 8, 10, 11, 16, 23, 25, 54, 73

P

Page, E.B. 143, 148, 208, 219
Panchapakesan, N. 186, 199
Parshall, C.G. vi, 129, 130, 131, 132, 137, 138, 148, 164, 170, 172, 181, 232, 242
Pashley, P.J. v, vi, 1, 28, 107, 129, 187, 232, 242, 280, 281
Patrick, R. 228, 243
Patsula, L.N. 86, 87, 98
Patz, R.J. 285, 287
Perlman, M. 137, 148
Petersen, N.S. 143, 148
Phillips, A. 224, 242
Piemme, T.E. 147
Pincetl, P.S. 147
Pine, J. 139, 140, 142, 148
Pommerich, M. 232, 242
Ponsoda, V. 166, 182
Popp, R.L. 138, 148
Potenza, M. 51
Powers, D.E. 221, 242

Q

Quardt, D. 133, 136, 140, 146

R

Raiffa, H. 295, 307, 308
Rao, C.R. 189, 190, 199
Rasch, G. 1, 25, 219, 290, 308

Reckase, M.D. 175, 181, 185, 199, 286, 287, 289, 291, 307, 308

Ree, M.J. 23

Reese, L.M. 34, 36, 41, 52, 77, 98, 151, 153, 155, 159, 162, 221, 242

Rettig, K. 206, 218

Revuelta, J. 166, 182

Rigdon, S.E. 184, 199

Ritter, J. 130, 131, 132, 137, 148

Robin, F. 166, 182

Robins, J. 224, 242

Rose, K.M. 147

Rosenbaum, P.R. 248, 269

Ross, L.P. 141, 143, 147

Roussos, L. 223, 226, 237, 238, 239, 242, 269

Rubin, D.B. 8, 24, 185, 186, 198, 199, 203, 218, 231, 241, 255, 269

S

Samejima, F. 6, 25

Sands, W.A. iv, viii, 25, 51, 148

Schaeffer,G.A. 77, 92, 98, 99

Scheiblechner, H.H. 186, 198

Schnipke, D.L. 20, 25, 34, 44, 52

Schweizer, D.A. 39, 51

Scott, E.L. 186, 199

Scrams, D.J. 43, 44, 52

Searle, S.R. 64, 73

Sebrechts, M.M. 132, 146

Segall, D.O. v, 2, 9, 25, 47, 51, 53, 54, 55, 56, 71, 73, 161, 162, 185, 232, 241

Shaeffer, G. 221, 242

Shavelson, R.J. 139, 140, 142, 148

Shea, J. 138, 148

Shealy, R. 223, 225, 226, 237, 242

Sheehan, K. 118, 123, 128, 289, 292, 294, 295, 308

Shepard, L.A. 221, 241

Sijtsma, K. 206, 207, 219

Simon, Th.A. viii

Singley, M.K. 135, 143, 146

Sireci, S.G. 246, 248, 269

Smith R.L. 292, 293, 294, 308

Smith, A.F.M. 56, 251, 269

Snijders, T.A.B. 202, 204, 206, 219

Soloway, E. 144, 146

Spray, J.A. 232, 242, 289, 291, 308

Steffen, M. vi, 75, 77, 78, 79, 84, 86, 87, 92, 98, 99, 135, 143, 146, 151, 162, 174, 181, 221, 242

Steinberg, L. 222, 242, 243, 269

Stern, H.S. 8, 24, 231, 241, 255, 269

Stewart, R. 130, 131, 132, 137, 148

Stocking, M.L. v, vi, viii, 9, 18, 25, 29, 40, 45, 46, 47, 51, 77, 98, 99, 126, 128, 149, 151, 162, 163, 165, 166, 168, 171, 172, 174, 180, 181, 182, 183, 184, 199, 240, 242

Stone, B. 135, 148

Stone, M.H. 203, 219

Stout, W.F. 223, 225, 226, 237, 242, 248, 269, 291

Straetmans, G.J.J.M. vi, 101, 107, 116

Swaminathan, H. 202, 218, 250, 269

Swanson, D.B. 117, 127, 128

Swanson, L. 29, 51, 99, 126, 128, 151, 162, 174, 182, 183, 199, 240, 242

Sweeney, S.F. 138, 148

Sylvey, S.D. 189, 190, 198

Sympson, J.B. vi, 23, 29, 45, 46, 51, 149, 162, 166, 168, 169, 170, 171, 172, 180, 182, 184, 199

T

Taggart, W.R. 138, 148

Tam, S.S. 54, 73

Tanner, M.A. 249, 269

Tatsuoka, M.M. 63, 73

Thayer, D.T. 222, 223, 224, 225, 227, 229, 230, 232, 233, 234, 235, 236, 241, 243, 244

Thissen, D. 10, 25, 184, 199, 222, 242, 243, 246, 248, 254, 267, 269

Thomasson, G.L. 166, 170, 173, 182, 240

Thompson, T. 171, 172, 180, 182

Timminga, E. 39, 43, 51

Trabin, T.E. 204, 219

Treder, D.W. 137, 148

Tsutakawa, R.K. 18, 19, 25, 184

Tversky, A. 307

Tyler, L. 137, 148

V

Vale, C.D. 54, 55, 72

van den Brink, W.P. 215, 219

van der Linden, W.J. v, vi, 1, 2, 9, 13, 15, 17, 18, 20, 21, 25, 27, 28, 30, 34, 36, 39, 41, 43, 44, 47, 50, 51, 52, 55, 73, 107, 110, 121, 125, 128, 149, 151, 152, 153, 155, 159, 162, 187, 198, 199, 280, 281, 287, 289, 294, 295, 300, 307, 308

van der Sluis, A. 298, 308

van Krimpen-Stoop, E.M.L.A. vii, 201, 204, 206, 216, 218, 219

Veerkamp, W.J.J. 14, 21, 25, 184, 192, 199, 207, 219, 281, 287

Veldkamp, B.P. vi, 39, 52, 149, 151, 153, 155, 156, 159, 162

Verhelst, N.D. 102, 103, 116, 295, 308

Verschoor, A.J. vi, 101

Vicino, F.L. 136, 148

Vispoel, W.P. 21, 23, 25, 137, 148

Vos, H.J. vii, 289, 293, 294, 308, 309

W

Wainer, H. v, vii, viii, 2, 20,
  25, 30, 52, 106, 107, 116,
  128, 162, 221, 222, 241, 242,
  243, 245, 246, 248, 250, 254,
  267, 268, 269, 271, 275, 279,
  286, 290, 299
Wald, A. 291, 309
Walpot, G. 138, 146
Wang, T. 21, 23, 25, 137, 148
Wang, X. 248, 268, 271, 286
Ward, W. 51
Warm, T.A. 5, 10, 21, 25, 212, 219
Waters, B.K. iv, viii, 25, 51, 148
Way, W.D. 84, 99, 151, 162, 174,
  181, 223, 243
Weiss, D.J. iv, v, viii, 21, 23,
  25, 101, 116, 165, 182, 204,
  219, 269, 289, 290, 291, 300,
  308, 309
Wenglinsky, H. 241, 243
Wetherill, G.B. 191, 199
Williams, E.A. 203, 218
Williams, V.S. 138, 148,
Wilson, M. 128, 185, 199, 286
Wingersky, M.S. 147, 182, 222,
  223, 228, 229, 243, 244, 254,
  269

Wolff, S. 143, 144, 147
Wolfowitz, J. 186, 199
Wong, W.H. 249, 269
Wood, R. 185, 199
Wright, B.D. 186, 199, 203, 219
Wu, P.-K. 17, 24, 185, 199

Y

Yen, W. 246, 248, 269
Ying, Z. 10, 13, 14, 15, 19, 20,
  21, 24, 34, 50, 190, 198

Z

Zaal, J.N. 308
Zack, J. 135
Zara, A.R. 29, 51, 166, 181, 183,
  199
Zhang, J. 248, 269
Zickar, M.J. 218, 219
Zieky, M. 224, 243
Zimowski, M.F. 185, 188, 198,
  199
Zwick, R. vii, 221, 222, 223, 225,
  227, 229, 230, 232, 233, 234,
  235, 236, 241, 243, 244

Subject Index

A

Aberrant score patterns, 204
Achievement test, 110, 112
Animation, 129, 139
Armed Services Vocational Aptitude
Battery (ASVAB) 43, 44, 55
ATA software, 122, 125
Attenuation paradox, 11
Audio, 137, 138
Automated test assembly (ATA),
122
Automatic scoring, 142, 143, 144

B

Bayes modal estimation, 6, 188
Bayesian
    approach, 4, 6, 174, 188,
        207, 255, 292
    decision theory, 290, 291
    estimation, 56
    procedures, 54, 55
    random effects model, 248
    sequential decision theory,
        292
BILOG, 7, 253, 254, 264, 275
Bilog-MG, 185
Branch-and-bound (BAB) method,
38
Branching item types, 142

C

Capitalization on change, 12,
    13, 20
CAT-ASVAB, 29
CATSIB, 223, 237, 238, 239
Cheating, 80
Classification table, 152, 153
Coaching schools, 88
College Board Exam, 1
Completion rates, 89
Component scores, 53
Conditional maximum likeli-
    hood estimates (CML),
103
Conditional multinomial expo-
    sure control, 170, 174
Constrained adaptive testing,
    47, 154
Constrained optimization
    problem, 125
Constrained sequential
    optimization, 29
Constraints, 36, 121, 124, 149,
    152
Constructed response items,
    133, 142
Content balance, 29
Content constraints, 88, 123
Content specifications, 121,
    125, 149

ConTEST, 39

Cross-validation, 20

Cumulative sum (CUSUM), 195,
212, 214, 217
statistic, 184
test, 184, 191, 192, 208

D

Decision accuracy, 109

Differential item functioning
(DIF), 188, 190, 221, 222,
223

Disclosed items, 101

E

Efficient score test, 189

Elementary functions, 298

EM algorithm, 18, 186, 272, 274

Empirical Bayes (EB), 18, 230

Empirical Bayes DIF method, 223

Empirical prior, 17

Essay scoring, 143

Ethnic group differences, 221

Expected A Posteriori (EAP)
estimator, 6, 7, 8, 15, 23,
158, 193, 271, 281

Expected information, 4, 15, 16

Exponential family, 5

F

Feedback, 140

Figureal response items, 132,
136, 141

Fisher method of scoring, 66,

Fisher's expected information,
4, 15

Fisher's information, 27, 55

Fisher scoring, 68

Flexi-level testing, 1

Full information factor analysis,
185

G

Gender group differences, 221

Gibbs sampler, 249, 251, 253,
254

Graduate Management Admission
Test (GMAT), 46, 157, 158

Graduate Record Examination
(GRE), 46, 75, 84, 231, 247,
262, 264, 275

Graphical modeling items, 136

Graphics, 129, 136

H

High-stakes adaptive testing, 164

I

Ignorability, 185
Information function, 9, 33
Information matrix, 188
Integer-programming, 151, 154
Interactive video assessment, 140
Interactivity, 139, 140, 141
Item-attributes, 29, 31, 36, 151
Item-exposure control, 20, 40,
    45, 126, 165, 166, 172
    control methods, 183
    control parameters, 167,
        168, 170, 175
    exposure rates, 149, 150,
        151, 161
Item format, 130
Item overlap, 79
Item pool development, 84
Item production, 150
Item writing process, 159, 161

K

Kullback-Leibler information, 13

L

Lagrange multiplier, 184
Lagrange multiplier statistic, 184
Lagrange multiplier test, 189
Law School Admission Test
    (LSAT), 1, 40, 48, 232
Likelihood function, 56, 57, 59

Likelihood-weighted information
    criterion, 14
Linear loss, 294
Linear programming (LP),
    30, 34, 123, 154
Linear test format, 2
Local dependence, 246, 247
Local independence, 190, 216, 248
LOGIST, 174, 228

M

Mantel-Haenszel (MH), 223, 230
Marginal maximum likelihood
    (MML), 103, 184, 185, 189,
    253, 254, 255, 256, 271,
    275
Markov chain Monte Carlo
    (MCMC) 19, 251, 252, 253,
    255, 271, 275
Mastery test, 120, 123, 289
MATHCAT software, 114
Maximum A Posteriori (MAP)
    estimator, 6, 7, 15, 23
Maximum information criterion,
    9, 10, 11, 21, 22, 23, 29,
    30, 32, 107
Maximum-likelihood (ML) esti-
    mator, 5, 9, 10, 20, 23, 55,
    56, 204
Minimax principle, 29
Minimax rules, 307
Multi-objective decision pro-
    blems, 156

Multiple choice, 131

Multiple objective decision making, 307

Multiple objective function, 125

Multiple response items, 131

Multi-stage testing, 30, 31, 32, 50, 118

N

National Assessment of Educational Progress (NAEP), 138

Network-flow programming, 34

Newton-Raphson method, 66, 68, 274

Number-correct scores, 47, 48

O

Objective function, 29, 30, 31, 36, 124, 149, 150, 155, 160

Observed information, 4, 15, 1(

OPLM model, 103

Ordered response items, 132

Owen's procedure, 3, 10, 54

P

Parallel tests, 121

Parameter drift, 183, 184, 188, 189

Partial credit, 133

Partial-credit scoring, 132

Person response function (PRF), 204

Person-fit, 201, 202

Person-fit statistics, 207

Placement test, 101, 106, 107

Posterior density, 59, 60

Posterior distribution, 4, 56, 63, 249

Posterior expected loss, 292

Posterior log-odds , 207

Posterior-weighted information criterion, 15

PRAXIS test, 46

Preknowledge, 163, 164, 201

Preposterior analysis, 16, 22

Pretest 2, 18, 183, 184

Prior density, 56, 57

Proportional scoring, 93, 94, 97

Q

Quality assurance, 118

R

Response-time constraints, 42

S

Scholastic Aptitude Test (SAT), 1, 247, 254

Scholastic Assessment Test (SAT-V), 262
Select-all-that-apply-formats, 104
Selected response items, 131
Sequential mastery testing, 290, 291, 292
Sequential probability ratio test (SPRT), 291
Shadow test, 29, 33, 34, 40, 47, 152, 154, 155
Short-answer item, 104
SIBTEST, 223, 225, 226, 237, 238, 239
Slutsky's theorems, 33
Sound, 129
Statistical Process Control, 207
Statistical quality control, 191
Sympson-Hetter method, 45, 46, 168, 169, 170, 171, 172, 173, 180, 184

T

Test assembly, 123
Test information, 54
Test information function, 9, 156

Test of English as a Foreign Language (TOEFL), 1, 275
Test security, 79, 126
TESTFACT, 185
Testlet, 30, 31, 32, 245, 246, 247, 298, 299
Testlet response theory (TRT), 247
Two-stage format, 101
Two-stage test, 102

U

Uniformly most powerful tests, 206
United States Medical Licensing ExaminationTM (USMLETM), 117
User interface, 136

V

Video, 129, 135, 138

W

Weighted deviation
    algorithm, 174
    method, 29, 151
Weighted likelihood estimator, 5, 10, 20